Fritz Oberhettinger

Tables of
Bessel Transforms

Springer-Verlag
Berlin Heidelberg New York 1972

Fritz Oberhettinger

Professor of Mathematics, Oregon State University, Corvallis, Oregon, U.S.A.

AMS Subject Classifications (1970): 33 A 40, 44 A 05, 44 A 20

ISBN-13: 978-3-540-05997-4   e-ISBN-13: 978-3-642-65462-6
DOI: 10.1007/ 978-3-642-65462-6

Library of Congress Catalog Card Number 72-88727.
Softcover reprint of the hardcover 1st edition 1972

Fritz Oberhettinger

Tables of
Bessel Transforms

Springer-Verlag
New York Heidelberg Berlin 1972

Fritz Oberhettinger

Professor of Mathematics, Oregon State University, Corvallis, Oregon, U.S.A.

AMS Subject Classifications (1970): 33 A 40, 44 A 05, 44 A 20

ISBN-13: 978-3-540-05997-4   e-ISBN-13: 978-3-642-65462-6
DOI: 10.1007/ 978-3-642-65462-6

For Joyce

Preface

This material represents a collection of integral transforms involving Bessel (or related) functions as kernel. The following types of inversion formulas have been singled out.

I. $\quad g(y) = \int_0^\infty f(x)(xy)^{\frac{1}{2}} J_\nu(xy) \, dx$

I'. $\quad f(x) = \int_0^\infty g(y)(xy)^{\frac{1}{2}} J_\nu(xy) \, dy$

II. $\quad g(y) = \int_0^\infty f(x)(xy)^{\frac{1}{2}} K_\nu(xy) \, dx$

II'. $\quad f(x) = \frac{1}{2\pi i} \int_{c-i\infty}^{c+i\infty} g(y)(xy)^{\frac{1}{2}} [I_\nu(xy) + I_{-\nu}(xy)] \, dy$

or also

II". $\quad f(x) = \frac{1}{\pi i} \int_{c-i\infty}^{c+i\infty} g(y)(xy)^{\frac{1}{2}} I_\nu(xy) \, dx$

III. $\quad g(y) + \int_0^\infty f(x)(xy)^{\frac{1}{2}} Y_\nu(xy) \, dx$

III'. $\quad f(x) = \int_0^\infty g(y)(xy)^{\frac{1}{2}} \mathbf{H}_\nu(xy) \, dy$

IV. $\quad g(y) = \int_0^\infty f(x)(xy)^{\frac{1}{2}} \mathbf{H}_\nu(xy) \, dx$

IV'. $\quad f(x) = \int_0^\infty g(y)(xy)^{\frac{1}{2}} Y_\nu(xy) \, dy$

V. $g(y) = \int_0^\infty f(x) K_{ix}(y) dx$

V'. $f(x) = 2\pi^{-2} x \sinh(\pi x) \int_0^\infty g(y) y^{-1} K_{ix}(y) dy$

VI. $g(y) = 2^{1-\mu} [\Gamma(\frac{1}{2}\mu + \frac{1}{2} - \frac{1}{2}\nu) \Gamma(\frac{1}{2}\mu + \frac{1}{2} + \frac{1}{2}\nu)]^{-1} \cdot$

$\cdot \int_0^\infty f(x) (xy)^{\frac{1}{2}} s_{\mu,\nu}(xy) dx$

VI'. $f(x) = 2^{1-\mu} [\Gamma(\frac{1}{2}\mu + \frac{1}{2} - \frac{1}{2}\nu) \Gamma(\frac{1}{2}\mu + \frac{1}{2} + \frac{1}{2}\nu)]^{-1} \cdot$

$\cdot \int_0^\infty g(y) (xy)^{\frac{1}{2}} [s_{\mu,\nu}(xy) - S_{\mu,\nu}(xy)] dy$

VII. $g(y) = \frac{1}{2} \int_0^\infty f(x) \lambda_0 [xy)^{\frac{1}{2}}] dx$

VII'. $f(x) = \frac{1}{2} \int_0^\infty g(y) \lambda_0 [(xy)^{\frac{1}{2}}] dy$

with $\lambda_0(z) = 2\pi^{-1} K_0(z) - Y_0(z)$.

(For notations and definitions see the appendix of
this book.)

The transform VII is also known as the divisor transform.

Greek letters denote complex parameters within the given
range of validity while latin letters signify positive real
numbers. A possible extension to complex values will in general
require a minor effort. In a few cases the expression for $g(y)$
is given only for a part of the internal $(0,\infty)$ for y. This
means that $g(y)$ cannot be given in a simple form for the

remaining part of y. Major contributions concerning integrals
involving Bessel functions as integrand (not necessarily of one
of the transform types I-VII) include the work by Y. L. Luke
(Integrals of Bessel functions, New York, McGraw-Hill, 1962,
419 p.) and A. Erdélyi et. al. (Tables of Integral Transforms,
Vol. 2. New York, McGraw-Hill 1954, 451 p.). Compared to the
latter (pp. 1-174) the material displayed here represents a
considerable extension. Large parts of it do not seem to have
been available before.

Oregon State University
Corvallis, Oregon 97331, U.S.A.
July 1971

Fritz Oberhettinger

Contents

Contents

Chapter I. Hankel Transforms

The representation of a given real function $f(t)$ of the real variable t by means of a double integral involving Bessel functions of order ν is known as Hankel's integral formula

$$f(x) = \int_0^\infty J_\nu(tx) \, t \, dt \int_0^\infty f(u) J_\nu(ut) \, u \, du$$

Equivalent with this is the pair of inversion formulas

(1) $$g(y;\nu) = \int_0^\infty f(x) (xy)^{\frac{1}{2}} J_\nu(xy) \, dx = h_\nu[f(x),y]$$

(2) $$f(x) = \int_0^\infty g(y;\nu) (xy)^{\frac{1}{2}} J_\nu(xy) \, dy$$

which represent the Hankel transform of a given function $f(x)$ and its inversion formula. The Hankel transform is self reciprocal and since

$$J_{\frac{1}{2}}(x) = (\tfrac{1}{2}\pi x)^{-\frac{1}{2}} \sin x \; ; \; J_{-\frac{1}{2}}(x) = (\tfrac{1}{2}\pi x)^{-\frac{1}{2}} \cos x$$

it is obvious that the Fourier sine transform $g_s(y)$, the Fourier cosine transform $g_c(y)$ and the exponential Fourier transform $g_e(y)$ of a function $f(x)$ are special cases of (1) and (2)

$$g_s(y) = (\tfrac{1}{2}\pi)^{-\frac{1}{2}} \int_0^\infty f(x) \sin(xy) \, dx = g(y,\tfrac{1}{2}) = h_{\frac{1}{2}}[f(x),y)$$

$$g_c(y) = (\tfrac{1}{2}\pi)^{-\frac{1}{2}} \int_0^\infty f(x) \cos(xy) \, dx = g(y,-\tfrac{1}{2}) = h_{-\frac{1}{2}}[f(x),y]$$

$$g_e(y) = \int_0^\infty f(x) e^{ixy} \, dx = \tfrac{1}{2}\{h_{-\frac{1}{2}}[f(x),y] + h_{-\frac{1}{2}}[f(-x),y]\} +$$

$$+ \tfrac{1}{2}i\{h_{\frac{1}{2}}[f(x),y] - h_{\frac{1}{2}}[f(-x),y]\}$$

The two dimensional Fourier transform of a given function $f(x,y)$ of two variables defined by

$$F(x,y) = \int_{-\infty}^{\infty} \int_{-\infty}^{\infty} f(x',y') e^{ixx'+iyy'} dx'dy'$$

leads if $f(x,y)$ is such that it depends on $\rho = (x^2+y^2)^{\frac{1}{2}}$ only i.e.

$$f(x,y) = f[(x^2+y^2)^{\frac{1}{2}}] = f(\rho) \qquad \text{to}$$

$$F(x,y) = F(\rho) = 2\pi\rho^{-\frac{1}{2}} \int_{0}^{\infty} \rho'^{\frac{1}{2}} f(\rho')(\rho\rho')^{\frac{1}{2}} J_0(\rho\rho') d\rho'$$

The integral occuring here is the Hankel transform of order zero of the function $\rho^{\frac{1}{2}} f(\rho)$. Similarly for the three dimensional Fourier transform of a function of three variables $f(x,y,z)$ such that

$$f(x,y,y) = f[(x^2+y^2+z^2)^{\frac{1}{2}}] = f(R)$$

$$F(x,y,z) = \int_{-\infty}^{\infty} \int_{-\infty}^{\infty} \int_{-\infty}^{\infty} f(x',y',x') e^{ixx'+iyy'+izz'} dx'dy'dz'$$

$$F(x,y,z) = F(R) = 4\pi R^{-1} \int_{R'=0}^{\infty} R' f(R') \sin(RR') dR'$$

The integral here represents the Fourier sine transform (or the Hankel transform of the order $\nu=\frac{1}{2}$) of the function $Rf(R)$.

In this connection it should be pointed out that in Poisson's summation formulas in one, two or three dimensions

$$F_1(x) = \sum_{n=-\infty}^{\infty} f_1(x+na) = \sum_{n=-\infty}^{\infty} A_n e^{-i2\pi n \frac{x}{a}}$$

$$F_2(x,y) = \sum_{n=-\infty}^{\infty} \sum_{m=-\infty}^{\infty} f_2(x+na,y+nb) =$$

$$= \sum_{n=-\infty}^{\infty} \sum_{m=-\infty}^{\infty} A_{nm} e^{-i2\pi n \frac{x}{a} - i2\pi m \frac{y}{b}}$$

$$F_3(x,y,z) = \sum_{n=-\infty}^{\infty} \sum_{m=-\infty}^{\infty} \sum_{k=-\infty}^{\infty} f_3(x+na, y+mb, z+kc)$$

$$= \sum_{n=-\infty}^{\infty} \sum_{m=-\infty}^{\infty} \sum_{k=-\infty}^{\infty} A_{nmk} e^{-i\pi n\frac{x}{a} -i2\pi m\frac{y}{b} -i2\pi k\frac{z}{c}}$$

the coefficients

$$A_n = \frac{1}{a}H_1(2\pi\frac{n}{a}), \quad A_{nm} = \frac{1}{ab}H_2(2\pi\frac{n}{a}, 2\pi\frac{m}{b})$$

$$A_{nmk} = \frac{1}{abc}H_3(2\pi\frac{n}{a}, 2\pi\frac{m}{b}, 2\pi\frac{k}{c})$$

of the Fourier series above are given by

$$H_1(u) = \int_{-\infty}^{\infty} f_1(x)e^{ixu}dx$$

$$H_2(u,v) = \int_{-\infty}^{\infty} \int_{-\infty}^{\infty} f_2(x,y)e^{ixu+iyv}dxdy$$

$$H_3(u,v,w) = \int_{-\infty}^{\infty} \int_{-\infty}^{\infty} \int_{-\infty}^{\infty} f_3(x,y,z)e^{iux+ivy+iwz}dxdydz$$

which are the Fourier transforms (in one, two or three dimensions) of the functions f_1, f_2, f_3 involved in the summation.

REFERENCES

Erdélyi, A. et.al., 1953: Higher transcendental functions, Vol. 2. McGraw-Hill, New York.

Erdélyi, A. et.al., 1954: Tables of integral transforms, 2 vols. McGraw-Hill, New York.

Oberhettinger, F., 1957: Tabellen zur Fourier Transformation.
                          Springer-Verlag, Berlin.

Sneddon, I. N., 1951: Fourier transforms.
                       Mc.Graw-Hill, New York.

Titchmarsh, E. C., 1937: Introduction to the theory of Fourier
                          integrals.
                          Oxford.

Watson, G. N., 1922: A treatise on the theory of Bessel
                      functions.
                      Cambridge.

1.1 General Formulas

	$f(x)$	$g(y;\nu) = \int_0^\infty f(x)(xy)^{\frac{1}{2}}J_\nu(xy)\,dx$
1.1	$\int_0^\infty g(y)(xy)^{\frac{1}{2}}J_\nu(xy)\,dy$	$g(y)$
1.2	$f(ax), \quad a>0$	$a^{-1}g(ya^{-1};\nu)$
1.3	$x^m f(x), m=0,1,2,\cdots$	$y^{\frac{1}{2}-\nu}\left(\dfrac{d}{y\,dy}\right)^m [y^{m+\nu-\frac{1}{2}}g(y;\nu+m)]$
1.4	$x^m f(x), m=0,1,2,\cdots$	$(-1)^m y^{\frac{1}{2}+\nu}\left(\dfrac{d}{y\,dy}\right)^m [y^{m-\nu-\frac{1}{2}}g(y;\nu-m)]$
1.5	$2\nu x^{-1}f(x)$	$yg(y;\nu-1)+yg(y;\nu+1)$
1.6	$x^{-\mu}f(x)$ $\mathrm{Re}\,\mu>0, \mathrm{Re}(\nu+1)>\mathrm{Re}\,\mu$	$2^{1-\mu}[\Gamma(\mu)]^{-1}y^{\frac{1}{2}-\nu}\,\cdot$ $\cdot \int_0^y \tau^{\nu-\mu+\frac{1}{2}}(y^2-\tau^2)^{\mu-1}g(\tau;\nu-\mu)\,d\tau$
1.7	$x^{-\mu}f(x)$ $\mathrm{Re}\,\mu>0, \mathrm{Re}\,\nu-\frac{3}{2}>\mathrm{Re}\,\mu$	$2^{1-\mu}[\Gamma(\mu)]^{-1}y^{\frac{1}{2}+\nu}\,\cdot$ $\cdot \int_y^\infty \tau^{-\nu-\mu+\frac{1}{2}}(\tau^2-y^2)^{\mu-1}g(\tau;\nu+\mu)\,d\tau$
1.8	$f'(x)$	$\frac{1}{2}\nu^{-1}[(\nu-\frac{1}{2})yg(y;\nu+1)-(\nu+\frac{1}{2})yg(y;\nu-1)]$

1.2 Transforms of Order Zero

	$f(x)$	$g(y) = \int_0^\infty f(x)(xy)^{\frac{1}{2}}J_0(xy)\,dx$
2.1	$x^{-\frac{1}{2}}$	$y^{-\frac{1}{2}}$
2.2	$x^{-\frac{1}{2}}$ $x < 1$ 0 $x > 1$	$y^{\frac{1}{2}}J_0(y)+\frac{1}{2}\pi y^{\frac{1}{2}}[J_1(y)\mathbf{H}_0(y)-J_0(y)\mathbf{H}_1(y)]$
2.3	0 $x < 1$ $x^{-\frac{1}{2}}$ $x > 1$	$y^{-\frac{1}{2}}[1-J_0(y)]+\frac{1}{2}\pi y^{\frac{1}{2}}[J_0(y)\mathbf{H}_1(y)-J_1(y)\mathbf{H}_0(y)]$
2.4	$x^{-\frac{1}{2}}(a+x)^{-1}$	$y^{\frac{1}{2}}[\mathbf{H}_0(ay)-Y_0(ay)]$
2.5	$x^{-\frac{1}{2}}(a^2+x^2)^{-\frac{1}{2}}\cdot$ $\cdot\,[a^2+x^2)^{\frac{1}{2}}+x]$	$y^{\frac{1}{2}}(1+e^{-ay})$
2.6	$x^{-\frac{1}{2}}(a^2+x^2)^{-\frac{1}{2}}\cdot$ $\cdot\,[(a^2+x^2)^{\frac{1}{2}}+x]^{-1}$	$2\pi^{3/2}a^{-2}y^{-\frac{1}{2}}\,(1-e^{-ay})$
2.7	$x^{\frac{1}{2}}(a^2-x^2)^{-\frac{1}{2}}\;x < a$ 0 $x > a$	$y^{-\frac{1}{2}}\sin(ay)$
2.8	$x^{-\frac{1}{2}}(a^2-x^2)^{-\frac{1}{2}}x < a$ 0 $x > a$	$\frac{1}{2}\pi y^{\frac{1}{2}}[J_0(\frac{1}{2}ay)]^2$

	$f(x)$	$g(y) = \int\limits_{0}^{\infty} f(x)\,(xy)^{\frac{1}{2}} J_0(xy)\,dx$
2.9	$x^{\frac{1}{2}}(a^2-x^2)^{\mu}, \quad x < a$ $0 \qquad\qquad x > a$ $\mathrm{Re}\ \mu > -1$	$2^{\mu} a^{\mu+1} y^{-\mu-\frac{1}{2}} \Gamma(\mu+1) J_{\mu+1}(ay)$
2.10	$0 \qquad\qquad x < a$ $x^{\frac{1}{2}}(x^2-a^2)^{-\frac{1}{2}} \quad x > a$	$y^{-\frac{1}{2}} \cos(ay)$
2.11	$0 \qquad\qquad x < a$ $x^{-\frac{1}{2}}(x^2-a^2)^{-\frac{1}{2}} \quad x > a$	$-\tfrac{1}{2}\pi y^{\frac{1}{2}} J_0(\tfrac{1}{2}ay) Y_0(\tfrac{1}{2}ay)$
2.12	$x^{\frac{1}{2}}(a^4+x^4)^{-1}$	$-a^{-2} y^{\frac{1}{2}} \mathrm{kei}_0(ay)$
2.13	$x^{5/2}(a^4+x^4)^{-1}$	$y^{\frac{1}{2}} \mathrm{ker}_0(ay)$
2.14	$x^{5/2}(a^4-x^4)^{-1}$ Cauchy principal value	$-\tfrac{1}{2}y^{\frac{1}{2}}[K_0(ay) - \tfrac{1}{2}\pi Y_0(ay)]$
2.15	$x^{-\frac{1}{2}}\dfrac{(a^2+x^2)^{\frac{1}{2}}-x}{(a^2+x^2)^{\frac{1}{2}}+x}$	$y^{-\frac{1}{2}}+2a^{-2}y^{-5/2}(aye^{-ay}+ e^{-ay}-1)$
2.16	$x^{\frac{1}{2}}(x^4+2a^2x^2+b^4)^{-\frac{1}{2}}$ $a > b$	$y^{\frac{1}{2}}I_0[2^{-\frac{1}{2}}y(a^2-b^2)^{\frac{1}{2}}] \cdot$ $\cdot\ K_0[2^{-\frac{1}{2}}y(a^2+b^2)^{\frac{1}{2}}]$

	$f(x)$	$g(y) = \int\limits_0^\infty f(x)(xy)^{\frac{1}{2}} J_0(xy)\,dx$
2.17	$x^{\frac{1}{2}}(x^4+2a^2x^2+b^4)^{-\frac{1}{2}}$ $b > a$	$y^{\frac{1}{2}} J_0[2^{-\frac{1}{2}}y(b^2-a^2)^{\frac{1}{2}}] \cdot$ $\cdot\, K_0[2^{-\frac{1}{2}}y(b^2+a^2)^{\frac{1}{2}}]$
2.18	$x^{\frac{1}{2}}(1+x^4)^{-\frac{1}{2}} \cdot$ $\cdot\, [x^2+(1+x^4)^{\frac{1}{2}}]^{-\nu}$ $\mathrm{Re}\ \nu > -\ ^{3/4}$	$y^{\frac{1}{2}} J_\nu(2^{-\frac{1}{2}}y) K_\nu(2^{-\frac{1}{2}}y)$
2.19	$x^{-\frac{1}{2}}e^{-ax}$	$y^{\frac{1}{2}}(a^2+y^2)^{-\frac{1}{2}}$
2.20	$x^{\frac{1}{2}}\, e^{-ax}$	$ay^{\frac{1}{2}}(a^2+y^2)^{-3/2}$
2.21	$x^{-3/2}(1-e^{-ax})$	$y^{\frac{1}{2}}\sinh(ay^{-1})$
2.22	$x^{-\frac{1}{2}}e^{-ax^2}$	$\frac{1}{2}\left(\frac{\pi}{a}\right)^{\frac{1}{2}} y^{\frac{1}{2}} e^{-\frac{y^2}{8a}} I_0\left(\frac{y^2}{8a}\right)$
2.23	$x^{\frac{1}{2}}\, e^{-ax^2}$	$(2a)^{-1}\, y^{\frac{1}{2}}\, e^{-\frac{y^2}{4a}}$
2.24	$x^{-3/2}e^{-\frac{a}{x}}$	$2y^{\frac{1}{2}}K_0[(2ay)^{\frac{1}{2}}] J_0[(2ay)^{\frac{1}{2}}]$
2.25	$x^{-1}\, e^{-ax^{\frac{1}{2}}}$	$\frac{1}{8}\,\pi a y^{-\frac{1}{2}}\{[J_{\frac{1}{4}}(\frac{a^2}{8y})]^2 + [Y_{\frac{1}{4}}(\frac{a^2}{8y})]^2\}$

	$f(x)$	$g(y) = \int\limits_{0}^{\infty} f(x)(xy)^{\frac{1}{2}} J_0(xy)\,dx$
2.26	$x^{\frac{1}{2}}\exp[-a(b^2+x^2)^{\frac{1}{2}}]$	$ay^{\frac{1}{2}}(a^2+y^2)^{-3/2}[1+b(a^2+y^2)^{\frac{1}{2}}] \cdot$ $\cdot\ \exp[-b(a^2+y^2)^{\frac{1}{2}}]$
2.27	$x^{\frac{1}{2}}(b^2+x^2)^{-\frac{1}{2}} \cdot$ $\cdot\ \exp[-a(b^2+x^2)^{\frac{1}{2}}]$	$y^{\frac{1}{2}}(a^2+y^2)^{-\frac{1}{2}}\exp[-b(a^2+y^2)^{\frac{1}{2}}]$
2.28	$x^{-\frac{1}{2}}\log x$	$-\ y^{-\frac{1}{2}}\log\ (2\gamma y)$
2.29	$x^{-\frac{1}{2}}(a^2+x^2)^{-\frac{1}{2}} \cdot$ $\cdot\ \log[x+(a^2+x^2)^{\frac{1}{2}}]$	$y^{\frac{1}{2}}[\tfrac{1}{2}K_0^2(\tfrac{1}{2}ay) + \log a\ I_0(\tfrac{1}{2}ay)K_0(\tfrac{1}{2}ay)]$
2.30	$x^{-\frac{1}{2}}(a^2+x^2)^{-\frac{1}{2}} \cdot$ $\cdot\log\ [\dfrac{(a^2+x^2)^{\frac{1}{2}}+x}{(a^2+x^2)^{\frac{1}{2}}-x}]$	$y^{\frac{1}{2}}\ K_0^2\ (\tfrac{1}{2}ay)$
2.31	$x^{\frac{1}{2}}(z^2-1)^{-\frac{1}{2}} \cdot$ $\cdot\ \log\ [z+(z^2-1)^{\frac{1}{2}}]$ $z = (2ab)^{-1}(a^2+b^2+y^2)$	$2aby^{\frac{1}{2}}\ K_0(ay)\ K_0(by)$
2.32	$x^{\frac{1}{2}}(a^4+x^4)^{-\frac{1}{2}} \cdot$ $\cdot\ \log[\dfrac{x^2+(a^4+x^4)^{\frac{1}{2}}}{a^2}]$	$-\ \tfrac{1}{2}\pi y^{\frac{1}{2}}\ Y_0(2^{-\frac{1}{2}}ay)\ K_0(2^{-\frac{1}{2}}ay)$

	$f(x)$	$g(y) = \int\limits_0^\infty f(x)(xy)^{\frac{1}{2}}J_0(xy)\,dx$	
2.33	$x^{\frac{1}{2}}\log(1+a^2x^{-2})$	$2y^{-\frac{1}{2}}[y^{-1}-aK_1(ay)]$	
2.34	$x^{\frac{1}{2}}\log[ax^{-1}+(1+a^2x^{-2})^{\frac{1}{2}}]$	$y^{-3/2}(1-e^{-ay})$	
2.35	$x^{-\frac{1}{2}}e^{-ax^2}\log x$	$2a^{-1}y^{\frac{1}{2}}e^{-\frac{y^2}{4a}}[\log(\frac{y}{2a}) - \frac{1}{2}\,\overline{Ei}\,(\frac{y^2}{4a})]$	
2.36	$x^{-\frac{1}{2}}\sin(ax)$	$y^{\frac{1}{2}}(a^2-y^2)^{-\frac{1}{2}}$	$y < a$
		0	$y > a$
2.37	$x^{-\frac{1}{2}}\cos(ax)$	0	$y < a$
		$y^{\frac{1}{2}}(y^2-a^2)^{-\frac{1}{2}}$	$y > a$
2.38	$x^{-3/2}\sin(ax)$	$\frac{1}{2}\pi y^{\frac{1}{2}}$	$y < a$
		$y^{\frac{1}{2}}\arcsin(ay^{-1})$	$y > a$
2.39	$x^{-3/2}[1-\cos(ax)]$	$y^{-\frac{1}{2}}\log[\frac{a}{y} + (\frac{a^2}{y^2} - 1)^{\frac{1}{2}}]$	$y < a$
		0	$y > a$
2.40	$x^{-\frac{1}{2}}(b^2+x^2)^{-1}\sin(ax)$	$y^{\frac{1}{2}}b^{-1}\sinh(ba)\,K_0(by)$	$y > a$
2.41	$x^{\frac{1}{2}}(b^2+x^2)^{-1}\sin(ax)$	$\frac{1}{2}\pi y^{\frac{1}{2}}e^{-ab}\,I_0(by)$	$y < a$

	$f(x)$	$g(y) = \int\limits_{0}^{\infty} f(x)(xy\,J_0(xy)\,dx$
2.42	$x^{-\frac{1}{2}}(b^2+x^2)^{-1}\cos(ax)$	$\frac{1}{2}b^{-1}\pi y^{\frac{1}{2}}e^{-ab}I_0(by)$ $\qquad y < a$
2.43	$x^{\frac{1}{2}}(b^2+x^2)^{-1}\cos(ax)$	$y^{\frac{1}{2}}\cosh(ba)\,K_0(by)$ $\qquad y > a$
2.44	$x^{\frac{1}{2}}\sin(ax^2)$	$\frac{1}{2}\,a^{-1}y^{\frac{1}{2}}\cos(\frac{y^2}{4a})$
2.45	$x^{\frac{1}{2}}\cos(ax^2)$	$\frac{1}{2}a^{-1}y^{\frac{1}{2}}\sin(\frac{y^2}{4a})$
2.46	$x^{-3/2}\sin(ax^2)$	$\frac{1}{2}y^{\frac{1}{2}}\text{si}(\frac{y^2}{4a^2})$
2.47	$x^{-3/2}[1-\cos(ax^2)]$	$-\frac{1}{2}y^{\frac{1}{2}}\text{Ci}(\frac{y^2}{4a^2})$
2.48	$x^{-1}\sin(ax^{\frac{1}{2}})$	$\frac{1}{4}\pi ay^{-\frac{1}{2}}J_{\frac{1}{4}}(\frac{a^2}{8y})[J_{\frac{1}{4}}(\frac{a^2}{8y}) - Y_{\frac{1}{4}}(\frac{a^2}{8y})]$
2.49	$x^{-1}\cos(ax^{\frac{1}{2}})$	$\frac{1}{8}\pi ay^{-\frac{1}{2}}\{Y_{\frac{1}{4}}(\frac{a^2}{8y})[Y_{\frac{1}{4}}(\frac{a^2}{8y})-J_{\frac{1}{4}}(\frac{a^2}{8y})] -$ $- J_{\frac{1}{4}}(\frac{a^2}{8y})[J_{\frac{1}{4}}(\frac{a^2}{8y}) + Y_{\frac{1}{4}}(\frac{a^2}{8y})]\}$
2.50	$x^{-1}e^{-ax^{\frac{1}{2}}}\sin(ax^{\frac{1}{2}})$	$\frac{1}{2}ay^{-\frac{1}{2}}I_{\frac{1}{4}}(\frac{a^2}{4y})\,K_{\frac{1}{4}}(\frac{a^2}{4y})$
2.51	$x^{-1}e^{-ax^{\frac{1}{2}}}\cos(ax^{\frac{1}{2}})$	$\frac{1}{2}ay^{-\frac{1}{2}}I_{-\frac{1}{4}}(\frac{a^2}{4y})\,K_{\frac{1}{4}}(\frac{a^2}{4y})$

	$f(x)$	$g(y) = \int_0^\infty f(x)(xy)^{1/2} J_0(xy)\, dx$
2.52	$x^{-1/2}\cos(ax)\log(bx)$	$-\tfrac{1}{2}\pi y^{1/2}(a^2-y^2)^{-1/2} \qquad y < a$ $-y^{1/2}(y^2-a^2)^{-1/2}[\gamma-\log(\tfrac{1}{2}by)+\log(y^2-a^2)]$ $\qquad\qquad y > a$
2.53	$x^{1/2}(b^2+x^2)^{-1/2}\sin[a(b^2+x^2)^{1/2}]$	$y^{1/2}(a^2-y^2)^{-1/2}\cos[b(a^2-y^2)^{1/2}] \quad y < a$ $0 \qquad\qquad y > a$
2.54	$x^{1/2}(b^2+x^2)^{-1/2}\cos[a(b^2+x^2)^{1/2}]$	$-y^{1/2}(a^2-y^2)^{-1/2}\sin[b(a^2-y^2)^{1/2}] \quad y < a$ $y^{1/2}(y^2-a^2)^{-1/2}\exp[-b(y^2-a^2)^{1/2}] \quad y > a$
2.55	$x^{1/2}(a^2-x^2)^{-1/2}\cos[b(a^2-x^2)^{1/2}]$ $\qquad x < a$ $0 \qquad x > a$	$y^{1/2}(b^2+y^2)^{-1/2}\sin[a(b^2+y^2)^{1/2}]$
2.56	$0 \qquad x < a$ $x^{1/2}(x^2-a^2)^{-1/2}\cos[b(x^2-a^2)^{1/2}]$ $\qquad x > a$	$0 \qquad\qquad y < b$ $y^{1/2}(y^2-b^2)^{-1/2}\cos[a(y^2-b^2)^{1/2}] \quad y > a$
2.57	$x^{1/2}\sin[b(a^2-x^2)^{1/2}]$ $\qquad x < a$ $0 \qquad x > a$	$ab(b^2+y^2)^{-1}\{a^{-1}(b^2+y^2)^{-1/2}\sin[a(b^2+y^2)^{1/2}]-$ $\quad -\cos[a(b^2+y^2)^{1/2}]\}$
2.58	$x^{-1/2}e^{-ax}\sinh(bx)$ $\qquad a \geq b$	$(aby)^{1/2}(uv)^{-1}(u-v)^{1/2}(u+v)^{-1/2}$ $u = [y^2+(b-a)^2]^{1/2},\ v=[y^2+(b+a)^2]^{1/2}$

	$f(x)$	$g(y(= \int_0^\infty f(x)(xy)^{\frac{1}{2}}J_0(xy)dx$
2.59	$x^{-\frac{1}{2}}e^{-ax}\cosh(bx)$ $a \geq b$	$(aby)^{\frac{1}{2}}(uv)^{-1}(u+v)^{\frac{1}{2}}(u-v)^{-\frac{1}{2}}$ $u = [y^2+(b-a)^2]^{\frac{1}{2}}, \ v = [y^2+(b-a)^2]^{\frac{1}{2}}$
2.60	$x^{-3/2}e^{-ax}\sinh(ax)$	$\tfrac{1}{2}y^{\frac{1}{2}}\log[\dfrac{2a}{y} + (1+\dfrac{4a^2}{y})^{\frac{1}{2}}]$
2.61	$x^{\frac{1}{2}}e^{-ax^{-1}}\sinh(ax^{-1})$	$2ay^{-\frac{1}{2}}J_1[2(ay)^{\frac{1}{2}}]K_1[2(ay)^{\frac{1}{2}}]$
2.62	$x^{\frac{1}{2}}P_n(1-2x^2) \quad x < 1$ $0 \qquad x > 1$ $n = 0,1,2,\cdots$	$y^{-\frac{1}{2}}J_{2n+1}(y)$
2.63	$x^{5/2}P_n(1-2x^2) \quad x < 1$ $0 \qquad x > 1$ $n = 0,1,2,\cdots$	$y^{-\frac{1}{2}}(2n+1)^{-1}[(n+1)J'_{2n+2}(y)-nJ'_{2n}(y)]$
2.64	$x^{\frac{1}{2}}(a^2+x^2)^{-\frac{1}{2}n-\frac{1}{2}} \cdot$ $\cdot P_n[a(a^2+x^2)^{-\frac{1}{2}}]$ $n = 0,1,2,\cdots$	$\dfrac{1}{n!}y^{n-\frac{1}{2}}e^{-ay}$

	$f(x)$	$g(y) = \int\limits_{0}^{\infty} f(x)\,(xy)^{\frac{1}{2}} J_0(xy)\,dx$
2.65	$x^{\frac{1}{2}}(1-x^2)^{-\frac{1}{2}}\sin[a(1-x^2)^{\frac{1}{2}}]\cdot$ $\cdot P_{2n+1}[(1-x^2)^{\frac{1}{2}}]\quad x<1$ $\qquad 0 \qquad x>1$ $n = 0,\ 1,\ 2,\cdots$	$(-1^n (\tfrac{1}{2}\pi)^{\frac{1}{2}} y^{\frac{1}{2}}(a^2+y^2)^{-\frac{1}{4}}\cdot$ $\cdot P_{2n+1}[a(a^2+y^2)^{-\frac{1}{2}}]J_{2n+3/2}[(a^2+y^2)^{\frac{1}{2}}]$
2.66	$x^{\frac{1}{2}}(1-x^2)^{-\frac{1}{2}}\cos[a(1-x^2)^{\frac{1}{2}}]$ $P_{2n}[(1-x^2)^{\frac{1}{2}}]\qquad x<1$ $\qquad 0 \qquad x>1$ $n = 0,1,2,\cdots$	$(-1)^n (\tfrac{1}{2}\pi)^{\frac{1}{2}} y^{\frac{1}{2}}(a^2+y^2)^{-\frac{1}{4}}\cdot$ $\cdot P_{2n}[a(a^2+y^2)^{-\frac{1}{2}}]J_{2n+\frac{1}{2}}[(a^2+y^2)^{\frac{1}{2}}]$
2.67	$x^{\frac{1}{2}}\exp(-a^2x^2)L_n(a^2x^2)$	$(n!)^{-1}2^{-2n-1}a^{-2n-2}y^{2n+\frac{1}{2}}\cdot$ $\cdot\exp(-\tfrac{1}{4}y^2 a^{-2})$
2.68	$x^{-\frac{1}{2}}\mathrm{Erf}(ax)$	$y^{-\frac{1}{2}}\mathrm{Erfc}(\tfrac{1}{2}ya^{-1})$
2.69	$x^{-\frac{1}{2}}\mathrm{Erfc}(ax)$	$y^{-\frac{1}{2}}\mathrm{Erf}(\tfrac{1}{2}ya^{-1})$
2.70	$x^{-\frac{1}{2}}\exp(a^2x^2)\cdot$ $\cdot\mathrm{Erfc}(ax)$	$2^{-\frac{1}{2}}\pi^{3/2}a^{-1}y^{\frac{1}{2}}\exp(\tfrac{1}{8}y^2 a^{-2})K_0(\tfrac{1}{8}y^2 a^{-2})$
2.71	$x^{\frac{1}{2}}\exp(a^2x^2)\cdot$ $\cdot\mathrm{Erfc}(ax)$	$a^{-1}(\pi y)^{-\frac{1}{2}}-\tfrac{1}{2}a^{-2}y^{\frac{1}{2}}\exp(\tfrac{1}{4}y^2 a^{-2})\mathrm{Erfc}(\tfrac{1}{2}ya^{-1})$

	$f(x)$	$g(y) = \int_0^\infty f(x)(xy)^{\frac{1}{2}}J_0(xy)\,dx$
2.72	$x^{-\frac{1}{2}}S(a^2x^2)$	$y^{-\frac{1}{2}}[\frac{1}{2}-C(\frac{1}{4}y^2a^{-2})]$
2.73	$x^{-\frac{1}{2}}C(a^2x^2)$	$y^{-\frac{1}{2}}[\frac{1}{2}-S(\frac{1}{4}y^2a^{-2})]$
2.74	$\frac{1}{2} - S(a^2x^2)$	$y^{-\frac{1}{2}}C(\frac{1}{4}y^2a^{-2})$
2.75	$\frac{1}{2} - C(a^2x^2)$	$y^{-\frac{1}{2}}S(\frac{1}{4}y^2a^{-2})$
2.76	$x^{\frac{1}{2}}Ei(-ax^2)$	$-2y^{-3/2}[1-\exp(-\frac{1}{4}y^2a^{-1})]$
2.77	$x^{\frac{1}{2}}\exp(ax^2)Ei(-ax^2)$	$(2a)^{-1}y^{\frac{1}{2}}\exp(\frac{1}{4}y^2a^{-1})Ei(-\frac{1}{4}y^2a^{-1})$
2.78	$x^{\frac{1}{2}}\exp(-ax^2) \cdot$ $\cdot L_k(ax^2)$ $\mathrm{Re}\ k > -\ ^{3/4}$	$2^{-2k-1}a^{-2-2k}[\Gamma(k+1)]^{-1}y^{2k+\frac{1}{2}} \cdot$ $\cdot \exp(-\frac{1}{4}y^2a^{-2})$
2.79	$x^{-\frac{1}{2}}si(ax)$	$-y^{-\frac{1}{2}}\arcsin(\frac{y}{a})$ $\quad y < a$ $0 \quad\quad\quad\quad\quad y > a$
2.80	$x^{\frac{1}{2}}si(ax^2)$	$-2y^{-3/2}\sin(\frac{x^2}{4a})$
2.81	$x^{\frac{1}{2}}Ci(ax^2)$	$2y^{-3/2}[1-\cos(\frac{x^2}{4a})]$

	$f(x)$	$g(y) = \int\limits_0^\infty f(x)(xy)^{\frac{1}{2}}J_0(xy)\,dx$
2.82	$x^{-\frac{1}{2}}\mathrm{Ci}(ax^2)$	$y^{-\frac{1}{2}}[\log(\frac{y^2x^2}{4a}) + \mathrm{Ci}(\frac{x^2}{4a})]$
2.83	$x^{\frac{1}{2}}(1+x^2)^{-\nu-1} \cdot$ $\cdot P_\nu[(1-x^2)(1+x^2)^{-1}]$ $\mathrm{Re}\ \nu > 0$	$[2^\nu\Gamma(\nu+1)]^{-2}\,y^{2\nu+\frac{1}{2}}K_0(y)$
2.84	$x^{\frac{1}{2}}\{p^\mu_{\nu-\frac{1}{2}}[(1+x^2a^{-2})^{\frac{1}{2}}]\}^2$ $\mathrm{Re}\ \mu<1,\ -\frac{1}{4} < \mathrm{Re}\ \nu < \frac{1}{4}$	$-i^{\pi-1}y^{-3/2}W_{\mu,\nu}(ay) \cdot$ $\cdot [W_{\mu,\nu}(aye^{i\pi}) - W_{\mu,\nu}(aye^{-i\pi})]$
2.85	$x^{\frac{1}{2}}p^\mu_{\nu-\frac{1}{2}}[(1+x^2a^{-2})^{\frac{1}{2}}] \cdot$ $\cdot p^{-\mu}_{\nu-\frac{1}{2}}[(1+x^2a^{-2})^{\frac{1}{2}}]$ $-\frac{1}{4}<\mathrm{Re}\ \nu < \frac{1}{4}$	$2\pi^{-1}y^{-3/2}\cos(\pi\nu)W_{\mu,\nu}(ay)W_{-\mu,\nu}(ay)$
2.86	$x^{\frac{1}{2}}p^\mu_{\nu-\frac{1}{2}}[(1+x^2a^{-2})^{\frac{1}{2}}] \cdot$ $\cdot q^\mu_{\nu-\frac{1}{2}}[(1+x^2a^{-2})^{\frac{1}{2}}]$ $\mathrm{Re}\ \nu>-\frac{1}{4},\ \mathrm{Re}\ \mu<1$	$e^{i\pi\mu}\frac{\Gamma(\frac{1}{2}+\nu-\mu)}{\Gamma(1+2\nu)}\,y^{-3/2} \cdot$ $\cdot W_{\mu,\nu}(ay)\,M_{-\mu,\nu}(ay)$

	$f(x)$	$g(y) = \int\limits_0^\infty f(x)(xy)^{\frac{1}{2}}J_0(xy)\,dx$
2.87	$J_0(ax)$	$2^{3/2}\pi^{-1}\dfrac{\Gamma(3/4)}{\Gamma(\frac{1}{4})}\left(\dfrac{y}{a}\right)^{\frac{1}{2}}(a^2-y^2)^{-\frac{1}{2}}\ \cdot$ $\cdot\ \mathbf{K}\{[\frac{1}{2}-\frac{1}{2}(1-y^2a^{-2})^{\frac{1}{2}}]^{\frac{1}{2}}\}\qquad y<a$ $2^{3/2}\pi^{-1}\dfrac{\Gamma(3/4)}{\Gamma(\frac{1}{4})}(y^2-a^2)^{-\frac{1}{2}}\ \cdot$ $\cdot\ \mathbf{K}\{[\frac{1}{2}-\frac{1}{2}(1-a^2y^{-2})^{\frac{1}{2}}]^{\frac{1}{2}}\}\qquad y>a$
2.88	$x^{-\frac{1}{2}}J_0(ax)$	$2(\pi a)^{-1}y^{\frac{1}{2}}\,\mathbf{K}\!\left(\dfrac{y}{a}\right)\qquad\qquad\quad y<a$ $2\pi^{-1}y^{-\frac{1}{2}}\,\mathbf{K}\!\left(\dfrac{a}{y}\right)\qquad\qquad\quad y>a$
2.89	$x^{-1}J_0(ax)$	$2^{\frac{1}{2}}\pi^{-1}\dfrac{\Gamma(\frac{1}{4})}{\Gamma(3/4)}(1-y^2a^{-2})^{\frac{1}{2}}$ $\cdot\ \mathbf{K}\{[\frac{1}{2}-\frac{1}{2}(1-y^2a^{-2})^{\frac{1}{2}}]^{\frac{1}{2}}\}\qquad y<a$ $2^{\frac{1}{2}}\pi^{-1}\dfrac{\Gamma(\frac{1}{4})}{\Gamma(3/4)}\mathbf{K}\{[\frac{1}{2}-\frac{1}{2}(1-a^2y^{-2})^{\frac{1}{2}}]^{\frac{1}{2}}\}$ $\qquad\qquad\qquad\qquad\qquad y>a$
2.90	$x^{-3/2}[1-J_0(ax)]$	$0\qquad\qquad y<a$ $y^{\frac{1}{2}}\log\left(\dfrac{a}{y}\right)\qquad y>a$
2.91	$x^{-\frac{1}{2}}e^{-bx}J_0(ax)$	$2\pi^{-1}y^{\frac{1}{2}}[(a+y)^2+b^2]^{-\frac{1}{2}}\ \cdot$ $\cdot\ \mathbf{K}\{2(ay)^{\frac{1}{2}}[(a+y)^2+b^2]^{-\frac{1}{2}}\}$

	$f(x)$	$g(y) = \int_0^\infty f(x)(xy)^{\frac{1}{2}}J_0(xy)dx$
2.92	$x^{-3/2}J_1(ax)$	$2\pi^{-1}y^{\frac{1}{2}}\mathbf{E}(\frac{y}{a})$ $\qquad$ $y < a$ $2(\pi a)^{-1}y^{3/2}[\mathbf{K}(\frac{a}{y}) - (1-a^2y^{-2})\mathbf{E}(\frac{a}{y})]$ $\quad$ $y > a$
2.93	$x^{-\frac{1}{2}}[J_0(\frac{1}{2}ax)]^2$	$4\pi^{-2}y^{-\frac{1}{2}}\mathbf{K}^2\{[\frac{1}{2}-\frac{1}{2}(1-a^2y^{-2})^{\frac{1}{2}}]^{\frac{1}{2}}\}$ $\quad$ $y > a$
2.94	$x^{\frac{1}{2}}J_0(ax)J_\nu(ax)$ $\mathrm{Re}\ \nu > -1$	$2\pi^{-1}y^{-\frac{1}{2}}(4a^2-y^2)^{-\frac{1}{2}}\cos[\nu\arccos(\frac{y}{2a})]$ $\qquad\qquad\qquad\qquad\qquad y < 2a$ $0 \qquad\qquad\qquad\qquad y > 2a$
2.95	$x^{\frac{1}{2}}J_0(ax)Y_0(ax)$	$0 \qquad\qquad\qquad y < 2a$ $-2\pi^{-1}y^{-\frac{1}{2}}(y^2-4a^2)^{-\frac{1}{2}} \qquad y > 2a$
2.96	$x^{\frac{1}{2}}J_0(ax)Y_0(bx)$ $a > b$	$-(\pi ab)^{-1}y^{\frac{1}{2}}(z_1^2-1)^{-\frac{1}{2}} \qquad y < a-b$ $0 \qquad\qquad a-b < y < a+b$ $-(\pi ab)^{-1}y^{\frac{1}{2}}(z_2^2-1)^{-\frac{1}{2}} \qquad y > a+b$ $z_{\frac{1}{2}} = \pm\frac{a^2+b^2-y^2}{2ab}$
2.97	$x^{\frac{1}{2}}J_0(bx)Y_0(ax)$ $a > b$	$(\pi ab)^{-1}y^{\frac{1}{2}}(z_1^2-1)^{-\frac{1}{2}} \qquad y < a-b$ $0 \qquad\qquad a-b < y < a+b$ $-(\pi ab)^{-1}y^{\frac{1}{2}}(z_2^2-1)^{-\frac{1}{2}} \qquad y > a+b$ $z_{\frac{1}{2}} = \pm\frac{a^2+b^2-y^2}{2ab}$

	$f(x)$	$g(y) = \int\limits_0^\infty f(x)(xy)^{\frac{1}{2}} J_0(xy)\,dx$
2.98	$x^{\frac{1}{2}}[Y_0(ax)]^2$	$4\pi^{-2} y^{-\frac{1}{2}} (4a^2-y^2)^{-\frac{1}{2}} \arcsin(1-\dfrac{y^2}{2a^2}\cdot)$ $\hspace{6cm} y < 2a$ $4\pi^{-2} y^{-\frac{1}{2}} (y^2-4a^2)^{-\frac{1}{2}}\cdot$ $\cdot\ \log[\dfrac{y^2}{2a^2}-1+\dfrac{y^2}{2a^2}(1-4\dfrac{a^2}{y^2})^{\frac{1}{2}}]\ \ y > 2a$
2.99	$x^{-1}[J_1(ax)]^2$	$\pi^{-1}\{\arccos(\dfrac{y}{2a})-\dfrac{y}{2a}(1-\dfrac{y^2}{4a^2})^{\frac{1}{2}}]$ $\hspace{5cm} y < 2a$ $0 \hspace{3.5cm} y > 2a$
2.100	$x^{\frac{1}{2}} J_\mu(ax) J_{-\mu}(ax)$	$2\pi^{-1} y^{-\frac{1}{2}} (4a^2-y^2)^{-\frac{1}{2}} \cos[\mu\arccos(\dfrac{y^2}{2a^2}-1)]$ $\hspace{5.5cm} y < 2a$ $0 \hspace{4cm} y > 2a$
2.101	$x^{\frac{1}{2}}[J_\mu(ax)]^2$ $\mathrm{Re}\ \mu > -1$	$2\pi^{-1} y^{-\frac{1}{2}} (4a^2-y^2)^{-\frac{1}{2}}\cdot$ $\cdot\ \cos[\mu\arccos(1-\tfrac{1}{2}y^2 a^{-2})] \hspace{1cm} y < 2a$ $-2\pi^{-1} y^{-\frac{1}{2}} \sin(\pi\mu)(y^2-4a^2)^{-\frac{1}{2}}\cdot$ $\cdot\ [\dfrac{y^2}{2a^2}-1+\dfrac{y^2}{2a^2}(1-4\dfrac{a^2}{y^2})^{\frac{1}{2}}]^{-\mu}\ \ y > 2a$

	$f(x)$	$g(y) = \int\limits_0^\infty f(x)(xy)^{\frac{1}{2}}J_0(xy)\,dx$
2.102	$x^{\frac{1}{2}}J_\mu(ax)Y_\mu(ax)$ $\mathrm{Re}\ \mu > -1$	$-2\pi^{-1}y^{-\frac{1}{2}}(4a^2-y^2)^{-\frac{1}{2}}\sin[\mu\arccos(1-\dfrac{y^2}{2a^2})]$ $\hspace{6cm} y < 2a$ $-2\pi^{-1}y^{-\frac{1}{2}}\cos(\pi\mu)(y^2-4a^2)^{-\frac{1}{2}}\ \cdot$ $\cdot\ [\dfrac{y^2}{2a^2} - 1 + \dfrac{y^2}{2a^2}(1-\dfrac{4a^2}{y^2})^{\frac{1}{2}}]^{-\mu}\quad y > 2a$
2.103	$x^{\frac{1}{2}}J_\mu(ax)J_\mu(bx)$ $\mathrm{Re}\ \mu > -1$	$0 \hspace{5cm} y < |a-b|$ $(\pi ab)^{-1}y^{\frac{1}{2}}(1-z_1^2)^{-\frac{1}{2}}\cos(\mu\arccos z_1)$ $\hspace{5cm} |a - b| < y < a+b$ $-(\pi ab)^{-1}y^{\frac{1}{2}}(z_2^2-1)^{-\frac{1}{2}}\sin(\pi\mu)\ \cdot$ $\cdot\ [z_2+(z_2^2-1)^{\frac{1}{2}}]^{-\mu} \hspace{2.5cm} y > a+b$ $z_{\substack{1\\2}} = \pm\ \dfrac{a^2+b^2-y^2}{2ab}$
2.104	$x^{\frac{1}{2}}\{[J_\mu(ax)]^2+[Y_\mu(ax)]^2\}$ $-1 < \mathrm{Re}\ \mu < 1$	$2\pi^{-1}\csc(\pi\mu)\,y^{-\frac{1}{2}}(y^2-4a^2)^{-\frac{1}{2}}\ \cdot$ $\cdot\ \{[\dfrac{y^2}{2a^2} -1+ \dfrac{y^2}{2a^2}(1-\dfrac{4a^2}{y^2})^{\frac{1}{2}}]^{\mu}\ -$ $-\ [\dfrac{y^2}{2a^2} - 1+\dfrac{y^2}{2a^2}(1-\dfrac{4a^2}{y^2})^{\frac{1}{2}}]^{-\mu}\}$

	$f(x)$	$g(y) = \int_0^\infty f(x)(xy)^{\frac{1}{2}} J_0(xy)\,dx$
2.105	$x^{\frac{1}{2}} J_\mu(bx) Y_\mu(ax)$ $\mathrm{Re}\ \mu > -1$ $a > b$	$(\pi ab)^{-1} y^{\frac{1}{2}} (z_1^2-1)^{-\frac{1}{2}} [z_1+(z_1^2-1)^{\frac{1}{2}}]^{-\mu}$ $\hspace{4cm} y < a-b$ $- (\pi ab)^{-1} y^{\frac{1}{2}} (1-z_1^2) \sin(\mu\arccos z_1)$ $\hspace{3cm} a-b < y < a+b$ $- (\pi ab)^{-1} y^{\frac{1}{2}} \cos(\pi\mu) (z_2^2-1)^{-\frac{1}{2}} [z_2+(z_2^2-1)^{\frac{1}{2}}]^{-\mu}$ $\hspace{4cm} y > a+b$ $z_{\underset{2}{1}} = \pm\dfrac{a^2+b^2-y^2}{2ab}$
2.106	$x^{\frac{1}{2}} J_\mu(ax) Y_\mu(bx)$ $\mathrm{Re}\ \mu > -1$ $a > b$	$- (\pi ab)^{-1} y^{\frac{1}{2}} (z_1^2-1)^{-\frac{1}{2}} [z_1+(z_1^2-1)^{\frac{1}{2}}]^{\mu}$ $\hspace{4cm} y < a-b$ For $z_{\underset{2}{1}}$; $a-k_y < a+b$; $\hspace{1.5cm} y > a+b$ the same values as before
2.107	$K_0(ax)$	$2\pi^{-1}\Gamma^2(\tfrac{3}{4}) y^{\frac{1}{2}} (a^2+y^2)^{-\frac{1}{2}} [a+(a^2+y^2)^{\frac{1}{2}}]^{-\frac{1}{2}} \cdot$ $\cdot\ \mathbf{K}\{[\dfrac{(a^2+y^2)^{\frac{1}{2}}-a}{(a^2+y^2)^{\frac{1}{2}}+a}]^{\frac{1}{2}}\}$ $= 2^{\frac{1}{2}}\pi^{-1}\Gamma^2(\tfrac{3}{4}) y^{\frac{1}{2}} (a^2+y^2)^{-\frac{1}{2}} [y+(a^2+y^2)^{\frac{1}{2}}]^{-\frac{1}{2}} \cdot$ $\cdot\ \mathbf{K}\{(2y)^{\frac{1}{2}} [(a^2+y^2)^{\frac{1}{2}}+y]^{-\frac{1}{2}}\}$

	$f(x)$	$g(y) = \int\limits_0^\infty f(x)(xy)^{\frac{1}{2}}J_0(xy)dx$
2.108	$x^{-1}K_0(ax)$	$\pi^{-1}\Gamma^2\left(\tfrac{1}{4}\right)y^{\frac{1}{2}}[a+(a^2+y^2)^{\frac{1}{2}}]^{-\frac{1}{2}}\cdot$ $\cdot \mathbf{K}\left\{\left[\dfrac{(a^2+y^2)^{\frac{1}{2}}-a}{(a^2+y^2)^{\frac{1}{2}}+a}\right]^{\frac{1}{2}}\right\}$ $= 2^{-\frac{1}{2}}\pi^{-1}\Gamma^2\left(\tfrac{1}{4}\right)y^{\frac{1}{2}}[y+(a^2+y^2)^{\frac{1}{2}}]^{-\frac{1}{2}}\cdot$ $\cdot \mathbf{K}\left\{(2y)^{\frac{1}{2}}[(a^2+y^2)^{\frac{1}{2}}+y]^{-\frac{1}{2}}\right\}$
2.109	$x^{-\frac{1}{2}}K_0(ax)$	$y^{\frac{1}{2}}(a^2+y^2)^{-\frac{1}{2}}\mathbf{K}[y(a^2+y^2)^{-\frac{1}{2}}]$
2.110	$x^{\frac{1}{2}}K_0(ax)$	$y^{\frac{1}{2}}(a^2+y^2)^{-1}$
2.111	$x^{\frac{1}{2}}J_0(ax)K_0(bx)$	$y^{\frac{1}{2}}(a^4+b^4+y^4-2a^2y^2+2a^2b^2+2b^2y^2)^{-\frac{1}{2}}$
2.112	$x^{3/2}J_1(ax(K_0(bx)$	$2ay^{\frac{1}{2}}(a^2+b^2-y^2)[a^2+b^2+y^2)^2-4a^2y^2]^{-3/2}$
2.113	$x^{\frac{1}{2}}I_0(ax)K_0(bx)$ $b>a$	$y^{\frac{1}{2}}(a^4+b^4+y^4-2a^2b^2+2a^2y^2+2b^2y^2)^{-\frac{1}{2}}$
2.114	$x^{3/2}I_0(ax)K_1(bx)$ $b>a$	$2y^{\frac{1}{2}}b(b^2+y^2-a^2([(a^2+b^2+y^2)^2-4a^2b^2]^{-3/2}$

	$f(x)$	$g(y) = \int\limits_0^\infty f(x)(xy)^{\frac{1}{2}} J_0(xy)\,dx$
2.115	$x^{\frac{1}{2}} I_1(ax) K_1(bx)$ $b > a$	$(2ab)^{-1} y^{\frac{1}{2}} \{(a^2+b^2+y^2) \cdot$ $\cdot \,[(a^2+b^2+y^2)^2 - 4a^2b^2]^{-\frac{1}{2}} - 1\}$
2.116	$x^{-\frac{1}{2}} I_0(ax) K_0(ax)$	$\pi^{-1} a^{-2} y^{\frac{1}{2}} [(4a^2+y^2)^{\frac{1}{2}} - y] \mathbf{K}\{2a[(4a^2+y^2)^{\frac{1}{2}}+y]^{-\frac{1}{2}}\} \cdot$ $\cdot\, \mathbf{K}\{(2y)^{\frac{1}{2}}[(4a^2+y^2)^{\frac{1}{2}}+y]^{-\frac{1}{2}}\} =$ $= 4\pi^{-1} y^{-3/2} [(4a^2+y^2)^{\frac{1}{2}} - 2a]\mathbf{K}\{y[(4a^2+y^2)^{\frac{1}{2}}+2a]^{-\frac{1}{2}}\}$ $\cdot\, \mathbf{K}\{2a^{\frac{1}{2}}[(4a^2+y^2)^{\frac{1}{2}}+2a]^{-\frac{1}{2}}\}$
2.117	$x^{-\frac{1}{2}}[K_0(\tfrac{1}{2}ax)]^2$	$2y^{\frac{1}{2}} a^{-2}[(a^2+y^2)^{\frac{1}{2}} - y]\mathbf{K}^2\{(2y)^{\frac{1}{2}}[(a^2+y^2)^{\frac{1}{2}}+y]^{-\frac{1}{2}}\}$ $= 4y^{-3/2}[(a^2+y^2)^{\frac{1}{2}} - a]\mathbf{K}^2\{y[(a^2+y^2)^{\frac{1}{2}}+a]^{-\frac{1}{2}}\}$
2.118	$x^{\frac{1}{2}} Y_0(ax) K_0(ax)$	$-2\pi^{-1} y^{\frac{1}{2}}(y^4+4a^4)^{-\frac{1}{2}} \log\left[\dfrac{y^2}{2a^2} + (1+\dfrac{y^2}{4a^4})^{\frac{1}{2}}\right]$
2.119	$x^{\frac{1}{2}} K_0(ax) K_0(bx)$	$\frac{1}{2}(ab)^{-1} y^{\frac{1}{2}}(z^2-1)^{-\frac{1}{2}} \log[z+(z^2-1)^{\frac{1}{2}}]$ $z = (2ab)^{-1}(a^2+b^2+y^2)$
2.120	$x^{\frac{1}{2}} K_\mu(ax) I_\mu(bx)$ $a > b,\ \mathrm{Re}\ \mu > -1$	$\frac{1}{2}(ab)^{-1} y^{\frac{1}{2}}(z^2-1)^{-\frac{1}{2}}[z+(z^2-1)^{\frac{1}{2}}]^{-\mu}$ $z = (2ab)^{-1}(a^2+b^2+y^2)$
2.121	$x^{\frac{1}{2}} K_\mu(ax) K_\mu(bx)$ $-1\ \ \mathrm{Re}\ \ 1$	$\frac{1}{4}\pi(ab)^{-1} \csc(\pi\mu) y^{\frac{1}{2}}(z^2-1)^{-\frac{1}{2}} \cdot$ $\cdot\, \{[z+(z^2-1)^{\frac{1}{2}}]^{\mu} - [z+(z^2-1)^{\frac{1}{2}}]^{-\mu}\}$

	$f(x)$	$g(y) = \int\limits_{0}^{\infty} f(x)(xy)^{\frac{1}{2}} J_0(xy)\,dx$
2.122	$x^{\frac{1}{2}}\exp(a^2x^2) \cdot K_0(a^2x^2)$	$\frac{1}{2}\pi^{-3/2}a^{-1}y^{-\frac{1}{2}}\exp(\frac{1}{8}y^2a^{-2})\,\mathrm{Ertc}(2^{-3/2}a^{-1}y)$
2.123	$x^{-\frac{1}{2}}I_{-\frac{1}{4}}(ax^{-1})K_{\frac{1}{4}}(ax^{-1})$	$a^{-\frac{1}{2}}y^{-1}e^{-2(ay)^{\frac{1}{2}}}\cos[2(ay)^{\frac{1}{2}}]$
2.124	$x^{-\frac{1}{2}}I_{\frac{1}{4}}(ax^{-1}(K_{\frac{1}{4}}(ax^{-1})$	$a^{-\frac{1}{2}}y^{-1}e^{-2(ay)^{\frac{1}{2}}}\sin[2(ay)^{\frac{1}{2}}]$
2.125	$x^{-\frac{1}{2}}[K_{\frac{1}{4}}(ax^{-1})]^2$	$\pi(2a)^{-\frac{1}{2}}y^{-1}e^{-2(ay)^{\frac{1}{2}}} \cdot \{\cos[2(ay)^{\frac{1}{2}}] - \sin[2(ay)^{\frac{1}{2}}]\}$
2.126	$x^{\frac{1}{2}}e^{-ax^2}I_0(ax^2)$	$(2\pi ay)^{-\frac{1}{2}}\exp(-\frac{y^2}{8a})$
2.127	$x^{\frac{1}{2}}[J_0(ax^2)]^2$	$-\frac{1}{4}a^{-1}y^{\frac{1}{2}}J_0(\frac{y^2}{16a})\,Y_0(\frac{y^2}{16a})$
2.128	$x^{\frac{1}{2}}J_0(ax^2)Y_0(ax^2)$	$-\frac{1}{4}a^{-1}y^{\frac{1}{2}}[J_0(\frac{y^2}{16a})]^2$
2.129	$x^{\frac{1}{2}}I_0(ax^2)K_0(ax^2)$	$\frac{1}{4}a^{-1}y^{\frac{1}{2}}K_0(\frac{y^2}{16a})\,I_0(\frac{y^2}{16a})$
2.130	$x^{\frac{1}{2}}J_0(ax^{\frac{1}{2}})K_0(ax^{\frac{1}{2}})$	$\frac{1}{2}y^{-3/2}e^{-\frac{1}{2}a^2y^{-1}}$

	$f(x)$	$g(y) = \int\limits_0^\infty f(x)(xy)^{\frac{1}{2}} J_0(xy)\,dx$
2.131	$x^{-\frac{1}{2}} J_\mu(a^2x^{-1}) J_{-\mu}(a^2x^{-1})$ $-\frac{1}{4} < \mathrm{Re}\ \mu < \frac{1}{4}$	$-i\csc(2\pi\mu) y^{-\frac{1}{2}} [e^{2\pi i\mu} J_{2\mu}(2ay^{\frac{1}{2}} e^{-i\frac{\pi}{4}}) \cdot$ $\cdot J_{-2\mu}(2ay^{\frac{1}{2}} e^{i\frac{\pi}{4}}) - e^{-2\pi i\mu} J_{2\mu}(2ay^{\frac{1}{2}} e^{-i\frac{\pi}{4}}) \cdot$ $\cdot J_{-2\mu}(2ay^{\frac{1}{2}} e^{-i\frac{\pi}{4}})]$
2.132	$x^{-\frac{1}{2}}[J_\mu^2(a^2x^{-1}) - J_{-\mu}^2(a^2x^{-1})]$ $-\frac{1}{4} < \mathrm{Re}\ \mu < \frac{1}{4}$	$\sec(\pi\mu) y^{-\frac{1}{2}}[J_{2\mu}(2ay^{\frac{1}{2}} e^{i\frac{\pi}{4}}) J_{2\mu}(2ay^{\frac{1}{2}} e^{i\frac{\pi}{4}}) -$ $-J_{-2\mu}(2ay^{\frac{1}{2}} e^{i\frac{\pi}{4}}) J_{-2\mu}(2ay^{\frac{1}{2}} e^{-i\frac{\pi}{4}})]$
2.133	$x^{-\frac{1}{2}} H_\mu^{(1)}(a^2x^{-1}) H_\mu^{(2)}(a^2x^{-1})$ $-\frac{1}{4} < \mathrm{Re}\ \mu < \frac{1}{4}$	$16\pi^{-2}\cos(\pi\mu) \cdot$ $\cdot y^{-\frac{1}{2}} K_{2\mu}(2ay^{\frac{1}{2}} e^{i\frac{\pi}{4}}) K_{2\mu}(2ay^{\frac{1}{2}} e^{-i\frac{\pi}{4}})$
2.134	$x^{-\frac{1}{2}} I_\mu(a^2x^{-1}) K_\mu(a^2x^{-1})$ $\mathrm{Re}\ \mu > -\frac{1}{4}$	$2y^{-\frac{1}{2}} J_{2\mu}(2ay^{\frac{1}{2}}) K_{2\mu}(2ay^{\frac{1}{2}})$
2.135	$x^{-\frac{1}{2}} K_\mu^2(a^2x^{-1})$ $-\frac{1}{4} < \mathrm{Re}\ \mu < \frac{1}{4}$	$-2\pi y^{-\frac{1}{2}} K_{2\mu}(2ay^{\frac{1}{2}}) \cdot$ $\cdot [\sin(\pi\mu) J_{2\mu}(2ay^{\frac{1}{2}}) + \cos(\pi\mu) Y_{2\mu}(2ay^{\frac{1}{2}})]$
2.136	$x^{-\frac{1}{2}} Y_0(ax^{\frac{1}{2}}) K_0(ax^{\frac{1}{2}})$	$-\frac{1}{2}\pi^{-1} y^{-\frac{1}{2}} K_0^2(\frac{1}{4}a^2y^{-1})$

	$f(x)$	$g(y) = \int\limits_0^\infty f(x)(xy)^{\frac{1}{2}}J_0(xy)dx$
2.137	$x^{-\frac{1}{2}}J_\mu(ax^{\frac{1}{2}})K_\mu(ax^{\frac{1}{2}})$ $\text{Re }\mu > -1$	$\frac{1}{2}y^{-\frac{1}{2}}I_{\frac{1}{2}\mu}(\frac{1}{4}a^2y^{-1})K_{\frac{1}{2}\mu}(\frac{1}{4}a^2y^{-1})$
2.138	$x^{-\frac{1}{2}}Y_\mu(ax^{\frac{1}{2}})K_\mu(ax^{\frac{1}{2}})$ $-1 < \text{Re }\mu < 1$	$-\frac{1}{2}y^{-\frac{1}{2}}\sec(\frac{1}{2}\pi\mu)K_{\frac{1}{2}\mu}(\frac{1}{4}a^2y^{-1})\ \cdot$ $\cdot\,[\pi^{-1}K_{\frac{1}{2}\mu}(\frac{1}{4}a^2y^{-1})+\sin(\frac{1}{2}\pi\mu)I_{\frac{1}{2}\mu}(\frac{1}{4}a^2y^{-1})]$
2.139	$x^{-\frac{1}{2}}K_\mu(ae^{i\frac{\pi}{4}}x^{\frac{1}{2}})\ \cdot$ $\cdot\ K_\mu(ae^{-i\frac{\pi}{4}}x^{\frac{1}{2}})$ $-1 < \text{Re }\mu < 1$	$\frac{1}{16}\pi^2\sec(\frac{1}{2}\pi\mu)\ H^{(1)}_{\frac{1}{2}\mu}(\frac{1}{4}a^2y^{-1})H^{(2)}_{\frac{1}{2}\mu}(\frac{1}{4}a^2y^{-1})$
2.140	$x^{-\frac{1}{2}}[\mathbf{H}_0(ax)-Y_0(ax)]$	$4\pi^{-1}y^{\frac{1}{2}}(a+y)^{-1}K[|a-y|(a+y)^{-1}]$
2.141	$x^{-\frac{1}{2}}[I_0(ax)-\mathbf{L}_0(ax)]$	$2\pi^{-1}y^{\frac{1}{2}}(a^2+y^2)^{-\frac{1}{2}}K[a(a^2+y^2)^{-\frac{1}{2}}]$
2.142	$x^{-1}D_\nu(a^{\frac{1}{2}}x^{\frac{1}{2}})D_{-\nu-1}(a^{\frac{1}{2}}x^{\frac{1}{2}})$	$2^{-3/2}\pi(\frac{y}{a})^{\frac{1}{2}}P^{\frac{1}{2}\nu+\frac{1}{4}}_{-\frac{1}{4}}[(1+4y^2a^{-2})^{\frac{1}{2}}]\ \cdot$ $\cdot\ P^{-\frac{1}{2}\nu-\frac{1}{4}}_{-\frac{1}{4}}[(1+4y^2a^{-2})^{\frac{1}{2}}]$ $= i2^{-3/2}(\frac{y}{a})^{-\frac{1}{2}}[\Gamma(-\frac{1}{2}\nu)\Gamma(\frac{1}{2}+\frac{1}{2}\nu)]^{-1}\ \cdot$ $\cdot\ q^{-\frac{1}{4}}_{-\frac{1}{2}\nu-3/4}[(1+\frac{1}{4}a^2y^{-2})^{\frac{1}{2}}]\ \cdot$ $\cdot\ q^{-\frac{1}{4}}_{\frac{1}{2}\nu-\frac{1}{4}}[(1+\frac{1}{4}a^2y^{-2})^{\frac{1}{2}}]$

	$f(x)$	$g(y) = \int\limits_{0}^{\infty} f(x)(xy)^{\frac{1}{2}}J_0(xy)dx$
2.143	$x^{-\frac{1}{2}}D_\nu(x)D_{\nu-1}(x)$	$\frac{1}{2}y^{-\frac{1}{2}}[D_\nu(y)D_{\nu-1}(-y)-D_\nu(-y)D_{\nu-1}(y)]$
2.144	$x^{-\frac{1}{2}}D_\nu(x)D_{\nu-1}(-x)$	$y^{-\frac{1}{2}}[D_\nu(y)D_{\nu-1}(y)+\frac{1}{2}D_\nu(y)D_{\nu-1}(-y) +$ $+ \frac{1}{2} D_\nu(-y)D_{\nu-1}(y)]$
2.145	$x^{-\frac{1}{2}}D_\nu(-x)D_{\nu-1}(x)$	$y^{\frac{1}{2}}[\frac{1}{2}D_\nu(y)D_{\nu-1}(-y)+\frac{1}{2}D_\nu(-y)D_{\nu-1}(y) -$ $- D_\nu(y)D_{\nu-1}(y)]$
2.146	$x^{-3/2}W_{k,\mu}(ax)M_{-k,\mu}(ax)$ $\mathrm{Re}\ \mu > -\frac{1}{2}, \mathrm{Re}\ k < \,^{3}/_{4}$	$e^{-ik\mu}\dfrac{\Gamma(1+2\mu)}{\Gamma(\frac{1}{2}+\mu+k)}\ y^{\frac{1}{2}} \cdot$ $\cdot p_{\mu-\frac{1}{2}}^{k}[(1+y^2a^{-2})^{\frac{1}{2}}]q_{\mu-\frac{1}{2}}^{k}[(1+y^2a^{-2})^{\frac{1}{2}}]$ $= e^{-i\pi\mu}\dfrac{\Gamma(1+2\mu)}{\Gamma(\frac{1}{2}-k+\mu)}\ ay^{-\frac{1}{2}} \cdot$ $\cdot p_{k-\frac{1}{2}}^{-\mu}[(1+a^2y^{-2})^{\frac{1}{2}}]q_{-k-\frac{1}{2}}^{\mu}[(1+a^2y^{-2})^{\frac{1}{2}}]$
2.147	$x^{-3/2}W_{k,\mu}(ax)W_{-k,\mu}(ax)$ $-\frac{1}{2} < \mathrm{Re}\ \mu < \frac{1}{2}$	$\frac{1}{2}\pi\cos(\pi\mu)y^{\frac{1}{2}} \cdot$ $\cdot p_{\mu-\frac{1}{2}}^{k}[(1+y^2a^{-2})^{\frac{1}{2}}]p_{\mu-\frac{1}{2}}^{-k}[(1+y^2a^{-2})^{\frac{1}{2}}]$ $= e^{i2\pi\mu}\cos(\pi\mu)[\Gamma(\frac{1}{2}-\mu+k)\Gamma(\frac{1}{2}-\mu-k)]^{-1}$ $\cdot ay^{-\frac{1}{2}}q_{-k-\frac{1}{2}}^{-\mu}[(1+a^2y^{-2})^{\frac{1}{2}}]q_{k-\frac{1}{2}}^{-\mu}[(1+a^2y^{-2})^{\frac{1}{2}}]$

1.3 Transforms of Order Unity

	$f(x)$	$g(y) = \int\limits_0^\infty f(x)(xy)^{\frac{1}{2}} J_1(xy)\,dx$
3.1	$x^{-\frac{1}{2}}$ $x < a$ 0 $x > a$	$y^{-\frac{1}{2}}[1-J_0(ay)]$
3.2	0 $x < a$ $x^{-\frac{1}{2}}$ $x > a$	$y^{-\frac{1}{2}}J_0(ay)$
3.3	$x^{-\frac{1}{2}}(a^2+x^2)^{-\frac{1}{2}}$	$a^{-1}y^{-\frac{1}{2}}(1-e^{-ay})$
3.4	$x^{-\frac{1}{2}}(a^2-x^2)^{-\frac{1}{2}}x < a$ 0 $x > a$	$a^{-1}y^{-\frac{1}{2}}[1-\cos(ay)]$
3.5	0 $x < a$ $x^{-\frac{1}{2}}(x^2-a^2)^{-\frac{1}{2}}x > a$	$a^{-1}y^{-\frac{1}{2}}\sin(ay)$
3.6	$x^{-\frac{1}{2}}e^{-ax}$	$y^{-\frac{1}{2}}[1-a(a^2+y^2)^{-\frac{1}{2}}]$
3.7	$x^{-\frac{1}{2}}e^{-ax^2}$	$y^{-\frac{1}{2}}[1-\exp(-\frac{y^2}{4a})]$
3.8	$x^{3/2}e^{-ax^2}$	$\tfrac{1}{4}a^{-2}y^{3/2}\exp(-\frac{y^2}{4a})$
3.9	$x^{-\frac{1}{2}}\exp[-a(b^2+x^2)^{\frac{1}{2}}]$	$y^{-\frac{1}{2}}\{e^{-ba}-a(a^2+y^2)^{-\frac{1}{2}}\exp[-b(a^2+y^2)^{\frac{1}{2}}]\}$

	$f(x)$	$g(y) = \int\limits_0^\infty f(x)(xy)^{\frac{1}{2}} J_1(xy)\,dx$
3.10	$x^{-\frac{1}{2}}(b^2+x^2)^{-\frac{1}{2}} \cdot$ $\cdot \exp[-a(b^2+x^2)^{\frac{1}{2}}]$	$y^{-\frac{1}{2}}b^{-1}[e^{-ba} - e^{-b(a^2+y^2)^{\frac{1}{2}}}]$
3.11	$x^{-3/2} e^{-\frac{a}{x}}$	$2y^{\frac{1}{2}} J_1[(2ay)^{\frac{1}{2}}] K_1(2ay)^{\frac{1}{2}}]$
3.12	$x^{-7/2} e^{-\frac{a}{x}}$	$2a^{-1}y^{3/2} J_2[(2ay)^{\frac{1}{2}}] K_2[(2ay)^{\frac{1}{2}}]$
3.13	$x^{-\frac{1}{2}}\log x$	$-y^{-\frac{1}{2}}\log(\tfrac{1}{2}\gamma y)$
3.14	$x^{-\frac{1}{2}}\log(a^2+x^2)$	$2y^{-\frac{1}{2}}[K_0(ay) + \log a]$
3.15	$x^{-\frac{1}{2}}\log[ax+(1+a^2x^2)^{\frac{1}{2}}]$	$y^{-\frac{1}{2}}I_0(\tfrac{y}{2a}) K_0(\tfrac{y}{2a})$
3.16	$x^{-\frac{1}{2}}\log(1+x^4)$	$4y^{-\frac{1}{2}}\ker_0 y$
3.17	$x^{-\frac{1}{2}}\sin(ax^2)$	$y^{-\frac{1}{2}}\sin(\tfrac{y^2}{4a})$
3.18	$x^{-\frac{1}{2}}\cos(ax^2)$	$2y^{-\frac{1}{2}}\sin(\tfrac{y^2}{8a})$
3.19	$x^{-\frac{1}{2}}\sin^2(ax^2)$	$y^{-\frac{1}{2}}\cos(\tfrac{y^2}{8a})$

	$f(x)$	$g(y) = \int\limits_0^\infty f(x)^{\frac{1}{2}} J_1(xy)\,dx$
3.20	$x^{-\frac{1}{2}}(a^2+x^2)^{-\frac{1}{2}} \cdot$ $\cdot \sin[b(a^2+x^2)^{\frac{1}{2}}]$	$a^{-1}y^{-\frac{1}{2}}\{\sin(ab)-\sin[a(b^2-y^2)^{\frac{1}{2}}]\}\,y < b$ $a^{-1}y^{-\frac{1}{2}}\sin(ab) \qquad\qquad y > b$
3.21	$x^{-\frac{1}{2}}(a^2+x^2)^{-\frac{1}{2}} \cdot$ $\cdot \cos[b(a^2+x^2)^{\frac{1}{2}}]$	$a^{-1}y^{-\frac{1}{2}}\{\cos(ab)-\cos[a(b^2-y^2)^{\frac{1}{2}}]\}\,y < b$ $a^{-1}y^{-\frac{1}{2}}\{\cos(ab)-\exp[-a(y^2-b^2)^{\frac{1}{2}}]\}\,y > b$
3.22	$x^{-\frac{1}{2}}(a^2-x^2)^{-\frac{1}{2}}\cos[b(a^2-x^2)^{\frac{1}{2}}]$ $\qquad\qquad x < a$ $0 \qquad\qquad x > a$	$a^{-1}y^{-\frac{1}{2}}\{\cos(ab)-\cos[a(b^2+y^2)^{\frac{1}{2}}]\}$
3.23	$0 \qquad\qquad x < a$ $x^{-\frac{1}{2}}\sin[b(x^2-a^2)^{\frac{1}{2}}]$ $\qquad\qquad x > a$	$0 \qquad\qquad\qquad y < b$ $by^{-\frac{1}{2}}(y^2-b^2)^{-\frac{1}{2}}\cos[a(y^2-b^2)^{\frac{1}{2}}] \quad y > b$
3.24	$0 \qquad\qquad x < a$ $x^{-\frac{1}{2}}(x^2-a^2)^{-\frac{1}{2}} \cdot$ $\cdot \cos[b(x^2-a^2)^{\frac{1}{2}}],\ x > a$	$0 \qquad\qquad\qquad y < b$ $a^{-1}y^{-\frac{1}{2}}\sin[a(y^2-b^2)^{\frac{1}{2}}] \qquad y > b$
3.25	$x^{-\frac{1}{2}}e^{-ax^2}\sin(bx^2)$	$y^{-\frac{1}{2}}\exp[-\tfrac{1}{4}ay^2(a^2+b^2)^{-1}] \cdot$ $\cdot \sin[\tfrac{1}{4}by^2(a^2+b^2)^{-1}]$

	$f(x)$	$g(y) = \int\limits_0^\infty f(x)(xy)^{\frac{1}{2}} J_1(xy)\,dx$
3.26	$x^{-\frac{1}{2}} e^{-ax^2} \sinh(bx^2)$ $a > b$	$y^{-\frac{1}{2}} \exp[-\tfrac{1}{4}ay^2(a^2-b^2)^{-1}] \cdot$ $\cdot \sinh[\tfrac{1}{4}by^2(a^2-b^2)^{-1}]$
3.27	$x^{-\frac{1}{2}} \arctan x^2$	$-2y^{\frac{1}{2}} \mathrm{kei}_0 y$
3.28	$x^{-3/2} P_n(1-2x^2) \quad x < 1$ $0 \qquad x > 1$	$(2n+1)^{-1} y^{-\frac{1}{2}} [(n+1)J_{2n+2}(y) - nJ_{2n}(y)]$
3.29	$x^{-\frac{1}{2}} [D_n(ax)]^2$	$(-1)^{n-1} y^{-\frac{1}{2}} [D_n(\tfrac{y}{a})]^2$
3.30	$x^{-\frac{1}{2}} \mathrm{si}(ax^2)$	$-y^{-\frac{1}{2}} [\tfrac{1}{2}\pi + \mathrm{si}(\tfrac{x^2}{4a})]$
3.31	$x^{-\frac{1}{2}} J_0(ax)$	$\begin{array}{ll} 0 & y < a \\ y^{-\frac{1}{2}} & y > a \end{array}$
3.32	$x^{-3/2} J_0(ax)$	$\begin{array}{ll} 2a\pi^{-1} y^{-\frac{1}{2}} [\mathbf{E}(\tfrac{y}{a}) - (1-y^2 a^{-2})\mathbf{K}(\tfrac{y}{a})] & y < a \\ 2\pi^{-1} y^{\frac{1}{2}} \mathbf{E}(\tfrac{a}{y}) & y > a \end{array}$
3.33	$x^{-5/2} [J_0(ax)-1]$	$\begin{array}{ll} -\tfrac{1}{4} y^{3/2} [1 + 2\log(\tfrac{a}{y})] & y < a \\ -\tfrac{1}{4} y^{\frac{1}{2}} a^2 & y > a \end{array}$

	$f(x)$	$g(y) = \int_0^\infty f(x)(xy)^{\frac{1}{2}}J_1(xy)\,dx$
3.34	$x^{-\frac{1}{2}}Y_0(ax)$	$-\pi^{-1}y^{-\frac{1}{2}}\log(1-y^2a^{-2})$ $\qquad\qquad y < a$
3.35	$x^{\frac{1}{2}}kei_0 x$	$-\tfrac{1}{2}y^{-\frac{1}{2}}\arctan(y^2)$
3.36	$x^{-\frac{1}{2}}ker_0 x$	$\tfrac{1}{4}y^{-\frac{1}{2}}\log(1+y^2)$
3.37	$x^{-\frac{1}{2}}J_0(ax)J_0(bx)$	$0 \qquad\qquad y < |a-b|$ $\pi^{-1}y^{-\frac{1}{2}}\arccos[(2ab)^{-1}(a^2+b^2-y^2)]$ $\qquad\qquad |a-b| < y < a+b$ $y^{-\frac{1}{2}} \qquad\qquad y > a+b$
3.38	$x^{\frac{1}{2}}J_1[(ax)^{\frac{1}{2}}]K_1[(ax)^{\frac{1}{2}}]$	$\tfrac{1}{2}y^{-3/2}e^{-\frac{a}{2y}}$
3.39	$x^{3/2}J_2[(ax)^{\frac{1}{2}}]K_2(ax)^{\frac{1}{2}}]$	$\tfrac{1}{4}ay^{-7/4}e^{-\frac{a}{2y}}$
3.40	$x^{-\frac{1}{2}}I_0(ax)K_0(ax)$	$y^{-\frac{1}{2}}\log[\tfrac{y}{2a} + (1+\tfrac{y^2}{4a^2})^{\frac{1}{2}}]$
3.41	$x^{-\frac{1}{2}}I_\mu(ax)K_\mu(ax)$ $Re\ \mu > -1$	$\tfrac{1}{2}y^{-\frac{1}{2}}\mu^{-1}\{1- [\tfrac{y}{2a} + (1+\tfrac{y^2}{4a^2})^{\frac{1}{2}}]\}$

Transforms of General Order

1.4 Algebraic Functions and Powers with Arbitray Index

	$f(x)$	$g(y) = \int\limits_0^\infty f(x)\,(xy)^{\frac{1}{2}} J_\nu(xy)\,dx$
4.1	$1, \quad 0 < x < 1$ $0, \quad\quad x > 1$ $\mathrm{Re}\ \nu > -\tfrac{3}{2}$	$2^{\frac{1}{2}} \dfrac{\Gamma(\frac{3}{4}+\frac{1}{2}\nu)}{\Gamma(\frac{1}{4}+\frac{1}{2}\nu)}\, y^{-1} + (\nu-\tfrac{1}{2})J_\nu(y)\ \cdot$ $\cdot\ S_{-\frac{1}{2},\nu-1}(y) - J_{\nu-1}(y)S_{\frac{1}{2},\nu}(y)$
4.2	$x^{-\frac{1}{2}}$ $\mathrm{Re}\ \nu > -1$	$y^{-\frac{1}{2}}$
4.3	$x^{\frac{1}{2}-\nu}, \ 0 < x < 1$ $0 \quad, \ x > 1$	$\dfrac{2^{1-\nu} y^{\nu-\frac{3}{2}}}{\Gamma(\nu)} - y^{-\frac{1}{2}} J_{\nu-1}(y)$
4.4	$x^{\nu-\frac{1}{2}}, \ 0 < x < 1$ $0 \quad, \quad x > 1$ $\mathrm{Re}\ \nu > -\tfrac{1}{2}$	$2^{\nu-1}\pi^{\frac{1}{2}}\,\Gamma(\tfrac{1}{2}+\nu)\ \cdot$ $\cdot\ y^{\frac{1}{2}-\nu}\,[J_\nu(y)\mathbf{H}_{\nu-1}(y) - \mathbf{H}_\nu(y)J_{\nu-1}(y)]$
4.5	$x^{\nu+\frac{1}{2}}, \ 0 < x < 1$ $0 \quad, \quad x > 1$ $\mathrm{Re}\ \nu > -1$	$y^{-\frac{1}{2}} J_{\nu+1}(y)$
4.6	x^μ $-\,\mathrm{Re}\ \nu-\tfrac{3}{2} < \mathrm{Re}\ \mu < 0$	$2^{\mu+\frac{1}{2}} y^{-\mu-1}\ \dfrac{\Gamma(\frac{1}{2}\mu+\frac{1}{2}\nu+\frac{3}{4})}{\Gamma(\frac{1}{2}\nu-\frac{1}{2}\mu+\frac{1}{4})}$

	$f(x)$	$g(y) = \int\limits_0^\infty f(x)(xy)^{\frac{1}{2}}J_\nu(xy)\,dx$
4.7	x^μ , $0 < x < 1$ 0 , $x > 1$ $\mathrm{Re}\,(\mu+\nu) > -\,{}^3\!/_2$	$y^{-\mu-1}[(\nu+\mu-\tfrac{1}{2})yJ_\nu(y)S_{\mu-\frac{1}{2},\nu-1}(y) -$ $-yJ_{\nu-1}(y)S_{\mu+\frac{1}{2},\nu}(y) +$ $+ 2^{\mu+\frac{1}{2}}\ \dfrac{\Gamma(\frac{1}{2}\mu+\frac{1}{2}\nu+\frac{3}{4})}{\Gamma(\frac{1}{2}\nu-\frac{1}{2}\mu+\frac{1}{4})}$
4.8	$x^{-\frac{1}{2}}(a+x)^{-1}$ $\mathrm{Re}\ \nu > -1$	$\pi\csc(\pi\nu)y^{\frac{1}{2}}[\mathbf{J}_\nu(ay) - J_\nu(ay)]$
4.9	$x^{\nu-\frac{1}{2}}(a+x)^{-1}$ $-\tfrac{1}{2} < \mathrm{Re}\ \nu < {}^3\!/_2$	$\tfrac{1}{2}\pi a^\nu \sec(\pi\nu)y^{\frac{1}{2}}[\mathbf{H}_\nu(ay) - Y_{-\nu}(ay)]$
4.10	$x^{\mu-\frac{1}{2}}(a+x)^{-1}$ $\mathrm{Re}\,(\mu+\nu) > -1$ $\mathrm{Re}\ \mu < {}^3\!/_2$	$(2a)^\mu y^{\frac{1}{2}}[\dfrac{\Gamma(\frac{1}{2}+\frac{1}{2}\mu+\frac{1}{2}\nu)}{\Gamma(\frac{1}{2}-\frac{1}{2}\mu+\frac{1}{2}\nu)}\ S_{-\mu,\nu}(ay) -$ $-\,2\,\dfrac{\Gamma(1+\frac{1}{2}\mu+\frac{1}{2}\nu)}{\Gamma(\frac{1}{2}\nu-\frac{1}{2}\mu)}\ S_{-\mu-1,\nu}(ay)\ ]$
4.11	$x^{-\frac{1}{2}}(a+x)^{-1}$ $\mathrm{Re}\ \nu > -1$	$\tfrac{1}{2}y^{\frac{1}{2}}a^{-1}\pi\{\sec(\tfrac{1}{2}\pi\nu)I_\nu(ay)+\csc(\pi\nu)[\mathbf{J}_\nu(iay) -$ $-\,\mathbf{J}_{-\nu}(iay)]\}$
4.12	$x^{\frac{1}{2}}(a^2+x^2)^{-1}$ $\mathrm{Re}\ \nu > -2$	$\tfrac{1}{2}\pi\csc(\pi\nu)y^{\frac{1}{2}}[\mathbf{J}_\nu(iay)+\mathbf{J}_\nu(-iay) -$ $-\,2\cos(\tfrac{1}{2}\pi\nu)I_\nu(ay)]$

	$f(x)$	$g(y) = \int\limits_{0}^{\infty} f(x)(xy)^{\frac{1}{2}}J_{\nu}(xy)\,dx$
4.13	$x^{\nu+\frac{1}{2}}(a^2+x^2)^{-1}$ $-1 \quad \mathrm{Re}\ \nu < {}^{3}/_{2}$	$a^{\nu}y^{\frac{1}{2}}K_{\nu}(ay)$
4.14	$x^{\nu-\frac{1}{2}}(a^2+x^2)^{-1}$ $-\frac{1}{2} < \mathrm{Re}\ \nu < {}^{5}/_{2}$	$\frac{1}{2}\pi a^{\nu-1}\ \sec(\pi\nu)y^{\frac{1}{2}}[I_{\nu}(ay)-\mathbf{L}_{-\nu}(ay)]$
4.15	$x^{-\nu-\frac{1}{2}}(a^2+x^2)^{-1}$ $\mathrm{Re}\ \nu >- \ {}^{5}/_{2}$	$\frac{1}{2}\pi a^{-\nu-1}y^{\frac{1}{2}}[I_{\nu}(ay)-\mathbf{L}_{\nu}(ay)]$
4.16	$x^{\mu-{}^{3}/_{2}}(a^2+x^2)^{-1}$ $-\mathrm{Re}\ \nu < \mathrm{Re}\ \mu < {}^{7}/_{2}$	$\frac{1}{2}\pi a^{\mu-2}\ \csc[\frac{1}{2}\pi(\nu+\mu)]y^{\frac{1}{2}}I_{\nu}(ay)\ -$ $-2^{\mu-3}\dfrac{\Gamma(\frac{1}{2}\nu+\frac{1}{2}\mu-1)}{\Gamma(\frac{1}{2}\nu-\frac{1}{2}\mu+2)}\ y^{\frac{1}{2}}a^{\mu-2}(2+\nu-\mu)(2-\nu-\mu)\ \cdot$ $\cdot\ e^{i\frac{\pi}{2}\mu}\ s_{1-\mu,\nu}(iay)$
4.17	$x^{-\frac{1}{2}}(a^2+x^2)^{-\frac{1}{2}}$ $\mathrm{Re}\ \nu > -1$	$y^{\frac{1}{2}}I_{\frac{1}{2}\nu}(\frac{1}{2}ay)\ K_{\frac{1}{2}\nu}(\frac{1}{2}ay)$
4.18	$x^{\nu+\frac{1}{2}}(a^2+x^2)^{-\frac{1}{2}}$ $-1 < \mathrm{Re}\ \nu < \frac{1}{2}$	$(\frac{1}{2}\pi)^{-\frac{1}{2}}\ a^{\nu+\frac{1}{2}}\ K_{\nu+\frac{1}{2}}(ay)$
4.19	$x^{\frac{1}{2}-\nu}(a^2+x^2)^{-\frac{1}{2}}$ $\mathrm{Re}\ \nu > -\frac{1}{2}$	$(\frac{1}{2}\pi)^{\frac{1}{2}}a^{\frac{1}{2}-\nu}[I_{\nu-\frac{1}{2}}(ay)\ -\ \mathbf{L}_{\nu-\frac{1}{2}}(ay)]$

	$f(x)$	$g(y) = \int\limits_0^\infty f(x)(xy)^{\frac{1}{2}}J_\nu(xy)\,dx$
4.20	$x^{-\nu-\frac{1}{2}}(a^2+x^2)^{-\frac{1}{2}-\nu}$ $\text{Re }\nu > -\frac{1}{2}$	$\dfrac{2^{-\nu}\pi^{\frac{1}{2}}a^{-2\nu}}{\Gamma(\frac{1}{2}+\nu)}\; y^{\frac{1}{2}+\nu}\; I_\nu(\tfrac{1}{2}ay)\; K_\nu(\tfrac{1}{2}ay)$
4.21	$x^{\nu+\frac{1}{2}}(a^2+x^2)^{-\nu-\frac{1}{2}}$ $\text{Re }\nu > -\frac{1}{2}$	$\dfrac{2^{-\nu}\pi^{\frac{1}{2}}y^{\nu-\frac{1}{2}}e^{-ay}}{\Gamma(\frac{1}{2}+\nu)}$
4.22	$x^{\nu+\frac{1}{2}}(a^2+x^2)^{-\nu-3/2}$ $\text{Re }\nu > -1$	$\dfrac{2^{-\nu-1}\pi^{\frac{1}{2}}y^{\nu+\frac{1}{2}}e^{-ay}}{a\Gamma(3/2+\nu)}$
4.23	$x^{\nu+\frac{1}{2}}(a^2+x^2)^{-\mu}$ $\text{Re }\nu > -1$ $\text{Re}(2\mu-\nu) > \frac{1}{2}$	$\dfrac{2^{1-\mu}a^{\nu-\mu+1}y^{\mu-\frac{1}{2}}}{\Gamma(\mu)}\; K_{\nu-\mu+1}(ay)$
4.24	$x^{\frac{1}{2}-\nu}(a^2+x^2)^{-\mu}$ $\text{Re}(\nu+2\mu) > \frac{1}{2}$	$a^{-\mu-\nu+1}\,y^{\mu-\frac{1}{2}}\,[2^{-\mu}\dfrac{\Gamma(1-\mu)}{(1-\nu)}\,I_{\nu+\mu-1}(ay)$ $-\dfrac{2^{1-\nu}}{\Gamma(\nu)}\,e^{-i\frac{\pi}{2}(\nu-\mu+1)}\;s_{-\mu+\nu,-\mu-\nu+1}(iay)$
4.25	$x^{\lambda-3/2}(a^2+x^2)^{-\mu-1}$ $-\text{Re}\nu<\text{Re}\lambda<2\text{Re}\mu+7/2$	$2^{-\nu-1}a^{\lambda+\nu-2-2\mu}y^{\frac{1}{2}+\nu}\dfrac{\Gamma(\frac{1}{2}\lambda+\frac{1}{2}\nu)\Gamma(1+\mu-\frac{1}{2}\lambda-\frac{1}{2}\nu)}{\Gamma(1+\mu)\Gamma(1+\nu)}\cdot$ $\cdot\,{}_1F_2(\tfrac{1}{2}\lambda+\tfrac{1}{2}\nu;\tfrac{1}{2}\lambda+\tfrac{1}{2}\nu-\mu,1+\nu;\tfrac{1}{4}a^2y^2) +$ $+\,2^{\lambda-2\mu-3}y^{2\mu-\lambda+52}\dfrac{\Gamma(\frac{1}{2}\lambda+\frac{1}{2}\nu-\mu-1)}{\Gamma(2+\mu+\frac{1}{2}\nu-\frac{1}{2}\lambda)}\cdot$ $\cdot\,{}_1F_2(1+\mu;2+\mu+\tfrac{1}{2}\nu-\tfrac{1}{2}\lambda,2+\mu-\tfrac{1}{2}\nu-\tfrac{1}{2}\lambda;\tfrac{1}{4}a^2y^2)$

	$f(x)$	$g(y) = \int_0^\infty f(x)(xy)^{\frac{1}{2}}J_\nu(xy)\,dx$
4.26	$x^{\nu+\frac{1}{2}}(x-a)^{-1}$ Principal value $-1 < \mathrm{Re}\ \nu < \frac{1}{2}$	$\frac{1}{2}\pi a^{\nu+1}\sec(\pi\nu)y^{\frac{1}{2}}\cdot$ $\cdot[J_\nu(ay)\sin(\pi\nu) - Y_\nu(ay)\cos(\pi\nu) -$ $-\mathbf{H}_{-\nu}(ay)] + 2^\nu\pi^{-\frac{1}{2}}\Gamma(\frac{1}{2}+\nu)y^{-\nu-\frac{1}{2}}$
4.27	$x^{\nu-\frac{1}{2}}(x-a)^{-1}$ Principal value $-\frac{1}{2} < \mathrm{Re}\ \nu < \frac{3}{2}$	$\frac{1}{2}\pi a^\nu \sec(\pi\nu)y^{\frac{1}{2}}\cdot$ $\cdot[J_\nu(ay)\sin(\pi\nu)-Y_\nu(ay)\cos(\pi\nu)-H_{-\nu}(ay)]$
4.28	$x^{-\frac{1}{2}}(x^2-a^2)^{-1}$ Principal value $\mathrm{Re}\ \nu > -1$	$\frac{1}{2}\pi a^{-1}y^{\frac{1}{2}}\{\tan(\frac{1}{2}\pi\nu)[J_\nu(ay)-\mathbf{J}_\nu(ay)]$ $+ \mathbf{E}_\nu(ay)\}$
4.29	$x^{\nu-\frac{1}{2}}(x^2-a^2)^{-1}$ Principal value $-\frac{1}{2} < \mathrm{Re}\ \nu < \frac{5}{2}$	$\frac{1}{2}\pi a^{\nu-1}y^{\frac{1}{2}}[J_\nu(ay)\tan(\pi\nu) -$ $-\mathbf{H}_{-\nu}(ay)\sec(\pi\nu)]$
4.30	$x^{\nu+\frac{1}{2}}(x^2-a^2)^{-1}$ Principal value $-1 < \mathrm{Re}\ \nu < \frac{3}{2}$	$-\frac{1}{2}\pi\ a^\nu y^{\frac{1}{2}}\ Y_\nu(ay)$
4.31	$x^{-\nu-\frac{1}{2}}(x^2-a^2)^{-1}$ Principal value $\mathrm{Re}\ \nu > -\frac{5}{2}$	$-\frac{1}{2}\ \pi\ a^{-\nu-1}\ y^{\frac{1}{2}}\ \mathbf{H}_\nu(ay)$

	$f(x)$	$g(y) = \int\limits_0^\infty f(x)(xy)^{\frac{1}{2}}J_\nu(xy)\,dx$
4.32	$x^{\mu-3/2}(x^2-a^2)^{-1}$ Principal value $\text{Re }\nu < \text{Re }\mu < 7/2$	$-\tfrac{1}{2}\pi a^{\mu-2}\ y^{\frac{1}{2}}Y_\nu(ay) -$ $-2^{\mu-1}a^{\mu-2}\dfrac{\Gamma(\frac{1}{2}\nu+\frac{1}{2}\mu)}{\Gamma(1+\frac{1}{2}\nu-\frac{1}{2}\mu)}\ y^{\frac{1}{2}}S_{1-\mu,\nu}(ay)$
4.33	$x^{-\frac{1}{2}}(a^2-x^2)^{-\frac{1}{2}},\ 0<x<a$ $0\qquad,x>a$ $\text{Re }\nu > -1$	$\tfrac{1}{2}\pi y^{\frac{1}{2}}[J_{\frac{1}{2}\nu}(\tfrac{1}{2}ay)]^2$
4.34	$x^{\frac{1}{2}-\nu}(a^2-x^2)^{-\frac{1}{2}},\ 0<x<a$ $0\qquad,x>a$	$(\tfrac{1}{2}\pi a)^{\frac{1}{2}}\ a^{-\nu}\mathbf{H}_{\nu-\frac{1}{2}}(ay)$
4.35	$x^{\nu+\frac{1}{2}}(a^2-x^2)^{-\nu-\frac{1}{2}},\ 0<x<a$ $0\qquad,x>a$ $-\tfrac{1}{2} < \text{Re }\nu < \tfrac{1}{2}$	$(\pi y)^{-\frac{1}{2}}(\tfrac{1}{2}y)^\nu\ \Gamma(\tfrac{1}{2}-\nu)\sin(ay)$
4.36	$x^{\nu+\frac{1}{2}}(a^2-x^2)^{-\nu-3/2},\ 0<x<a$ $0\qquad,x>a$ $-1 < \text{Re }\nu < -\tfrac{1}{2}$	$\pi^{-\frac{1}{2}}2^{-\nu-1}a^{-1}\Gamma(-\tfrac{1}{2}-\nu)y^{\nu+\frac{1}{2}}\cos(ay)$
4.37	$x^{\frac{1}{2}-\nu}(a^2-x^2)^\mu,\ 0<x<a$ $0\qquad,x>a$ $\text{Re }\mu > -1$	$\dfrac{2^{1-\nu}a^{\mu-\nu+1}}{\Gamma(\nu)}\ y^{-\mu-\frac{1}{2}}\ S_{\nu+\mu,\mu-\nu+1}(ay)$

	$f(x)$	$g(y) = \int\limits_{0}^{\infty} f(x)(xy)^{\frac{1}{2}}J_{\nu}(xy)\,dx$
4.38	$x^{\nu+\frac{1}{2}}(a^2-x^2)^{\mu}, 0<x<a$ $0 \qquad , x > a$ Re $\nu>-1$, Re $\mu>-1$	$2^{\mu}\Gamma(\mu+1)a^{\nu+\mu+1}y^{-\mu-\frac{1}{2}} J_{\nu+\mu+1}(ay)$
4.39	$x^{\mu-\frac{1}{2}}(a^2-x^2)^{\lambda}, 0<x<a$ $0 \qquad , x > a$ Re $\lambda>-1$, Re$(\mu+\nu)>-1$	$\dfrac{2^{-\nu-1}a^{2\lambda+\mu+\nu+1}}{\Gamma(\nu+1)} B(\lambda+1, \tfrac{1}{2}\mu+\tfrac{1}{2}\nu+\tfrac{1}{2})\cdot$ $\cdot y^{\nu+\frac{1}{2}}{}_1F_2(\tfrac{1}{2}+\tfrac{1}{2}\mu+\tfrac{1}{2}\nu; \nu+1, {}^3/_2+{}^3/_2\mu+{}^3/_2\nu+\lambda; -\tfrac{1}{4}a^2y^2)$
4.40	$0 \qquad , 0 < x < a$ $x^{-\frac{1}{2}}(x^2-a^2)^{-\frac{1}{2}}$	$-\tfrac{1}{2}\pi y^{\frac{1}{2}} J_{\frac{1}{2}\nu}(\tfrac{1}{2}ay)\, Y_{\frac{1}{2}\nu}(\tfrac{1}{2}ay)$
4.41	$0 \qquad , 0 < x < a$ $x^{-\nu-\frac{1}{2}}(x^2-a^2)^{-\nu-\frac{1}{2}}, x>a$ $-\tfrac{1}{2} <$ Re $\nu < \tfrac{1}{2}$	$-\pi^{\frac{1}{2}}2^{-\nu-1}a^{-2\nu}\Gamma(\tfrac{1}{2}-\nu)y^{\nu+\frac{1}{2}}J_{\nu}(\tfrac{1}{2}ay)Y_{\nu}(\tfrac{1}{2}ay)$
4.42	$0 \qquad , 0 < x < a$ $x^{\frac{1}{2}-\nu}(x^2-a^2)^{\nu-\frac{1}{2}}, x > a$ $-\tfrac{1}{2} <$ Re $\nu < \tfrac{1}{2}$	$\pi^{-\frac{1}{2}}2^{-\nu}\Gamma(\tfrac{1}{2}+\nu)y^{-\nu-\frac{1}{2}}\cos(ay)$
4.43	$0 \qquad , 0 < x < a$ $x^{\frac{1}{2}-\nu}(x^2-a^2)^{\nu-{}^3/_2}, x>a$ $\tfrac{1}{2} <$ Re $\nu < {}^5/_2$	$\pi^{-\frac{1}{2}}2^{-\nu-1}a^{-1}\,\Gamma(\nu-\tfrac{1}{2})y^{\frac{1}{2}-\nu}\sin(ay)$

	$f(x)$	$g(y) = \int\limits_{0}^{\infty} f(x)(xy)^{\frac{1}{2}} J_\nu(xy)dx$
4.44	$\begin{array}{l} 0 \quad,\ 0 < x < a \\ x^{\nu-\frac{1}{2}}(x^2-a^2)^{\nu-\frac{1}{2}},\ x>a \\ -\frac{1}{2} < \text{Re } \nu < \frac{1}{2} \end{array}$	$\pi^{\frac{1}{2}}2^{\nu-2}\ \Gamma(\frac{1}{2}+\nu)a^{2\nu}y^{\frac{1}{2}-\nu}$ $\cdot\ [J_\nu(\frac{1}{2}ay)Y_{-\nu}(\frac{1}{2}ay) + Y_\nu(\frac{1}{2}ay)J_{-\nu}(\frac{1}{2}ay)]$
4.45	$\begin{array}{l} 0 \quad,\ 0 < x < a \\ x^{\frac{1}{2}-\nu}(x^2-a^2)^{\mu},\ x > a \\ \text{Re } \mu>-1,\ \text{Re}(\nu-2\mu)>\frac{1}{2} \end{array}$	$2^\mu a^{1+\mu-\nu}\ \Gamma(1+\mu)y^{-\mu-\frac{1}{2}}\ J_{\nu-\mu-1}(ay)$
4.46	$\begin{array}{l} 0 \quad,\ 0 < x < a \\ x^{\frac{1}{2}+\nu}(x^2-a^2)^{\mu},\ x > a \\ \text{Re } \mu>-1,\ \text{Re}(\nu+2\mu)<-\frac{1}{2} \end{array}$	$2^\mu a^{1+\mu+\nu}\ \Gamma(1+\mu)y^{-\mu-\frac{1}{2}}\ \cdot$ $\cdot\ [\sin(\pi\mu)Y_{\mu+\nu+1}(ay)-\cos(\pi\mu)J_{\mu+\nu+1}(ay)]$
4.47	$\begin{array}{l} 0 \quad,\ 0 < x < a \\ x^{\mu-\frac{1}{2}}(x^2-a^2)^{\lambda},\ x > a \\ \text{Re } \mu>-1,\ \text{Re}(\mu+2\lambda)<\frac{1}{2} \end{array}$	$y^{-2\lambda-\mu-1}\ 2^{\mu+2\lambda}\ \dfrac{\Gamma(\lambda+\frac{1}{2}+\frac{1}{2}\nu+\frac{1}{2}\mu)}{\Gamma(\frac{1}{2}\nu-\frac{1}{2}\mu+\frac{1}{2}-\lambda)}\ \cdot$ $\cdot {}_1F_2(-\lambda;\frac{1}{2}\nu+\frac{1}{2}-\frac{1}{2}\mu-\lambda,\frac{1}{2}-\frac{1}{2}\nu-\frac{1}{2}\mu-\lambda;\ -\frac{1}{4}a^2y^2)$ $+ \dfrac{2^{-\nu-1}a^{2\lambda+\nu+\mu+1}}{\Gamma(1+\nu)}\ B(1+\lambda,-\lambda-\frac{1}{2}\nu-\frac{1}{2}\mu-\frac{1}{2})$ $\cdot y^\nu {}_1F_2(\frac{1}{2}+\frac{1}{2}\nu+\frac{1}{2}\mu;\frac{3}{2}+\frac{1}{2}\nu+\frac{1}{2}\mu+\lambda;\nu+1;-\frac{1}{4}a^2y^2)$
4.48	$x^{\rho-\frac{3}{2}}(4a^4+x^4)^{-\mu-1}$	Watson, 1944, p. 435 Bessel functions

	$f(x)$	$g(y) = \int\limits_0^\infty f(x)(xy)^{\frac{1}{2}}J_\nu(xy)\,dx$
4.49	$x^{\nu+5/2}(4a^4+x^4)^{-\nu-\frac{1}{2}}$ $\mathrm{Re}\ \nu > 1/6$	$\dfrac{\pi^{\frac{1}{2}}2^{1-3\nu}a^{2-2\nu}}{\Gamma(\frac{1}{2}+\nu)}\,y^{\nu+\frac{1}{2}}J_{\nu-1}(ay)K_{\nu-1}(ay)$
4.50	$x^{\nu+\frac{1}{2}}(4a^4+x^4)^{-\nu-\frac{1}{2}}$ $\mathrm{Re}\ \nu > -\frac{1}{2}$	$\dfrac{\pi^{\frac{1}{2}}2^{-3\nu}a^{-2\nu}}{\Gamma(\frac{1}{2}+\nu)}\,y^{\nu+\frac{1}{2}}J_\nu(ay)K_\nu(ay)$
4.51	$x^{\mu-\frac{1}{2}}(1-2ax+a^2)^{-\frac{1}{2}}$ $,\ 0<x<1$ $0\quad,\ x>1$	see Bose, S.K., 1946: Bull. Calcutta Math. Soc., 38, 177-180.
4.52	$x^{\nu+\frac{1}{2}}(x^4\pm2a^2x^2+b^4)^{-\frac{1}{2}}\cdot$ $\cdot\ [b^2+x^2+$ $+(x^4\pm2a^2x^2+b^4)^{\frac{1}{2}}]^{-2\nu}$ $0<a<b$ $\mathrm{Re}\ \nu > -\frac{1}{2}$	$(b^2\mp a^2)^{-\nu}2^\nu y^{\frac{1}{2}}\cdot$ $\cdot\ K_\nu[\frac{1}{2}b^2\pm\frac{1}{2}a^2)^{\frac{1}{2}}y]J_\nu[(\frac{1}{2}b^2\pm\frac{1}{2}a^2)$
4.53	$x^{-\frac{1}{2}}(a^2+x^2)^{-\frac{1}{2}}\cdot$ $\cdot\ [(x^2+a^2)^{\frac{1}{2}}+x]^{\nu-1}$ $-1<\mathrm{Re}\ \nu<5/2$	$2\pi^{-\frac{1}{2}}a^{\nu-3/2}\sinh(\frac{1}{2}ay)K_{\nu-\frac{1}{2}}(\frac{1}{2}ay)$

	$f(x)$	$g(y) = \int\limits_0^\infty f(x)(xy)^{\frac{1}{2}}J_\nu(xy)\,dx$
4.54	$x^{-\frac{1}{2}}(x^2+a^2)^{-\frac{1}{2}} \cdot$ $\cdot\,[(x^2+a^2)^{\frac{1}{2}}+x]^{1-\nu}$ $\mathrm{Re}\ \nu > -\tfrac{1}{2}$	$\pi^{\frac{1}{2}}a^{\frac{1}{2}-\nu}e^{-\frac{1}{2}ay}\,I_{\nu-\frac{1}{2}}(\tfrac{1}{2}ay)$
4.55	$x^{-\frac{1}{2}}(x^2+a^2)^{-\frac{1}{2}} \cdot$ $\cdot\,[(x^2+a^2)^{\frac{1}{2}}\pm x]^{\mu}$ $\mathrm{Re}\ \nu > -1,\ \mathrm{Re}\ \mu < {}^{3}/_{2}$	$a^{\mu}y^{\frac{1}{2}}I_{\frac{1}{2}\nu\mp\frac{1}{2}\mu}(\tfrac{1}{2}ay)K_{\frac{1}{2}\nu\pm\frac{1}{2}\mu}(\tfrac{1}{2}ay)$
4.56	$x^{\frac{1}{2}-\nu}(x^2+a^2)^{-\frac{1}{2}} \cdot$ $\cdot\,[(x^2+a^2)^{\frac{1}{2}}-a]^{\nu}$ $\mathrm{Re}\ \nu > -1$	$y^{-\frac{1}{2}}e^{-ay}$
4.57	$x^{-\mu-\frac{1}{2}}(x^2+a^2)^{-\frac{1}{2}}$ $[(x^2+a^2)^{\frac{1}{2}}+a]^{\mu}$ $\mathrm{Re}(\nu-\mu) > -1$	$\dfrac{\Gamma(\frac{1}{2}+\frac{1}{2}\nu-\frac{1}{2}\mu)}{a\Gamma(1+\nu)}\,y^{-\frac{1}{2}}W_{\frac{1}{2}\mu,\frac{1}{2}\nu}(ay)\ \cdot$ $\cdot\,M_{-\frac{1}{2}\mu,\frac{1}{2}\nu}(ay)$
4.58	$x^{-\frac{1}{2}}(a^2-x^2)^{-\frac{1}{2}} \cdot$ $\cdot\,\{[x+i(a^2-x^2)^{\frac{1}{2}}]^{\mu}+$ $+\,[x-i(a^2-x^2)^{\frac{1}{2}}]^{\mu}\}$ $,\ 0<x<a$ $\qquad 0\quad,\ x > a$ $\mathrm{Re}(\mu+\nu) > -1$	$\pi a^{\mu}y^{\frac{1}{2}}J_{\frac{1}{2}\nu+\frac{1}{2}\mu}(\tfrac{1}{2}ay)J_{\frac{1}{2}\nu-\frac{1}{2}\mu}(\tfrac{1}{2}ay)$

	$f(x)$	$g(y) = \int\limits_0^\infty f(x)(xy)^{\frac{1}{2}}J_\nu(xy)\,dx$
4.59	$x^{-2\mu-\frac{1}{2}}(a^2-x^2)^{-\frac{1}{2}} \cdot$ $\cdot \{[a+(a^2-x^2)^{\frac{1}{2}}]^{2\mu} +$ $+ [a-(a^2-x^2)^{\frac{1}{2}}]^{2\mu}\}$ $,0<x<a$ $0 \qquad ,x > a$ $2\,\mathrm{Re}\ \mu < 1 + \mathrm{Re}\ \nu$	$\dfrac{a^\nu B(\frac{1}{2}+\frac{1}{2}\nu+\mu,\frac{1}{2}+\frac{1}{2}\nu-\mu)}{\Gamma(1+\nu)}\ y^{\nu+\frac{1}{2}} \cdot$ $\cdot\ _1F_1(\frac{1}{2}+\frac{1}{2}\nu-\mu;\ \nu+1;\ -iay) \cdot$ $\cdot\ _1F_1(\frac{1}{2}+\frac{1}{2}\nu-\mu;\ \nu+1;\ iay) =$ $= a^{-1}[\Gamma(1+\nu)]^{-1}B(\frac{1}{2}+\frac{1}{2}\nu+\mu,\ \frac{1}{2}+\frac{1}{2}\nu-\mu)y^{-\frac{1}{2}} \cdot$ $\cdot\ M_{\mu,\frac{1}{2}\nu}(iay)M_{\mu,\frac{1}{2}\nu}(-iay)$
4.60	$0 \qquad ,0 < x < a$ $x^{-\frac{1}{2}}(x^2-a^2)^{-\frac{1}{2}} \cdot$ $\cdot \{[x + (x^2-a^2)^{\frac{1}{2}}]^{\nu+1} +$ $+ [x-(x^2-a^2)^{\frac{1}{2}}]^{\nu+1}\}$ $\qquad\qquad x > a$ $\mathrm{Re}\ \nu < \frac{1}{2}$	$-\pi^{\frac{1}{2}}a^{\nu-3/2}[\sin(\frac{1}{2}ay)J_{\nu+\frac{1}{2}}(\frac{1}{2}ay) +$ $+ \cos(\frac{1}{2}ay)Y_{\nu+\frac{1}{2}}(\frac{1}{2}ay)]$
4.61	$0 \qquad ,0 < x < a$ $x^{-\frac{1}{2}}(x^2-a^2)^{-\frac{1}{2}} \cdot$ $\cdot \{[x+(x^2-a^2)^{\frac{1}{2}}]^{\nu-1} +$ $+ [x-(x^2-a^2)^{\frac{1}{2}}]^{\nu-1}\}$ $\qquad , \quad x > a$ $\mathrm{Re}\ \nu < 5/2$	$\pi^{\frac{1}{2}}a^{\nu-3/2}[\cos(\frac{1}{2}ay)J_{\nu-\frac{1}{2}}(\frac{1}{2}ay) -$ $- \sin(\frac{1}{2}ay)Y_{\nu-\frac{1}{2}}(\frac{1}{2}ay)]$

	$f(x)$	$g(y) = \int\limits_0^\infty f(x)(xy)^{\frac{1}{2}} J_\nu(xy)\,dx$
4.62	$0, \quad 0 < x < a$ $x^{-\frac{1}{2}}(x^2-a^2)^{-\frac{1}{2}} \cdot$ $\cdot \{[x+(x^2-a^2)^{\frac{1}{2}}]^\mu +$ $+ [x-(x^2-a^2)^{\frac{1}{2}}]^\mu\}$ $x > a$ $\mathrm{Re}\ \mu < {}^{3\!/2}$	$-\tfrac{1}{2}\pi y^{\frac{1}{2}} a^\mu [J_{\frac{1}{2}\mu+\frac{1}{2}\nu}(\tfrac{1}{2}ay) \cdot$ $\cdot\ Y_{\frac{1}{2}\nu-\frac{1}{2}\mu}(\tfrac{1}{2}ay) +$ $+ J_{\frac{1}{2}\nu-\frac{1}{2}\mu}(\tfrac{1}{2}ay) Y_{\frac{1}{2}\nu+\frac{1}{2}\mu}(\tfrac{1}{2}ay)]$

1.5 Exponential and Logarithmic Functions

	$f(x)$	$g(y) = \int\limits_{0}^{\infty} f(x)\,(xy)^{\frac{1}{2}} J_{\nu}(xy)\,dx$
5.1	$x^{-\frac{1}{2}} e^{-\alpha x}$ $\mathrm{Re}\ \nu > -1,\ \mathrm{Re}\ \alpha > 0$	$y^{\frac{1}{2}+\nu} (\alpha^2+y^2)^{-\frac{1}{2}} [\alpha+(\alpha^2+y^2)^{\frac{1}{2}}]^{-\nu}$
5.2	$x^{-\frac{3}{2}} e^{-\alpha x}$ $\mathrm{Re}\ \nu > 0,\ \mathrm{Re}\ \alpha > 0$	$\nu^{-1} y^{\frac{1}{2}+\nu} [\alpha+(\alpha^2+y^2)^{\frac{1}{2}}]^{-\nu}$
5.3	$x^{m+\frac{1}{2}} e^{-\alpha x}$ $\mathrm{Re}\ \nu > -m-2,\ \mathrm{Re}\ \alpha > 0$ $m=0,1,2,\cdots$	$(-1)^{m+1} y^{\frac{1}{2}+\nu} \dfrac{d^{m+1}}{d\alpha^{m+1}} \{ (\alpha^2+y^2)^{-\frac{1}{2}} \cdot$ $\qquad \cdot\ [\alpha+(\alpha^2+y^2)^{\frac{1}{2}}]^{-\nu} \}$
5.4	$x^{\nu+\frac{1}{2}} e^{-\alpha x}$ $\mathrm{Re}\ \nu > -1,\ \mathrm{Re}\ \alpha > 0$	$2^{\nu+1} \pi^{-\frac{1}{2}} \Gamma(\nu+\tfrac{3}{2})\, \alpha y^{\frac{1}{2}+\nu} (\alpha^2+y^2)^{-\nu-\frac{3}{2}}$
5.5	$x^{\nu-\frac{1}{2}} e^{-\alpha x}$ $\mathrm{Re}\ \nu > -\tfrac{1}{2},\ \mathrm{Re}\ \alpha > 0$	$2^{\nu} \pi^{-\frac{1}{2}} \Gamma(\tfrac{1}{2}+\nu)\, y^{\nu+\frac{1}{2}} (\alpha^2+y^2)^{-\nu-\frac{1}{2}}$
5.6	$x^{\mu-\frac{1}{2}} e^{-\alpha x}$ $\mathrm{Re}(\mu+\nu) > -1,\ \mathrm{Re}\ \alpha > 0$	$y^{\frac{1}{2}} (\alpha^2+y^2)^{-\frac{1}{2}\mu-\frac{1}{2}} \Gamma(\nu+\mu+1)\, P_{\mu}^{-\nu} [\alpha(\alpha^2+y^2)^{-\frac{1}{2}}]$
5.7	$x^{-\frac{1}{2}} e^{-\alpha x^2}$ $\mathrm{Re}\ \nu > -1,\ \mathrm{Re}\ \alpha > 0$	$\tfrac{1}{2}(\pi \tfrac{y}{\alpha})^{\frac{1}{2}} \exp\left(-\dfrac{y^2}{8\alpha}\right) I_{\frac{1}{2}\nu}\left(\dfrac{y^2}{8\alpha}\right)$

	$f(x)$	$g(y) = \int\limits_0^\infty f(x)(xy)^{\frac{1}{2}}J_\nu(xy)\,dx$
5.8	$x^{\frac{1}{2}}e^{-\alpha x^2}$ $\mathrm{Re}\ \nu > -2,\ \mathrm{Re}\ \alpha > 0$	$\frac{1}{8}\pi^{\frac{1}{2}}\alpha^{-3/2}y^{3/2}\exp\left(-\frac{y^2}{8\alpha}\right)\cdot$ $\cdot\left[I_{\frac{1}{2}\nu-\frac{1}{2}}\left(\frac{y^2}{8\alpha}\right) - I_{\frac{1}{2}\nu+\frac{1}{2}}\left(\frac{y^2}{8\alpha}\right)\right]$
5.9	$x^{\nu+\frac{1}{2}}e^{-\alpha x^2}$ $\mathrm{Re}\ \nu > -1,\ \mathrm{Re}\ \alpha > 0$	$(2\alpha)^{-\nu-1}y^{\nu+\frac{1}{2}}\exp\left(-\frac{y^2}{4\alpha}\right)$
5.10	$x^{\nu-3/2}e^{-\alpha x^2}$ $\mathrm{Re}\ \nu > 0,\ \mathrm{Re}\ \alpha > 0$	$2^{\nu-1}y^{\frac{1}{2}-\nu}\,\gamma\left(\nu,\ \frac{y^2}{4\alpha}\right)$
5.11	$x^{\nu+\frac{1}{2}}e^{\pm i\alpha x^2}$ $-1 < \mathrm{Re}\ \nu < \frac{1}{2},\ \alpha > 0$	$(2\alpha)^{-\nu-1}y^{\nu+\frac{1}{2}}\exp\left[\pm i\left(\frac{\nu+1}{2}\pi - \frac{y^2}{4\alpha}\right)\right]$
5.12	$x^{2n+\nu+\frac{1}{2}}e^{-\frac{1}{4}x^2}$ $\mathrm{Re}\ \nu > -1-2n, n=0,1,2,\cdot$	$2^{2n+\nu+1}\,n!\,y^{\nu+\frac{1}{2}}e^{-y^2}L_n^\nu(y^2)$
5.13	$x^{\mu-\frac{1}{2}}e^{-\alpha x^2}$ $\mathrm{Re}(\mu+\nu) > -1, \mathrm{Re}\ \alpha > 0$	$\frac{\Gamma(\frac{1}{2}+\frac{1}{2}\nu+\frac{1}{2}\mu)}{\Gamma(1+\nu)}\alpha^{-\frac{1}{2}\mu}y^{-\frac{1}{2}}\exp\left(-\frac{y^2}{8\alpha}\right)\cdot$ $\cdot M_{\frac{1}{2}\mu,\frac{1}{2}\nu}\left(\frac{y^2}{4\alpha}\right)$
5.14	$x^{-3/2}e^{-\frac{\alpha}{x}},\ \mathrm{Re}\ \alpha > 0$	$2y^{\frac{1}{2}}J_\nu\left[(2\alpha y)^{\frac{1}{2}}\right]K_\nu\left[(2\alpha y)^{\frac{1}{2}}\right]$

	$f(x)$	$g(y) = \int\limits_{0}^{\infty} f(x)\,(xy)^{\frac{1}{2}} J_{\nu}(xy)\,dx$
5.15	$x^{-3/2} e^{-\frac{\alpha}{x}-\beta x}$ Re $\alpha>0$, Re $\beta>0$	$2y^{\frac{1}{2}} J_{\nu}\{(2\alpha)^{\frac{1}{2}}[(\beta^2+y^2)^{\frac{1}{2}}-\beta]^{\frac{1}{2}}\}\cdot$ $\cdot\, K_{\nu}\{(2\alpha)^{\frac{1}{2}}[(\beta^2+y^2)^{\frac{1}{2}}+\beta]^{\frac{1}{2}}\}$
5.16	$x^{-1} e^{-(\alpha x)^{\frac{1}{2}}}$ Re $\nu>-\frac{1}{2}$, Re $\alpha>0$	$\pi^{-\frac{1}{2}} 2^{\frac{1}{2}} \Gamma(\frac{1}{2}+\nu) D_{-\nu-\frac{1}{2}}[e^{i\frac{\pi}{4}}(\frac{2y}{\alpha})^{-\frac{1}{2}}] D_{-\nu-\frac{1}{2}}[e^{-i\frac{\pi}{4}}(2\frac{y}{a})^{-\frac{1}{2}}]$
5.17	$x^{\nu+\frac{1}{2}} e^{\alpha(1-x^2)}$, $x<1$ 0 , $x>1$ Re $\nu>-\frac{1}{2}$	$(2i\alpha)^{-\nu-1} y^{\frac{1}{2}+\nu}[U_{\nu+1}(2i\alpha,y)-iU_{\nu+2}(2i\alpha,y)]$
5.18	$x^{\nu+\frac{1}{2}}\exp[-\alpha(b^2+x^2)^{\frac{1}{2}}]$ Re $\nu>-1$, Re $\alpha>0$	$(\frac{1}{2}\pi)^{-\frac{1}{2}}\alpha b^{\nu+3/2} y^{\frac{1}{2}+\nu}(y^2+\alpha^2)^{-\nu-3/4}\cdot$ $\cdot\, K_{\nu+3/2}[b(y^2+\alpha^2)^{\frac{1}{2}}]$
5.19	$x^{-\frac{1}{2}}(b^2+x^2)^{-\frac{1}{2}}\cdot$ $\cdot\exp[-\alpha(b^2+x^2)^{\frac{1}{2}}]$ Re $\nu>-1$, Re $\alpha>0$	$y^{\frac{1}{2}} I_{\frac{1}{2}\nu}\{\frac{1}{2}b[(\alpha^2+y^2)^{\frac{1}{2}}-\alpha]\} K_{\frac{1}{2}\nu}\{\frac{1}{2}b[(\alpha^2+y^2)^{\frac{1}{2}}+\alpha]\}$
5.20	$x^{\nu+\frac{1}{2}}(b^2+x^2)^{-\frac{1}{2}}\cdot$ $\cdot\exp[-\alpha(b^2+x^2)^{\frac{1}{2}}]$ Re $\nu>-1$, Re $\alpha>0$	$(\frac{1}{2}\pi)^{-\frac{1}{2}} b^{\nu+\frac{1}{2}} y^{\nu+\frac{1}{2}}(\alpha^2+y^2)^{-\frac{1}{2}\nu-\frac{1}{4}}\cdot$ $\cdot\, K_{\nu+\frac{1}{2}}[b(\alpha^2+y^2)^{\frac{1}{2}}]$

	$f(x)$	$g(y) = \int\limits_0^\infty f(x)(xy)^{\frac{1}{2}} J_\nu(xy)\,dx$
5.21	$x^{\frac{1}{2}-\nu}(b^2+x^2)^{-\frac{1}{2}} \cdot$ $\cdot [(x^2+b^2)^{\frac{1}{2}}-b]^\nu \cdot$ $\cdot \exp[-\alpha(x^2+b^2)^{\frac{1}{2}}]$ $\mathrm{Re}\ \nu>-1,\ \mathrm{Re}\ \alpha>0$	$y^{\nu+\frac{1}{2}}[\alpha+(\alpha^2+y^2)^{\frac{1}{2}}]^{-\nu}(\alpha^2+y^2)^{-\frac{1}{2}} \cdot$ $\cdot \exp[-b(\alpha^2+y^2)^{\frac{1}{2}}]$
5.22	$x^{\tau-\frac{1}{2}}(b^2+x^2)^{-\frac{1}{2}} \cdot$ $\cdot [(b^2+x^2)^{\frac{1}{2}}+b]^{-\tau} \cdot$ $\cdot \exp[-\alpha(b^2+x^2)^{\frac{1}{2}}]$ $\mathrm{Re}(\nu+\tau) > -1$ $\mathrm{Re}\ \alpha > 0$	$\dfrac{\Gamma(\frac{1}{2}+\frac{1}{2}\nu+\frac{1}{2}\tau)}{b\Gamma(1+\nu)}\ y^{-\frac{1}{2}} M_{\frac{1}{2}\tau,\frac{1}{2}\nu}\{b[(\alpha^2+y^2)^{\frac{1}{2}}-\alpha]\} \cdot$ $\cdot W_{-\frac{1}{2}\tau,\frac{1}{2}\nu}\{b[(\alpha^2+y^2)^{\frac{1}{2}}+\alpha]\}$

1.6 Trigonometric and Inverse Trigonometric Functions

	$f(x)$	$g(y) = \int_0^\infty f(x)(xy)^{\frac{1}{2}}J_\nu(xy)\,dx$
6.1	$x^{-\frac{1}{2}}\sin(ax)$ $\mathrm{Re}\ \nu > -2$	$\cos(\tfrac{1}{2}\pi\nu)y^{\nu+\frac{1}{2}}(a^2-y^2)^{-\frac{1}{2}}[a+(a^2-y^2)^{\frac{1}{2}}]^{-\nu},y<a$ $y^{\frac{1}{2}}(y^2-a^2)^{-\frac{1}{2}}\sin[\nu\arcsin(\tfrac{a}{y})]\ ,\qquad y>a$
6.2	$x^{-\frac{1}{2}}\cos(ax)$ $\mathrm{Re}\ \nu > -1$	$-\sin(\tfrac{1}{2}\pi\nu)y^{\nu+\frac{1}{2}}(a^2-y^2)^{-\frac{1}{2}}[a+(a^2-y^2)^{\frac{1}{2}}]^{-\nu},y<a$ $y^{\frac{1}{2}}(y^2-a^2)^{-\frac{1}{2}}\cos[\nu\arcsin(\tfrac{a}{y})]\ ,\qquad y>a$
6.3	$x^{-3/2}\sin(ax)$ $\mathrm{Re}\ \nu > -1$	$\nu^{-1}\sin(\tfrac{1}{2}\pi\nu)y^{\nu+\frac{1}{2}}[a+(a^2-y^2)^{\frac{1}{2}}]^{-\nu},\qquad y<a$ $\nu^{-1}y^{\frac{1}{2}}\sin[\nu\arcsin(\tfrac{a}{y})]\qquad\qquad ,\qquad y>a$
6.4	$x^{-3/2}\cos(ax)$ $\mathrm{Re}\ \nu > 0$	$\nu^{-1}\cos(\tfrac{1}{2}\pi\nu)y^{\nu+\frac{1}{2}}[a+(a^2-y^2)^{\frac{1}{2}}]^{-\nu},\qquad y<a$ $\nu^{-1}y^{\frac{1}{2}}\cos[\nu\arcsin(\tfrac{a}{y})]\qquad\qquad ,\qquad y>a$
6.5	$x^{\nu-\frac{1}{2}}\sin(ax)$ $-1 < \mathrm{Re}\ \nu < \frac{1}{2}$	$\pi^{\frac{1}{2}}[\Gamma(\tfrac{1}{2}-\nu)]^{-1}2^\nu y^{\nu+\frac{1}{2}}(a^2-y^2)^{-\nu-\frac{1}{2}},\qquad y<a$ $0\qquad\qquad\qquad\qquad\qquad ,\qquad y>a$
6.6	$x^{\nu-\frac{1}{2}}\cos(ax)$ $-\frac{1}{2} < \mathrm{Re}\ \nu < \frac{1}{2}$	$-2^\nu\pi^{-\frac{1}{2}}\sin(\pi\nu)\Gamma(\tfrac{1}{2}+\nu)y^{\nu+\frac{1}{2}}(a^2-y^2)^{-\nu-\frac{1}{2}},y<a$ $2^\nu\pi^{-\frac{1}{2}}\Gamma(\tfrac{1}{2}+\nu)y^{\nu+\frac{1}{2}}(y^2-a^2)^{-\nu-\frac{1}{2}},\qquad y\quad a$
6.7	$x^{\nu+\frac{1}{2}}\sin(ax)$ $-3/2 < \mathrm{Re}\ \nu < -\frac{1}{2}$	$-2^{1+\nu}\pi^{-\frac{1}{2}}a\,\sin(\pi\nu)\Gamma(3/2+\nu)y^{\nu+\frac{1}{2}}(a^2-y^2)^{-\nu-3/2}$ $\qquad\qquad\qquad\qquad\qquad\qquad\qquad y>a$ $-2^{1+\nu}\pi^{-\frac{1}{2}}a\Gamma(3/2+\nu)y^{\nu+\frac{1}{2}}(y^2-a^2)^{-\nu-3/2}$

	$f(x)$	$g(y) = \int\limits_{0}^{\infty} f(x)(xy)^{\frac{1}{2}} J_{\nu}(xy)\,dx$
6.8	$x^{\nu+\frac{1}{2}}\cos(ax)$ $-1 < \mathrm{Re}\ \nu < -\frac{1}{2}$	$2^{1+\nu}\pi^{\frac{1}{2}}a[\Gamma(-\frac{1}{2}-\nu)]^{-1}y^{\nu+\frac{1}{2}}(a^2-y^2)^{-\nu-3/2}\ ,y<a$ $\qquad\qquad 0 \qquad\qquad\qquad ,\quad y>a$
6.9	$x^{\frac{1}{2}-\nu}\sin(ax)$ $\mathrm{Re}\ \nu > \frac{1}{2}$	$\qquad\qquad 0 \qquad\qquad\qquad y < a$ $2^{1-\nu}\pi^{\frac{1}{2}}a[\Gamma(\nu-\frac{1}{2})]^{-1}y^{\frac{1}{2}+\nu}(y^2-a^2)^{\nu-3/2}\ ,\ y > a$
6.10	$x^{\mu-3/2}\sin(ax)$ $1-\mathrm{Re}\ \nu < \mathrm{Re}\ \mu < 3/2$	$2^{-\nu}a^{-\nu-\mu}\sin[\frac{\pi}{2}(\nu+\mu)]\ \dfrac{\Gamma(\nu+\mu)}{\Gamma(\nu+1)}\ y^{\nu+\frac{1}{2}}\ \cdot$ $\cdot\ {}_2F_1(\frac{1}{2}+\frac{1}{2}\nu+\frac{1}{2}\mu,\frac{1}{2}\nu+\frac{1}{2}\mu;\nu+1;\frac{y^2}{a^2})\qquad,\ y < a$ $2^{\mu}a\ \dfrac{\Gamma(\frac{1}{2}+\frac{1}{2}\nu+\frac{1}{2}\mu)}{\Gamma(\frac{1}{2}+\frac{1}{2}\nu-\frac{1}{2}\mu)}\ y^{-\mu-\frac{1}{2}}\ \cdot$ $\cdot\ {}_2F_1(\frac{1}{2}+\frac{1}{2}\mu+\frac{1}{2}\nu,\frac{1}{2}+\frac{1}{2}\mu-\frac{1}{2}\nu;\frac{3}{2};\frac{a^2}{y^2})\qquad,\ y > a$
6.11	$x^{\mu-3/2}\cos(ax)$ $-\mathrm{Re}\ \nu < \mathrm{Re}\ \mu < 3/2$	$2^{-\nu}a^{-\nu-\mu}\cos[\frac{\pi}{2}(\nu+\mu)]\ \dfrac{\Gamma(\nu+\mu)}{\Gamma(\nu+1)}\ y^{\nu+\frac{1}{2}}\ \cdot$ $\cdot\ {}_2F_1(\frac{1}{2}\nu+\frac{1}{2}\mu,\frac{1}{2}+\frac{1}{2}\nu+\frac{1}{2}\mu;\nu+1;\frac{y^2}{a^2})\qquad,\quad y < a$ $2^{\mu-1}\ \dfrac{\Gamma(\frac{1}{2}\nu+\frac{1}{2}\mu)}{\Gamma(1+\frac{1}{2}\nu-\frac{1}{2}\mu)}\ y^{\frac{1}{2}-\nu}\ \cdot$ $\cdot\ {}_2F_1(\frac{1}{2}\nu+\frac{1}{2}\mu,\frac{1}{2}\mu-\frac{1}{2}\nu;\frac{1}{2};\frac{a^2}{y^2})\qquad,\quad y > a$
6.12	$x^{\nu-\frac{1}{2}}(b^2+x^2)^{-1}\sin(ax)$ $-1 < \mathrm{Re}\ \nu < 3/2$	$b^{\nu-1}\sinh(ba)y^{\frac{1}{2}}K_{\nu}(by)\qquad\qquad\qquad y \geqq a$

	$f(x)$	$g(y) = \int\limits_0^\infty f(x)(xy)^{\frac{1}{2}}J_\nu(xy)\,dx$
6.13	$x^{\frac{1}{2}-\nu}(b^2+x^2)^{-1}\sin(ax)$ Re $\nu > -\frac{1}{2}$	$\frac{1}{2}\pi b^{-\nu}e^{-ab}y^{\frac{1}{2}}I_\nu(by)$ $\qquad\qquad y \leq a$
6.14	$x^{\nu+\frac{1}{2}}(b^2+x^2)^{-1}\cos(ax)$ $1- <$ Re $\nu < \frac{1}{2}$	$b^\nu\cosh(ab)y^{\frac{1}{2}}K_\nu(by)$ $\qquad\qquad y \geq a$
6.15	$x^{-\nu-\frac{1}{2}}(b^2+x^2)^{-1}\cos(ax)$ Re $\nu > -\,^{5/2}$	$\frac{1}{2}\pi b^{-\nu-1}e^{-ab}y^{\frac{1}{2}}I_\nu(by)$ $\qquad\qquad y \leq a$
6.16	$x^{\nu+2n-\frac{1}{2}}(b^2+x^2)^{-1}\cdot$ $\cdot\,\sin(ax)$ Re$(\nu+n) > -1$ Re$(\nu+2n) < \,^{5/2}$ $n = 0,1,2,\cdots$	$(-1)^n b^{\nu+2n-\frac{1}{2}}\sinh(ab)y^{\frac{1}{2}}K_\nu(by)$ $\quad y \geq a$
6.17	$x^{2n+\frac{1}{2}-\nu}(b^2+x^2)^{-1}\cdot$ $\cdot\,\sin(ax)$ Re $\nu > 2n-\,^{3/2},n=-1,0,1,$	$(-1)^n\frac{1}{2}\pi b^{2n-\nu}e^{-ab}y^{\frac{1}{2}}I_\nu(by)$ $\qquad y \leq a$
6.18	$x^{2n-\frac{1}{2}-\nu}(b^2+x^2)^{-1}\cdot$ $\cdot\,\cos(ax)$ Re $\nu > 2n-\,^{5/2}$ $n = 0,1,2,\cdots$	$(-1)^n\frac{1}{2}\pi b^{2n-\nu-1}e^{-ab}y^{\frac{1}{2}}I_\nu(by)$ $\qquad y \leq a$

	$f(x)$	$g(y) = \int\limits_0^\infty f(x)(xy)^{\frac{1}{2}} J_\nu(xy)\,dx$
6.19	$x^{\nu+2n+\frac{1}{2}}(b^2+x^2)^{-1}\cdot$ $\cdot\cos(ax)$ $-1 < \mathrm{Re}(\nu+n) < 3/2-n$ $n = 0,1,2,\cdots$	$(-1)^n b^{\nu+2n}\cosh(ab)\,y^{\frac{1}{2}}K_\nu(by) \qquad y \geq a$
6.20	$x^{-\frac{1}{2}}\sin(ax^2)$ $\mathrm{Re}\,\nu > -3$	$-\tfrac{1}{2}\left(\tfrac{\pi y}{a}\right)^{\frac{1}{2}}\sin\left(\tfrac{y^2}{8a} - \tfrac{\nu+1}{4}\pi\right)J_{\frac{1}{2}\nu}\left(\tfrac{y^2}{8a}\right)$
6.21	$x^{-\frac{1}{2}}\cos(ax^2)$ $\mathrm{Re}\,\nu > -1$	$\tfrac{1}{2}\left(\tfrac{\pi y}{a}\right)^{\frac{1}{2}}\cos\left(\tfrac{y^2}{8a} - \tfrac{\nu+1}{4}\pi\right)J_{\frac{1}{2}\nu}\left(\tfrac{y^2}{8a}\right)$
6.22	$x^{\frac{1}{2}}\sin(ax^2)$ $\mathrm{Re}\,\nu > -4$	$\tfrac{1}{8}\pi^{\frac{1}{2}}\left(\tfrac{y}{a}\right)^{3/2}\left[\cos\left(\tfrac{y^2}{8a} - \nu\tfrac{\pi}{4}\right)J_{\frac{1}{2}\nu-\frac{1}{2}}\left(\tfrac{y^2}{8a}\right) - \sin\left(\tfrac{y^2}{8a} - \nu\tfrac{\pi}{4}\right)J_{\frac{1}{2}\nu+\frac{1}{2}}\left(\tfrac{y^2}{8a}\right)\right] +$
6.23	$x^{\frac{1}{2}}\cos(ax^2)$ $\mathrm{Re}\,\nu > -2$	$\tfrac{1}{8}\pi^{\frac{1}{2}}\left(\tfrac{y}{a}\right)^{3\,2}\left[\cos\left(\tfrac{y^2}{8a} - \tfrac{\pi}{4}\right)J_{\frac{1}{2}\nu+\frac{1}{2}}\left(\tfrac{y^2}{8a}\right) + \sin\left(\tfrac{y^2}{8a} - \nu\tfrac{\pi}{4}\right)J_{\frac{1}{2}\nu-\frac{1}{2}}\left(\tfrac{y^2}{8a}\right)\right]$
6.24	$x^{\nu+\frac{1}{2}}\sin(ax^2)$ $-2 < \mathrm{Re}\,\nu < \frac{1}{2}$	$(2a)^{-\nu-1}y^{\nu+\frac{1}{2}}\cos\left(\tfrac{y^2}{4a} - \tfrac{1}{2}\pi\nu\right)$

	$f(x)$	$g(y)=\int\limits_{0}^{\infty} f(x)(xy)^{\frac{1}{2}}J_{\nu}(xy)\,dx$
6.25	$x^{\nu+\frac{1}{2}}\cos(ax^2)$ $-1 < \mathrm{Re}\ \nu < \frac{1}{2}$	$(2a)^{-\nu-1}y^{\nu+\frac{1}{2}}\sin(\frac{y^2}{4a} - \frac{1}{2}\pi\nu)$
6.26	$x^{\nu+\frac{1}{2}}\sin(ax^2)$, $\ x<b$ $0 \qquad\quad$, $x>b$ $\mathrm{Re}\ \nu > -2$	$(2a)^{-\nu-1}y^{\nu+\frac{1}{2}}[U_{\nu+1}(2a^2b,by)\sin(ab^2) -$ $- U_{\nu+2}(2ab^2,by)\cos(ab^2)]$
6.27	$x^{\nu+\frac{1}{2}}\cos(ax^2)$, $\ x<b$ $0 \qquad\quad$,$x>b$ $\mathrm{Re}\ \nu > -1$	$(2a)^{-\nu-1}y^{\nu+\frac{1}{2}}[U_{\nu+2}(2a^2b,by)\sin(ab^2) +$ $+ U_{\nu+1}(2a^2b,by)\cos(ab^2)]$
6.28	$x^{\nu+\frac{1}{2}}\sin[b(a^2-x^2)]$ $\qquad\qquad x<a$ $0 \qquad\quad x>a$ $\mathrm{Re}\ \nu > -1$	$(2b)^{-\nu-1}y^{\nu+\frac{1}{2}}U_{\nu+2}(2a^2b,ay)$
6.29	$x^{\nu+\frac{1}{2}}\cos[b(a^2-x^2)]$ $\qquad\qquad x<a$ $0 \qquad\quad x>a$ $\mathrm{Re}\ \nu > -1$	$(2b)^{-\nu-1}y^{\nu+\frac{1}{2}}U_{\nu+1}(2a^2b,ay)$

	$f(x)$	$g(y) = \int\limits_{0}^{\infty} f(x)(xy)^{\frac{1}{2}}J_{\nu}(xy)\,dx$
6.30	$\begin{array}{ll} 0 & x<a \\ x^{\nu+\frac{1}{2}}\sin[b(x^2-a^2)] & \\ & x>a \\ \text{Re } \nu > \frac{1}{2} & \end{array}$	$(2b)^{-\nu-1}y^{\nu+\frac{1}{2}}V_{-\nu}(2a^2b,ay)$
6.31	$\begin{array}{ll} 0 & x<a \\ x^{\nu+\frac{1}{2}}\cos[b(x^2-a^2)] & \\ & x>a \\ \text{Re } \nu > \frac{1}{2} & \end{array}$	$-(2b)^{-\nu-1}y^{\nu+\frac{1}{2}}V_{1-\nu}(2a^2b,ay)$
6.32	$x^{-3/2}\sin(ax)\sin(bx^{-1})$ Re $\nu > -2$	$\frac{1}{2}\pi y^{\frac{1}{2}}J_{\nu}(cb^{\frac{1}{2}})[J_{\nu}(db^{\frac{1}{2}})\sin(\frac{1}{2}\pi\nu) +$ $+ Y_{\nu}(db^{\frac{1}{2}})\cos(\frac{1}{2}\pi\nu)] +$ $+ y^{\frac{1}{2}}I_{\nu}(cb^{\frac{1}{2}})K_{\nu}(db^{\frac{1}{2}})\cos(\frac{1}{2}\pi\nu)$ $\begin{array}{l} c \\ d \end{array} = (a+y)^{\frac{1}{2}}\pm(a-y)^{\frac{1}{2}}, \qquad y < a$
6.33	$x^{-3/2}\sin(ax)\cos(bx^{-1})$ Re $\nu > -2$	$\frac{1}{2}\pi y^{\frac{1}{2}}J_{\nu}(cb^{\frac{1}{2}})[J_{\nu}(db^{\frac{1}{2}})\cos(\frac{1}{2}\pi\nu) -$ $- Y_{\nu}(db^{\frac{1}{2}})\sin(\frac{1}{2}\pi\nu)]+y^{\frac{1}{2}}I_{\nu}(cb^{\frac{1}{2}})K_{\nu}(db^{\frac{1}{2}})\cdot$ $\cdot \sin(\frac{1}{2}\pi\nu)$ $\begin{array}{l} c \\ d \end{array} = (a+y)^{\frac{1}{2}}\pm(a-y)^{\frac{1}{2}}, \qquad y < a$

	$f(x)$	$g(y) = \int_0^\infty f(x)\,(xy)^{\frac{1}{2}} J_\nu(xy)\,dx$
6.34	$x^{-3/2}\cos(ax)\sin(bx^{-1})$ $\mathrm{Re}\ \nu > -1$	$\frac{1}{2}\pi y^{\frac{1}{2}} J_\nu(cb^{\frac{1}{2}})\,[J_\nu(db^{\frac{1}{2}})\cos(\frac{1}{2}\pi\nu) -$ $-\ Y_\nu(db^{\frac{1}{2}})\sin(\frac{1}{2}\pi\nu)] -$ $-\ y^{\frac{1}{2}}\sin(\frac{1}{2}\pi\nu)\,I_\nu(cb^{\frac{1}{2}})K_\nu(db^{\frac{1}{2}})]$ ${}^c_d = (a+y)^{\frac{1}{2}}\pm(a-y)^{\frac{1}{2}}$, $y < a$
6.35	$x^{-3/2}\cos(ax)\cos(bx^{-1})$ $\mathrm{Re}\ \nu > -1$	$-\frac{1}{2}\pi y^{\frac{1}{2}} J_\nu(cb^{\frac{1}{2}})\,[J_\nu(db^{\frac{1}{2}})\sin(\frac{1}{2}\pi\nu) +$ $+\ Y_\nu(db^{\frac{1}{2}})\ \cos(\frac{1}{2}\pi\nu)] +$ $+\ y^{\frac{1}{2}}\cos(\frac{1}{2}\pi\nu)\,I_\nu(cb^{\frac{1}{2}})K_\nu(db^{\frac{1}{2}})$ ${}^c_d = (a+y^{\frac{1}{2}}\pm(a-y)^{\frac{1}{2}}$, $y < a$
6.36	$x^{-3/2}\sin(ax-bx^{-1})$ $\mathrm{Re}\ \nu > -1$	$2\,\sin(\frac{1}{2}\pi\nu)y^{\frac{1}{2}}I_\nu(cb^{\frac{1}{2}})K_\nu(db^{\frac{1}{2}})$ ${}^c_d = (a+y^{\frac{1}{2}}\pm(a-y)^{\frac{1}{2}}$, $y < a$
6.37	$x^{-3/2}\sin(ax+bx^{-1})$ $\mathrm{Re}\ \nu > -1$	$y^{\frac{1}{2}} J_\nu(cb^{\frac{1}{2}})\,[J_\nu(db^{\frac{1}{2}})\cos(\frac{1}{2}\pi\nu) -$ $-\ Y_\nu(db^{\frac{1}{2}})\sin(\frac{1}{2}\pi\nu)]$ ${}^c_d = (a+y)^{\frac{1}{2}}\pm(a-y)^{\frac{1}{2}}$, $y < a$

	$f(x)$	$g(y) = \int\limits_0^\infty f(x)(xy)^{\frac{1}{2}}J_\nu(xy)\,dx$
6.38	$x^{-3/2}\cos(ax-bx^{-1})$ $\mathrm{Re}\ \nu > -1$	$2\cos(\tfrac{1}{2}\pi\nu)\,y^{\frac{1}{2}}Y_\nu(cb^{\frac{1}{2}})K_\nu(db^{\frac{1}{2}})$ $\genfrac{}{}{0pt}{}{c}{d} = (a+y)^{\frac{1}{2}}\pm(a-y)^{\frac{1}{2}}$, $y < a$
6.39	$x^{-3/2}\cos(ax-bx^{-1})$ $\mathrm{Re}\ \nu > -1$	$-\pi y^{\frac{1}{2}}J_\nu(cb^{\frac{1}{2}})[J_\nu(db^{\frac{1}{2}})\sin(\tfrac{1}{2}\pi\nu) +$ $+\ Y_\nu(db^{\frac{1}{2}})\cos(\tfrac{1}{2}\pi\nu)]$ $\genfrac{}{}{0pt}{}{c}{d} = (a+y)^{\frac{1}{2}}\pm(a-y)^{\frac{1}{2}}$, $y < a$
6.40	$x^{-\frac{1}{2}}(b^2+x^2)^{-\frac{1}{2}} \cdot$ $\cdot \cos[a(b^2+x^2)^{\frac{1}{2}}]$ $\mathrm{Re}\ \nu > -1$	$-\tfrac{1}{2}\pi y^{\frac{1}{2}}J_{\frac{1}{2}\nu}\{\tfrac{1}{2}b[a-(a^2-y^2)^{\frac{1}{2}}]\} \cdot$ $\cdot Y_{-\frac{1}{2}\nu}\{\tfrac{1}{2}b[a+(a^2-y^2)^{\frac{1}{2}}]\}$, $y < a$
6.41	$x^{-\frac{1}{2}}(b^2+x^2)^{-\frac{1}{2}} \cdot$ $\cdot \sin[a(b^2+x^2)^{\frac{1}{2}}]$ $\mathrm{Re}\ \nu > -1$	$\tfrac{1}{2}\pi y^{\frac{1}{2}}J_{\frac{1}{2}\nu}\{\tfrac{1}{2}b[a-(a^2-y^2)^{\frac{1}{2}}]\} \cdot$ $\cdot J_{-\frac{1}{2}\nu}\{\tfrac{1}{2}b[a+(a^2-y^2)^{\frac{1}{2}}]$, $y < a$

	$f(x)$	$g(y) = \int\limits_{0}^{\infty} f(x)(xy)^{\frac{1}{2}} J_{\nu}(xy)\,dx$
6.42	$x^{\nu+\frac{1}{2}}\sin[a(b^2+x^2)^{\frac{1}{2}}]$ $-1 < \operatorname{Re}\nu < -\tfrac{1}{2}$	$(\tfrac{1}{2}\pi)^{\frac{1}{2}} ab^{\nu+3/2} y^{\nu+\frac{1}{2}} (a^2-y^2)^{-\frac{1}{2}\nu-3/4}\cdot$ $\cdot\{\sin(\pi\nu)J_{\nu+3/2}[b(a^2-y^2)^{\frac{1}{2}}] +$ $+\cos(\pi\nu)Y_{\nu+3/2}[b(a^2-y^2)^{\frac{1}{2}}]\},\; y < a$ $-(\tfrac{1}{2}\pi)^{-\frac{1}{2}} ab^{\nu+3/2} y^{\nu+\frac{1}{2}} (y^2-a^2)^{-\frac{1}{2}\nu-3/4}\cdot$ $\cdot K_{\nu+3/2}[b(y^2-a^2)^{\frac{1}{2}}]\;,\qquad y > a$
6.43	$x^{\nu+\frac{1}{2}}\cos[a(b^2+x^2)^{\frac{1}{2}}]$ $-1 < \operatorname{Re}\nu < -\tfrac{1}{2}$	$(\tfrac{1}{2}\pi)^{\frac{1}{2}} ab^{\nu+3/2} y^{\nu+\frac{1}{2}} (a^2-y^2)^{-\frac{1}{2}\nu-3/4}\cdot$ $\cdot\{\cos(\pi\nu)J_{\nu+3/2}[b(a^2-y^2)^{\frac{1}{2}}] -$ $-\sin(\pi\nu)Y_{\nu+3/2}[b(a^2-y^2)^{\frac{1}{2}}]\},\; y < a$ $0\qquad,\quad y > a$
	$x^{\nu+\frac{1}{2}}(b^2+x^2)^{-\frac{1}{2}}\cdot$ $\cdot\sin[a(b^2+x^2)^{\frac{1}{2}}]$ $-1 < \operatorname{Re}\nu < \tfrac{1}{2}$	$(\tfrac{1}{2}\pi)^{\frac{1}{2}} b^{\frac{1}{2}+\nu} y^{\frac{1}{2}+\nu} (a^2-y^2)^{-\frac{1}{4}-\frac{1}{2}\nu}\cdot$ $\cdot J_{-\nu-\frac{1}{2}}[b(a^2-y^2)^{\frac{1}{2}}]\qquad y < a$ $0\qquad y > a$

	$f(x)$	$g(y) = \int_0^\infty f(x)(xy)^{\frac{1}{2}} J_\nu(xy)\,dx$
6.44	$x^{\nu+\frac{1}{2}}(b^2+x^2)^{-\frac{1}{2}} \cdot$ $\cdot \sin[a(b^2+x^2)^{\frac{1}{2}}]$ $-1 < \text{Re } \nu < \frac{1}{2}$	$(\frac{1}{2}\pi)^{\frac{1}{2}} b^{\frac{1}{2}+\nu} y^{\frac{1}{2}+\nu}(a^2-y^2)^{-\frac{1}{4}-\frac{1}{2}\nu} \cdot$ $\cdot J_{-\nu-\frac{1}{2}}[b(a^2-y^2)^{\frac{1}{2}}]$ $y < a$ 0 $y > a$
6.45	$x^{\nu+\frac{1}{2}}(b^2+x^2)^{-\frac{1}{2}} \cdot$ $\cdot \cos[a(b^2+x^2)^{\frac{1}{2}}]$ $-1 < \text{Re } \nu < \frac{1}{2}$	$-(\frac{1}{2}\pi)^{\frac{1}{2}} b^{\frac{1}{2}+\nu} y^{\frac{1}{2}+\nu}(a^2-y^2)^{-\frac{1}{4}-\frac{1}{2}\nu} \cdot$ $\cdot Y_{-\nu-\frac{1}{2}}[b(a^2-y^2)^{\frac{1}{2}}]$ $y < a$ $(\frac{1}{2}\pi)^{-\frac{1}{2}} b^{\frac{1}{2}+\nu} y^{\frac{1}{2}+\nu}(y^2-a^2)^{-\frac{1}{4}-\frac{1}{2}\nu} \cdot$ $\cdot K_{\nu+\frac{1}{2}}[b(y^2-a^2)^{\frac{1}{2}}]$ $y > a$
6.46	$x^{\nu+\frac{1}{2}}(k^2+x^2)^{-1} \cdot$ $\cdot \cos[a(b^2+x^2)^{\frac{1}{2}}]$ $-1 < \text{Re } \nu < 3/2$	$k^\nu y^{\frac{1}{2}}\cosh[a(k^2-b^2)^{\frac{1}{2}}] K_\nu(ky)$ $y > a$
6.47	$x^{\nu+\frac{1}{2}}(k^2+x^2)^{-1} \cdot$ $\cdot (b^2+x^2)^{-\frac{1}{2}} \cdot$ $\cdot \sin[a(b^2+x^2)^{\frac{1}{2}}]$ $-1 < \text{Re } \nu < 5/2$	$k^\nu y^{\frac{1}{2}}(k^2-b^2)^{-\frac{1}{2}}\sinh[a(k^2-b^2)^{\frac{1}{2}}] \cdot$ $\cdot K_\nu(ky)$ $y > a$

	$f(x)$	$g(y) = \int\limits_{0}^{\infty} f(x)(xy)^{\frac{1}{2}} J_{\nu}(xy)\,dx$
6.48	$x^{-\frac{1}{2}}(a^2-x^2)^{-\frac{1}{2}} \cdot$ $\cdot \cos[b(a^2-x^2)^{\frac{1}{2}}], x<a$ $0 \quad , \quad x>a$ $\mathrm{Re}\ \nu > -1$	$\frac{1}{2}\pi y^{\frac{1}{2}} J_{\frac{1}{2}\nu}\{\frac{1}{2}a[(b^2+y^2)^{\frac{1}{2}}+b]\} \cdot$ $\cdot J_{\frac{1}{2}\nu}\{\frac{1}{2}a[(b^2+y^2)^{\frac{1}{2}}-b]\}$
6.49	$x^{\nu+\frac{1}{2}}(a^2-x^2)^{-\frac{1}{2}} \cdot$ $\cdot \cos[b(a^2-x^2)^{\frac{1}{2}}], x<a$ $0 \quad , \quad x>a$ $\mathrm{Re}\ \nu > -1$	$(\frac{1}{2}a\pi)^{\frac{1}{2}}a^{\nu}y^{\nu+\frac{1}{2}}(b^2+y^2)^{-\frac{1}{2}\nu-\frac{1}{4}} \cdot$ $\cdot J_{\nu+\frac{1}{2}}[a(b^2+y^2)^{\frac{1}{2}}]$
6.50	$x^{-\frac{1}{2}}(a^2-x^2)^{-\frac{1}{2}} \cdot$ $\cdot \sin[b(a^2-x^2)^{\frac{1}{2}}]\ x<a$ $-x^{-\frac{1}{2}}(x^2-a^2)^{-\frac{1}{2}} \cdot$ $\cdot \exp[-b(x^2-a^2)^{\frac{1}{2}}]$ $\qquad\qquad x>a$ $\mathrm{Re}\ \nu > -1$	$\frac{1}{2}\pi y^{\frac{1}{2}} J_{\frac{1}{2}\nu}\{\frac{1}{2}a[(b^2+y^2)^{\frac{1}{2}}-b]\} \cdot$ $\cdot Y_{\frac{1}{2}\nu}\{\frac{1}{2}a[(b^2+y^2)^{\frac{1}{2}}+b]\}$

	$f(x)$	$g(y) = \int\limits_{0}^{\infty} f(x)(xy)^{\frac{1}{2}} J_{\nu}(xy)\,dx$
6.51	$x^{\nu+\frac{1}{2}}(a^2-x^2)^{-\frac{1}{2}} \cdot$ $\cdot \sin[b(a^2-x^2)^{\frac{1}{2}}]$ $x < a$ $-x^{\nu+\frac{1}{2}}(x^2-a^2)^{-\frac{1}{2}} \cdot$ $\cdot \exp[-b(x^2-a^2)^{\frac{1}{2}}]$ $x > a$ $\mathrm{Re}\ \nu > -1$	$2^{-\frac{1}{2}} \pi^{\frac{1}{2}} a^{\nu+\frac{1}{2}}(b^2+y^2)^{-\frac{1}{2}\nu-\frac{1}{4}} y^{\nu+\frac{1}{2}} \cdot$ $\cdot\ Y_{\nu+\frac{1}{2}}[a(b^2+y^2)^{\frac{1}{2}}]$
6.52	$0 \qquad x < a$ $x^{\frac{1}{2}-\nu}(x^2-a^2)^{-\frac{1}{2}} \cdot$ $\cdot \cos[b(x^2-a^2)^{\frac{1}{2}}]$ $x > a$ $\mathrm{Re}\ \nu > -\frac{1}{2}$	$(\tfrac{1}{2}\pi a)^{\frac{1}{2}} a^{-\nu} y^{\frac{1}{2}-\nu}(y^2-b^2)^{\frac{1}{2}\nu-\frac{1}{4}} J_{\nu-\frac{1}{2}}[a(y^2-b^2)^{\frac{1}{2}}]$ $y > b$ $0 \qquad y < b$
	$0 \qquad x < a$ $x^{\frac{1}{2}+\nu}\sin[b(x^2-a^2)^{\frac{1}{2}}]$ $x > a$ $\mathrm{Re}\ \nu < -\frac{1}{2}$	$-(\tfrac{1}{2}\pi)^{-\frac{1}{2}} a^{\nu+3/2}\sin(\pi\nu) y^{\nu+\frac{1}{2}}(b^2-y^2)^{-\frac{1}{2}\nu-3/4} \cdot$ $\cdot\ K_{\nu+3/2}[a(b^2-y^2)^{\frac{1}{2}}] \qquad y < b$ $(\tfrac{1}{2}\pi)^{\frac{1}{2}} a^{\nu+3/2} b y^{\nu+\frac{1}{2}}(y^2-b^2)^{-\frac{1}{2}\nu-3/4} \cdot$ $\cdot\ Y_{\nu+3/2}[a(y^2-b^2)^{\frac{1}{2}}] \qquad y > b$

	$f(x)$	$g(y) = \int_0^\infty f(x)(xy)^{\frac{1}{2}} J_\nu(xy)\,dx$
6.53	$0 \qquad x < a$ $x^{\frac{1}{2}+\nu}\sin[b(x^2-a^2)^{\frac{1}{2}}]$ $x > a$ $\mathrm{Re}\ \nu < -\tfrac{1}{2}$	$-(\tfrac{1}{2}\pi)^{-\frac{1}{2}} a^{\nu+3/2}\sin(\pi\nu) y^{\nu+\frac{1}{2}}(b^2-y^2)^{-\frac{1}{2}\nu-3/4}\ \cdot$ $\cdot\ K_{\nu+3/2}[a(b^2-y^2)^{\frac{1}{2}}] \qquad y < b$ $(\tfrac{1}{2}\pi)^{\frac{1}{2}} a^{\nu+3/2} b y^{\nu+\frac{1}{2}}(y^2-b^2)^{-\frac{1}{2}\nu-3/4}\ \cdot$ $\cdot\ Y_{\nu+3/2}[a(y^2-b^2)^{\frac{1}{2}}] \qquad,\qquad y > b$
6.54	$0 \qquad x < a$ $x^{\frac{1}{2}-\nu}\sin[b(x^2-a^2)^{\frac{1}{2}}]$ $x > a$ $\mathrm{Re}\ \nu > \tfrac{1}{2}$	$0 \qquad y < b$ $(\tfrac{1}{2}\pi)^{\frac{1}{2}} a^{3/2-\nu} b y^{\frac{1}{2}-\nu}(y^2-b^2)^{\frac{1}{2}\nu-3/4}\ \cdot$ $\cdot\ J_{\nu-3/2}[a(y^2-b^2)^{\frac{1}{2}}] \qquad,\qquad y > b$
6.55	$0 \qquad x < a$ $x^{-\frac{1}{2}}(x^2-a^2)^{-\frac{1}{2}}\ \cdot$ $\cdot \cos[b(x^2-a^2)^{\frac{1}{2}}], x>a$	$\cos(\tfrac{1}{2}\pi\nu) y^{\frac{1}{2}} I_{\frac{1}{2}\nu}\{\tfrac{1}{2}a[b-(b^2-y^2)^{\frac{1}{2}}]\}\ \cdot$ $\cdot\ K_{\frac{1}{2}\nu}\{\tfrac{1}{2}a[b+(b^2-y^2)^{\frac{1}{2}}]\} \qquad,\qquad y < b$
6.56	$0 \qquad x < a$ $x^{\frac{1}{2}-\nu}(k^2+x^2)^{-1}\ \cdot$ $\cdot\ (x^2-a^2)^{-\frac{1}{2}}\ \cdot$ $\cdot \cos[b(x^2-a^2)^{\frac{1}{2}}]\ x>a$ $\mathrm{Re}\ \nu > -5/2$	$\tfrac{1}{2}\pi y^{\frac{1}{2}} k^{-\nu}(a^2+k^2)^{-\frac{1}{2}}\exp[-b(a^2+k^2)^{\frac{1}{2}}]\ \cdot$ $\cdot\ I_\nu(ky) \qquad,\qquad y < b$

	$f(x)$	$g(y = \int\limits_0^\infty f(x)(xy)^{\frac{1}{2}}J_\nu(xy)dx$
6.57	$\begin{array}{ll}0 & x < a\end{array}$ $x^{\frac{1}{2}-\nu}(k^2+x^2)^{-1}\cdot$ $\cdot\sin[b(x^2-a^2)^{\frac{1}{2}}]$ x>a $\operatorname{Re}\nu > -3/2$	$\frac{1}{2}\pi y^{\frac{1}{2}}k^{-\nu}\exp[-b(a^2+k^2)^{\frac{1}{2}}]I_\nu(ky)\quad y > b$
6.58	$x^{\frac{1}{2}-\nu}(b^2+x^2)^{-\frac{1}{2}}\cdot$ $\cdot[(b^2+x^2)^{\frac{1}{2}}-b]^\nu\cdot$ $\cdot\sin[a(b^2+x^2)^{\frac{1}{2}}]$ $\operatorname{Re}\nu > -1$	$y^{\nu+\frac{1}{2}}[a+(a^2-y^2)^{\frac{1}{2}}]^{-\nu}(a^2-y^2)^{-\frac{1}{2}}\cdot$ $\cdot\cos[b(a^2-y^2)^{\frac{1}{2}}+\frac{1}{2}\pi\nu]\quad y < a$ $y^{-\frac{1}{2}}(y^2-a^2)^{-\frac{1}{2}}\exp[-b(y^2-a^2)^{\frac{1}{2}}]\cdot$ $\cdot\sin[\nu\arcsin(\frac{a}{y})]\quad y > a$
6.59	$x^{\frac{1}{2}-\nu}(b^2+x^2)^{-\frac{1}{2}}\cdot$ $\cdot[(b^2+x^2)^{\frac{1}{2}}-b]^\nu\cdot$ $\cdot\cos[a(b^2+x^2)^{\frac{1}{2}}]$ $\operatorname{Re}\nu > -1$	$-y^{\nu+\frac{1}{2}}[a+(a^2-y^2)^{\frac{1}{2}}]^{-\nu}(a^2-y^2)^{-\frac{1}{2}}\cdot$ $\cdot\sin[b(a^2-y^2)^{\frac{1}{2}}+\frac{1}{2}\pi\nu]\quad y < a$ $y^{-\frac{1}{2}}(y^2-a^2)^{-\frac{1}{2}}\exp[-b(y^2-a^2)^{\frac{1}{2}}]\cdot$ $\cdot\cos[\nu\arcsin(\frac{a}{y})]\quad,\quad y > a$

	$f(x)$	$g(y) = \int\limits_{0}^{\infty} f(x)(xy)^{\frac{1}{2}} J_{\nu}(xy)\,dx$
6.60	$x^{-1} e^{-ax^{\frac{1}{2}}} \cdot$ $\cdot \cos[ax^{\frac{1}{2}} - \frac{1}{2}\pi(\nu - \frac{1}{2})]$ $\mathrm{Re}\ \nu > -\frac{1}{2}$	$D_{\nu-\frac{1}{2}}(ay^{-\frac{1}{2}})\ D_{-\nu-\frac{1}{2}}(ay^{-\frac{1}{2}})$
6.61	$x^{-\frac{1}{2}}(a^2-x^2)^{-\frac{1}{2}} \cdot$ $\cdot \cos[\mu \arccos(\frac{x}{a})]$ $\mathrm{Re}(\nu+\mu) > -1$	$\frac{1}{2}\pi y^{\frac{1}{2}} J_{\frac{1}{2}\nu+\frac{1}{2}\mu}(\frac{1}{2}ay)\, J_{\frac{1}{2}\nu-\frac{1}{2}\mu}(\frac{1}{2}ay)$

1.7 Orthogonal Polynomials

	$f(x)$	$g(y) = \int_0^\infty f(x)(xy)^{\frac{1}{2}} J_\nu(xy)\,dx$
7.1	$x^{-\frac{1}{2}}(a^2-x^2)^{-\frac{1}{2}} T_n(xa^{-1})$ $x < a$ $0 \quad x > a$ $\mathrm{Re}\,(\nu+n) > -1$	$\frac{1}{2}\pi y^{\frac{1}{2}} J_{\frac{1}{2}\nu+\frac{1}{2}n}(\frac{1}{2}ay) J_{\frac{1}{2}\nu-\frac{1}{2}n}(\frac{1}{2}ay)$
7.2	$x^{\nu-\frac{1}{2}}\exp(-\frac{1}{2}a^2x^2)\ \cdot$ $\cdot\ \mathrm{He}_n(x)$ $\mathrm{Re}\,\nu > -\frac{1}{2}$	$2^{\frac{1}{2}\nu+\frac{1}{4}n}\,a^{-\nu-\frac{1}{2}n}\,\Gamma(\frac{1}{2}+\nu)\,[\Gamma(1+\nu-\frac{1}{2}n)]^{-1}\ \cdot$ $\cdot\ y^{\frac{1}{2}n-\frac{1}{2}}\exp(-\frac{1}{4}y^2a^{-2})\ \cdot$ $\cdot\ M_{\frac{1}{2}\nu+\frac{1}{4}n,\ \frac{1}{2}\nu-\frac{1}{4}n}(\frac{1}{2}y^2a^{-2})$
7.3	$x^{\nu+\frac{1}{2}}\exp(-\frac{1}{2}a^2x^2)\ \cdot$ $\cdot\ \mathrm{He}_n(x)$ $\mathrm{Re}\,\nu > -1$	$2^{\frac{1}{2}\nu+3/4+\frac{1}{4}n}a^{-\nu-\frac{1}{2}-\frac{1}{2}n}\Gamma(3/2+\nu)\,[\Gamma(\frac{3}{2}+\nu-\frac{1}{2}n)]^{-1}\ \cdot$ $\cdot\ y^{\frac{1}{2}n-\frac{1}{2}}\exp(-\frac{1}{4}y^2a^{-2})\ \cdot$ $\cdot\ M_{\frac{1}{2}\nu+3/4+\frac{1}{4}n,\ \frac{1}{2}\nu+\frac{1}{4}-\frac{1}{2}n}(\frac{1}{2}y^2a^{-2})$
7.4	$x^{\frac{1}{2}-\nu}\exp(-a^2x^2)\quad\cdot$ $\cdot\ L_n(a^2x^2)$ $\mathrm{Re}\,(\nu+n) > 0$	$[\Gamma(1+n+\nu)]^{-1}2^{-n}a^{\nu-n-1}y^{n-\frac{1}{2}}\ \cdot$ $\cdot\ \exp(-\frac{1}{8}y^2a^{-2})\ \cdot$ $\cdot\ M_{\frac{1}{2}n-\frac{1}{2}\nu+\frac{1}{2},\ \frac{1}{2}n+\frac{1}{2}\nu}(\frac{1}{4}y^2a^{-2})$

	$f(x)$	$g(y) = \int_0^\infty f(x)(xy)^{\frac{1}{2}}J_\nu(xy)\,dx$
7.5	$x^{\frac{1}{2}+\nu}\exp(-a^2x^2)\cdot$ $\cdot L_n(a^2x^2)$ $\mathrm{Re}(2n-\nu) > -\tfrac{3}{2}$ $\mathrm{Re}\ \nu > -1$	$(n!)^{-1}2^{-n}a^{-n-\nu-1}y^{n-\frac{1}{2}}\exp(-\tfrac{1}{8}y^2a^{-2})\cdot$ $\cdot W_{\frac{1}{2}n+\frac{1}{2}+\frac{1}{2}\nu,\,\frac{1}{2}n-\frac{1}{2}\nu}(\tfrac{1}{8}y^2a^{-2})$
7.6	$x^{\nu+\frac{1}{2}}\exp(-a^2x^2)\cdot$ $\cdot L_n^\nu(a^2x^2)$ $\mathrm{Re}\ \nu > -1$	$(n!)^{-1}2^{-2n-\nu-1}a^{-2\nu-2n-2}\cdot$ $\cdot y^{2n+\nu+\frac{1}{2}}\exp(-\tfrac{1}{4}y^2a^{-2})$
7.7	$x^{\nu+\frac{1}{2}}\exp(-\tfrac{1}{2}a^2x^2)\cdot$ $\cdot L_n^\nu(a^2x^2)$ $\mathrm{Re}\ \nu > -1$	$(-1)^n a^{-2\nu-2}\exp(-\tfrac{1}{2}y^2a^{-2})\cdot$ $\cdot y^{\nu+\frac{1}{2}}L_n^\nu(y^2a^{-2})$
7.8	$x^{2n+\nu+\frac{1}{2}}\exp(-a^2x^2)\cdot$ $\cdot L_n^{\nu+n}(a^2x^2)$ $\mathrm{Re}\ \nu > -1$	$a^{-4n-2\nu-2}2^{-2n-\nu-1}\exp(-\tfrac{1}{4}y^2a^{-2})\cdot$ $\cdot L_n^{\nu+n}(\tfrac{1}{4}y^2a^{-2})$

	$f(x)$	$g(y) = \int\limits_0^\infty f(x)(xy)^{\frac{1}{2}}J_\nu(xy)\,dx$
7.9	$x^{\nu+\frac{1}{2}}\exp(-bx^2)\cdot$ $\cdot L_n^\nu(ax^2)$ $\mathrm{Re}\ \nu > 0$	$2^{-\nu-1}b^{-\nu-n-1}(b-a)^n y^{\nu+\frac{1}{2}}\exp(-\tfrac{1}{4}y^2 b^{-1})\cdot$ $\cdot L_n^\nu[\tfrac{1}{4}ay^2 b^{-1}(a-b)^{-1}]$
7.10	$x^{\nu+\frac{1}{2}}\exp(-ax^2)\cdot$ $\cdot[L_n^{\frac{1}{2}\nu}(ax^2)]^2$ $\mathrm{Re}\ \nu > -1$	$(2a)^{-\nu-1}y^{\nu+\frac{1}{2}}\exp(-\tfrac{1}{4}y^2 a^{-1})\cdot$ $\cdot[L_n^{\frac{1}{2}\nu}(\tfrac{1}{4}y^2 a^{-1})]^2$
7.11	$x^{\nu+\frac{1}{2}}\exp(-bx^2)\cdot$ $\cdot[L_n^{\frac{1}{2}\nu}(ax^2)]^2$ $\mathrm{Re}\ \nu > -1$	$(\pi n!)^{-1}\Gamma(n+1+\tfrac{1}{2}\nu)(2b)^{-\nu-1}y^{\nu+\frac{1}{2}}\cdot$ $\cdot\exp(-\tfrac{1}{4}y^2 b^{-1})\sum\limits_{m=0}^n \dfrac{(-1)^m\Gamma(n-m+\tfrac{1}{2})\Gamma(m+\tfrac{1}{2})}{\Gamma(m+1+\tfrac{1}{2}\nu)(n-m)!}\cdot$ $\cdot(1-2ab^{-1})^{2m}L_{2m}^\nu[\tfrac{1}{2}y^2 a(2ab-b^2)^{-1}]$
7.12	$x^{\nu+\frac{1}{2}}\exp(-ax^2)\cdot$ $\cdot L_m^{\nu-\mu}(ax^2)L_n^\mu(ax^2)$ $\mathrm{Re}\ \nu > -1$	$(-1)^{m+n}(2a)^{-\nu-1}y^{\nu+\frac{1}{2}}\exp(-\tfrac{1}{4}y^2 a^{-1})\cdot$ $\cdot L_n^{\mu-m+n}(\tfrac{1}{4}y^2 a^{-1})L_m^{\nu-\mu+m-n}(\tfrac{1}{4}y^2 a^{-1})$
7.13	$\begin{array}{ll}0 & x < a\end{array}$ $x^{2n+\frac{1}{2}-\nu}(x^2-a^2)^{\nu-2n-\frac{1}{2}}\cdot$ $\cdot C_{2n}^{\nu-2n}(ax^{-1})\quad x > a$ $2n-\tfrac{1}{2} < \mathrm{Re}\ \nu < 2n+\tfrac{1}{2}$	$(-1)^n 2^{2n-\nu+1}\Gamma(2\nu-2n)\cdot$ $\cdot[(2n)!\,\Gamma(\nu-2n)]^{-1}y^{2n-\nu-\frac{1}{2}}\cos(ay)$

	$f(x)$	$g(y) = \int_0^\infty f(x)(xy)^{\frac{1}{2}} J_\nu(xy)\,dx$
7.14	$0 \qquad x < a$ $x^{2n+3/2-\nu}(x^2-a^2)^{\nu-2n-3/2} \cdot$ $\cdot\ C_{2n+1}^{\nu-2n-1}(ax^{-1})$ $x > a$ $2n+\tfrac{1}{2} < \operatorname{Re}\ \nu < 2n+{3/2}$	$(-1)^n 2^{2n-\nu+2}\Gamma(2\nu-2n-1)\ \cdot$ $\cdot\ [(2n+1)!\,\Gamma(\nu-2n-1)]^{-1} y^{2n-\nu+\frac{1}{2}}\sin(ay)$
7.15	$x^{\nu+\frac{1}{2}}(a^2-x^2)^{-\frac{1}{2}} \cdot$ $\cdot\ \sin[b(a^2-x^2)^{\frac{1}{2}}] \cdot$ $\cdot C_{2n+1}^{\nu+\frac{1}{2}}[(1-x^2a^{-2})^{\frac{1}{2}}]$ $x < a$ $0 \qquad x > a$ $\operatorname{Re}\ \nu > -\tfrac{1}{2}$	$(-1)^n(\tfrac{1}{2}\pi)^{\frac{1}{2}}(ay)^{\nu+\frac{1}{2}}(b^2+y^2)^{-\frac{1}{2}\nu-\frac{1}{4}} \cdot$ $\cdot\ C_{2n+1}^{\nu+\frac{1}{2}}[b(b^2+y^2)^{-\frac{1}{2}}] \cdot$ $\cdot\ J_{\nu+3/2+2n}[a(b^2+y^2)^{\frac{1}{2}}]$
7.16	$x^{\nu+\frac{1}{2}}(a^2-x^2)^{-\frac{1}{2}}$ $\cdot\ \cos[b(a^2-x^2)^{\frac{1}{2}}] \cdot$ $\cdot C_{2n}^{\nu+\frac{1}{2}}[(1-x^2a^{-2})^{\frac{1}{2}}]$ $x < a$ $0 \qquad x > a$ $\operatorname{Re}\ \nu > -\tfrac{1}{2}$	$(-1)^n(\tfrac{1}{2}\pi)^{\frac{1}{2}}(ay)^{\nu+\frac{1}{2}}(b^2+y^2)^{-\frac{1}{2}\nu-\frac{1}{4}} \cdot$ $\cdot\ C_{2n}^{\nu+\frac{1}{2}}[b(b^2+y^2)^{-\frac{1}{2}}] \cdot$ $\cdot\ J_{\nu+\frac{1}{2}+2n}[a(b^2+y^2)^{\frac{1}{2}}]$

1.8 Miscellaneous Functions

	$f(x)$	$g(y) = \int_0^\infty f(x)(xy)^{\frac{1}{2}} J_\nu(xy)\,dx$
8.1	$x^{\nu-\frac{1}{2}} \mathrm{Erfc}(ax)$ $\mathrm{Re}\ \nu > -\tfrac{1}{2}$	$\pi^{-\frac{1}{2}} 2^\nu y^{-\nu-\frac{1}{2}} \gamma(\nu+\tfrac{1}{2}, \tfrac{1}{4}y^2 a^{-1})$
8.2	$x^{\nu+\frac{1}{2}} \mathrm{Ei}(-ax^2)$ $\mathrm{Re}\ \nu > -1$	$-2^{\nu+1} y^{-\nu-3/2} \gamma(\nu+1, \tfrac{1}{4}y^2 a^{-1})$
8.3	$x^{\nu+\frac{1}{2}} e^{ax^2} \mathrm{Ei}(-ax^2)$ $-1 < \mathrm{Re}\ \nu < 3/2$	$-\Gamma(1+\nu)(2a)^{-\nu-1} y^{\nu+\frac{1}{2}} \exp(\tfrac{1}{4}y^2 a^{-1}) \cdot$ $\cdot\ \Gamma(-\nu, \tfrac{1}{4}y^2 a^{-1})$
8.4	$x^{\nu-\frac{1}{2}} \exp(a^2 x^2) \cdot$ $\cdot\ \mathrm{Erfc}(ax)$ $-\tfrac{1}{2} < \mathrm{Re}\ \nu < 3/2$	$\pi^{-\frac{1}{2}} a^{-\nu} \Gamma(\tfrac{1}{2}+\nu) y^{-\frac{1}{2}} \exp(\tfrac{1}{8}y^2 a^{-2}) \cdot$ $\cdot\ W_{-\nu,\nu}(\tfrac{1}{4}y^2 a^{-2})$
8.5	$x^{\frac{1}{2}+\nu} \mathrm{Erfc}(ax)$ $\mathrm{Re}\ \nu > -1$	$a^{-\nu} \Gamma(3/2+\nu) [\Gamma(\nu+2)]^{-1} y^{-3/2} \cdot$ $\cdot \exp(-\tfrac{1}{8}y^2 a^{-2}) M_{\frac{1}{2}\nu+\frac{1}{2},\ \frac{1}{2}\nu+\frac{1}{2}}(\tfrac{1}{4}y^2 a^{-2})$
8.6	$x^{-\nu-\frac{1}{2}} \gamma(\nu+\tfrac{1}{2}, ax^2)$ $\mathrm{Re}\ \nu > -\tfrac{1}{2}$	$2^{-\nu} \pi^{\frac{1}{2}} y^{\nu-\frac{1}{2}} \mathrm{Erfc}(\tfrac{1}{2}y a^{-1})$
8.7	$x^{-\nu-3/2} \gamma(\nu+1, ax^2)$ $\mathrm{Re}\ \nu > -1$	$-2^{-\nu-1} y^{\nu+\frac{1}{2}} \mathrm{Ei}(-\tfrac{1}{4}y^2 a^{-1})$

	$f(x)$	$g(y) = \int\limits_0^\infty f(x)(xy)^{\frac{1}{2}} J_\nu(xy)\,dx$
8.8	$x^{\nu+\frac{1}{2}} \exp(ax^2)\,\Gamma(-\nu, ax^2)$ $-1 < \mathrm{Re} < \tfrac{3}{2}$	$-[\Gamma(1+\nu)]^{-1}(2a)^{\nu+1} y^{\nu+\frac{1}{2}} \exp(ax^2)\,\mathrm{Ei}(-ax^2)$
8.9	$x^{\frac{1}{2}-\nu}\exp(-a^2x^2)\cdot$ $\cdot\, L_{-\nu}(a^2x^2)$ $\mathrm{Re}\ \nu < \tfrac{1}{2}$	$(2a^2)^{\nu-1} y^{\frac{1}{2}-\nu}\exp(-\tfrac{1}{4}y^2a^{-2})\cdot$ $\cdot\, L_{-\nu}(\tfrac{1}{4}y^2a^{-2})$
8.10	$x^{\frac{1}{2}+\nu}\exp(-a^2x^2)\cdot$ $\cdot\, L_{\nu-\frac{1}{2}}(a^2x^2)$ $\mathrm{Re}\ \nu > -\tfrac{1}{2}$	$[\Gamma(\tfrac{1}{2}+\nu)]^{-1}2^{-2\nu}a^{-2\nu-1}y^{\nu-\frac{1}{2}}\cdot$ $\cdot\, \exp(-\tfrac{1}{8}y^2a^{-2})\,D_{2\nu}(2^{-\frac{1}{2}}a^{-1}y)$
8.11	$x^{\frac{1}{2}+\nu}\exp(-a^2x^2)\cdot$ $\cdot\, L_{\nu+\frac{1}{2}}(a^2x^2)$ $\mathrm{Re}\ \nu > -1$	$\cdot\,[\Gamma(\tfrac{3}{2}+\nu)]^{-1}2^{-2\nu-\frac{3}{2}}y^{\nu+\frac{1}{2}}\exp(-\tfrac{1}{8}y^2a^{-2})\cdot$ $\cdot\, D_{2\nu+1}(2^{-\frac{1}{2}}a^{-1}y)$
8.12	$x^{\frac{1}{2}-\nu}\exp(-a^2x^2)\cdot$ $\cdot\, L_{\nu-1}(a^2x^2)$ $\mathrm{Re}\ \nu > 0$	$\pi^{\frac{1}{2}}[\Gamma(\nu)]^{-1}2^{-\nu}a^{-1}y^{\nu-\frac{1}{2}}\exp(-\tfrac{1}{8}y^2a^{-2})\cdot$ $\cdot\, I_{\nu-\frac{1}{2}}(\tfrac{1}{8}y^2a^{-2})$

	$f(x)$	$g(y) = \int\limits_0^\infty f(x)(xy)^{\frac{1}{2}} J_\nu(xy)\,dx$
8.13	$x^{\frac{1}{2}+\nu} \exp(-a^2x^2) \cdot$ $\cdot\ L_{-\nu-1}(a^2x^2)$ $-1 < \mathrm{Re}\ \nu < -\tfrac{1}{6}$	$\pi^{-\frac{1}{2}}[\Gamma(-\nu)]^{-1} 2^\nu a^{-1} y^{-\nu-\frac{1}{2}} \exp(-\tfrac{1}{8}y^2a^{-2}) \cdot$ $\cdot\ K_{\nu+\frac{1}{2}}(\tfrac{1}{8}y^2a^{-2})$
8.14	$x^{\frac{1}{2}-\nu} \exp(-a^2x^2) \cdot$ $\cdot\ L_k(a^2x^2)$ $\mathrm{Re}(\nu+k) > 0$	$[\Gamma(\nu+k+1)]^{-1} 2^{-k} a^{\nu-k-1} y^{k-\frac{1}{2}} \cdot$ $\cdot\ \exp(-\tfrac{1}{8}y^2a^{-2}) \cdot$ $\cdot\ M_{\frac{1}{2}k-\frac{1}{2}\nu+\frac{1}{2},\ \frac{1}{2}k+\frac{1}{2}\nu}(\tfrac{1}{4}y^2a^{-2})$
8.15	$x^{\frac{1}{2}+\nu} \exp(-a^2x^2) \cdot$ $\cdot\ L_k(a^2x^2)$ $\mathrm{Re}(2k-\nu) > -\tfrac{3}{2}$ $\mathrm{Re}\ \nu > -1$	$[\Gamma(1+k)]^{-1} 2^{-k} a^{-k-\nu-1} y^{k-\frac{1}{2}} \cdot$ $\cdot\ \exp(-\tfrac{1}{8}y^2a^{-2}) \cdot$ $\cdot\ W_{\frac{1}{2}k+\frac{1}{2}+\frac{1}{2}\nu,\ \frac{1}{2}k-\frac{1}{2}\nu}(\tfrac{1}{8}y^2a^{-2})$

1.9 Legendre Functions

	$f(x)$	$g(y) = \int\limits_{0}^{\infty} f(x)(xy)^{\frac{1}{2}}J_{\nu}(xy)\,dx$
9.1	$x^{-\mu}(a^2-x^2)^{-\frac{1}{2}\mu}P^{\mu}_{\nu-\frac{1}{2}}(\frac{x}{a})$ $x < a$ $0 \quad x > a$ $\text{Re } \mu < 1, \text{Re}(\nu-\mu) > -\frac{3}{2}$	$(\tfrac{1}{2}\pi)^{\frac{1}{2}}a^{1-\mu}y^{\mu}J_{\frac{1}{2}-\mu}(\tfrac{1}{2}ay)J_{\nu}(\tfrac{1}{2}ay)$
9.2	$x^{\nu-\frac{1}{2}}(a^2-x^2)^{\frac{1}{2}\nu+\frac{1}{4}}\cdot$ $\cdot P^{-\nu-\frac{1}{2}}_{\mu}(2a^2x^{-2}-1)$ $x < a$ $0 \quad x > a$ $-\frac{3}{2}-\text{Re }\nu < \text{Re }\mu < \frac{1}{2}+\text{Re }\nu$	$(2\pi)^{-\frac{1}{2}}[\Gamma(\tfrac{3}{2}+\nu)]^{-2}\Gamma(\tfrac{3}{2}+\mu+\nu)\Gamma(\tfrac{1}{2}+\nu-\mu)\cdot$ $\cdot a^{3\nu+\frac{3}{2}}(2y)^{\nu+\frac{1}{2}}{}_1F_1(\nu+\mu+\tfrac{3}{2};2\nu+2;iay)\cdot$ $\cdot {}_1F_1(\nu+\mu+\tfrac{3}{2};2\nu+2;-iay)$ $= (2\pi)^{-\frac{1}{2}}[\Gamma(\tfrac{3}{2}+\nu)]^{-2}\Gamma(\tfrac{3}{2}+\mu+\nu)\Gamma(\tfrac{1}{2}+\nu-\mu)\cdot$ $\cdot 2^{\nu+\frac{1}{2}}a^{\nu-\frac{1}{2}}y^{-\nu-\frac{3}{2}}\cdot$ $\cdot M_{-\frac{1}{2}-\mu,\frac{1}{2}+\nu}(iay)M_{-\frac{1}{2}-\mu,\frac{1}{2}+\nu}(-iay)$
9.3	$(a^2-x^2)^{-\frac{1}{2}\nu-\frac{1}{4}}\cdot$ $\cdot P^{\frac{1}{2}+\nu}_{\mu}(2x^2a^{-2}-1)$ $x < a$ $0 \quad x > a$ $-1 < \text{Re }\nu < \frac{1}{2}$	$\pi^{\frac{1}{2}}2^{-\nu-1}ay^{\nu+\frac{1}{2}}J_{\frac{1}{2}-\mu}(ay)J_{-\frac{1}{2}+\mu}(ay)$

	$f(x)$	$g(y) = \int\limits_0^\infty f(x)(xy)^{\frac{1}{2}}J_\nu(xy)dx$
9.4	$\begin{array}{l} \quad 0 \quad x < a \\ (x^2-a^2)^{\frac{1}{2}\nu-\frac{1}{4}} \cdot \\ \cdot P_\mu^{\frac{1}{2}-\nu}(2x^2a^{-2}-1) \quad x>a \\ \quad \mathrm{Re}\ \nu > -\frac{1}{2} \\ \mathrm{Re}\ \nu -^3\!/\!2<2\mathrm{Re}\ \mu<\frac{1}{2}-\mathrm{Re}\ \nu \end{array}$	$\begin{array}{l} -\pi^{\frac{1}{2}}\sec(\pi\mu)2^{\nu-2}ay^{\frac{1}{2}-\nu}\cdot \\ \cdot\{[J_{\nu+\frac{1}{2}}(\frac{1}{2}ay)]^2-[J_{-\mu-\frac{1}{2}}(\frac{1}{2}ay)]^2\} \end{array}$
9.5	$\begin{array}{l} \quad 0 \quad x < a \\ (x^2-a^2)^{-\frac{1}{2}\mu}P_{\nu-\frac{1}{2}}^\mu(\frac{a}{x}) \\ \quad x > a \\ 0 < \mathrm{Re}\ \mu < 1, \mathrm{Re}\ \nu>-1 \end{array}$	$(\tfrac{1}{2}\pi)^{-\frac{1}{2}}y^{\mu-1}\cos[ay+\frac{\pi}{2}(\nu+\mu-\frac{1}{2})]$
9.6	$\begin{array}{l} \quad 0 \quad x < a \\ x^{-\mu}(x^2-a^2)^{-\frac{1}{2}\mu}P_{\nu-\frac{1}{2}}^\mu(\frac{x}{a}) \\ \quad x > a \\ -\frac{1}{4}<\mathrm{Re}\mu < 1, \\ |\mathrm{Re}\ \nu|<\frac{1}{2}+2\ \mathrm{Re}\ \mu \end{array}$	$\begin{array}{l} -2^{-3\!/\!2}\pi^{\frac{1}{2}}a^{1-\mu}y^\mu\cdot \\ \cdot\ [J_{\mu-\frac{1}{2}}(\frac{1}{2}ay)Y_\nu(\frac{1}{2}ay)+ \\ +\ Y_{\mu-\frac{1}{2}}(\frac{1}{2}ay)J_\nu(\frac{1}{2}ay)] \end{array}$
9.7	$\begin{array}{l} \quad 0 \quad x < a \\ x^{\nu-\frac{1}{2}}(x^2-a^2)^{\frac{1}{2}\mu}P_{\mu+1}^\mu(\frac{x}{a}) \\ \quad x > a \\ \mathrm{Re}\ \nu<^5\!/\!2, \mathrm{Re}(2\mu+\nu)<-\frac{1}{2} \end{array}$	$\begin{array}{l} \pi^{-\frac{1}{2}}2^{\nu+\mu+1}a^\nu\frac{\Gamma(\frac{1}{2}+\mu)}{\Gamma(-\mu)}\cdot \\ \cdot\ y^{-\mu-\frac{1}{2}}S_{\mu-\nu+1,\mu+\nu+1}(ay) \end{array}$

	$f(x)$	$g(y) = \int\limits_{0}^{\infty} f(x)(xy)^{\frac{1}{2}}J_{\nu}(xy)\,dx$
9.8	$(x^2-a^2)^{-\frac{1}{2}\nu-\frac{1}{4}}$ · · $P_{\mu-\frac{1}{2}}^{-\nu-\frac{1}{2}}(2x^2a^{-2}-1)$ Re $\nu < \frac{1}{2}$, Re $\nu > -2$ Re$(\nu\pm\mu) > -2$	$2^{-\nu-2}\pi^{3/2}a[\Gamma(1+\nu-\mu)\Gamma(1+\nu+\mu)]^{-1}$ · · $y^{\nu+\frac{1}{2}}\{[J_{\mu}(\tfrac{1}{2}ay)]^2 + [Y_{\mu}(\tfrac{1}{2}ay)]^2\}$
9.9	$x^{\frac{1}{2}}(a^2+x^2)^{-\frac{1}{2}\mu}$ · · $P_{\mu-1}^{-\nu}[a(a^2+x^2)^{-\frac{1}{2}}]$ Re $\nu > -1$, Re $\mu > \frac{1}{2}$	$[\Gamma(\mu+\nu)]^{-1}y^{\mu-3/2}e^{-ay}$
9.10	$x^{\frac{1}{2}}(a^2+x^2)^{\frac{1}{2}\mu-\frac{1}{2}}$ · · $P_{\frac{1}{2}\mu-\frac{1}{2}}^{-\nu}\left(\dfrac{a^2-x^2}{a^2+x^2}\right)$ Re $\mu < 1$, Re$(2\nu-\mu) > -1$	$2^{1+\mu}[\Gamma(\tfrac{1}{2}+\nu-\tfrac{1}{2}\mu)\Gamma(\tfrac{1}{2}-\tfrac{1}{2}\mu)]^{-1}$ · · $y^{-\mu-\frac{1}{2}}K_{\nu}(ay)$
9.11	$(|a^2-x^2|)^{\frac{1}{2}\mu-\frac{1}{2}}$ · · $P_{\frac{1}{2}\mu-\frac{1}{2}}^{\nu}\left(\dfrac{a^2+x^2}{|a^2-x^2|}\right)$ Re $\mu > -1$, Re$(\mu-2\nu) < 1$	$2^{\mu}\,\dfrac{\Gamma(\tfrac{1}{2}+\tfrac{1}{2}\mu)}{\Gamma(\tfrac{1}{2}+\nu-\tfrac{1}{2}\mu)}\,y^{-\mu-\frac{1}{2}}J_{\nu}(ay)$

	$f(x)$	$g(y) = \int\limits_0^\infty f(x)(xy)^{\frac{1}{2}}J_\nu(xy)\,dx$
9.12	$x^{\frac{1}{2}}(a^2+x^2)^{\frac{1}{2}\nu}$ $\cdot$ $\cdot\; P_{\frac{1}{2}\mu-\frac{1}{2}}^{-\nu}(1+2x^2a^{-2})$ $-1 < \mathrm{Re}\;\mu < 1$	$2^{1+\nu}\pi^{-1}a\,\cos(\tfrac{1}{2}\pi\mu)\,y^{-\nu-\frac{1}{2}}K_\mu(ay)$
9.13	$x^{\frac{1}{2}}(a^2+x^2)^{-\frac{1}{2}\nu}$ $\cdot$ $\cdot\; P_{\frac{1}{2}\mu-\frac{1}{2}}^{-\nu}(1+2x^2a^{-2})$ $\mathrm{Re}(2\nu\pm\mu) > -1$	$2^{1-\nu}a\,[\Gamma(\tfrac{1}{2}+\nu+\tfrac{1}{2}\mu)\,\Gamma(\tfrac{1}{2}+\nu-\tfrac{1}{2}\mu)]^{-1}$ $\cdot$ $\cdot\; y^{\nu-\frac{1}{2}}K_\mu(ay)$
9.14	$x^{\nu+\frac{1}{2}}(a^2+x^2)^{\frac{1}{2}\nu}$ $\cdot$ $\cdot\; P_\nu\!\left[\dfrac{2a^2+x^2}{2a(a^2+x^2)^{\frac{1}{2}}}\right]$ $-1 < \mathrm{Re}\;\nu < 0$	$(2a)^{\nu+1}[\pi\Gamma(-\nu)]^{-1}$ $\cdot$ $\cdot\; y^{-\nu-\frac{1}{2}}[K_{\nu+\frac{1}{2}}(\tfrac{1}{2}ay)]^2$
9.15	$x^{\nu+\frac{1}{2}}P_\nu\!\left[\dfrac{x^2+2a^2}{2a(x^2+a^2)^{\frac{1}{2}}}\right]$ $0 < \mathrm{Re}\;\nu < {}^3\!/_2$	$(2a)^{\nu+1}[\Gamma(-\nu)]^{-1}y^{-\nu-\frac{1}{2}}$ $\cdot$ $\cdot\; I_{-\nu-\frac{1}{2}}(\tfrac{1}{2}ay)\,K_{\nu+\frac{1}{2}}(\tfrac{1}{2}ay)$

	$f(x)$	$g(y) = \int_0^\infty f(x)(xy)^{\frac{1}{2}}J_\nu(xy)\,dx$
9.16	$x^{\nu+\frac{1}{2}}[(a+2b^2x^2)^2-1]^{-\frac{1}{2}\nu-\frac{1}{4}} \cdot$ $\cdot\, p_{\mu-\frac{1}{2}}^{-\nu-\frac{1}{2}}(a+2b^2x^2)$ $\operatorname{Re}\nu>-1, \operatorname{Re}(\nu\pm\mu)>-1$ $a \geq 1$	$\pi^{-\frac{1}{2}}2^{-2\nu-\frac{1}{2}}b^{-2\nu-2}[\Gamma(1+\nu+\mu)\Gamma(1+\nu-\mu)]^{-1} \cdot$ $\cdot\, y^{\nu+\frac{1}{2}}K_\mu\{2^{-3/2}b^{-1}y[(a+1)^{\frac{1}{2}}+(a-1)^{\frac{1}{2}}]\} \cdot$ $\cdot\, K_\mu\{2^{-3/2}b^{-1}y[(a+1)^{\frac{1}{2}}-(a-1)^{\frac{1}{2}}]\}$
9.17	$x^{\frac{1}{2}}(a^2+x^2)^{-\frac{1}{2}} \cdot$ $\cdot\, p_\mu^{-\nu}[(1+x^2a^{-2})^{\frac{1}{2}}]$ $\operatorname{Re}(\nu-\mu)>-3/2,$ $\operatorname{Re}(\nu+\mu) > -\frac{1}{2}$	$2^{\frac{1}{2}-\nu}a^{\frac{1}{2}}[\Gamma(1+\frac{1}{2}\nu+\frac{1}{2}\mu)\Gamma(\frac{1}{2}+\frac{1}{2}\nu-\frac{1}{2}\mu)]^{-1} \cdot$ $\cdot\, K_{\frac{1}{2}+\mu}(ay)$
9.18	$x^{\frac{1}{2}}p_{\mu-\frac{1}{2}}^{-\nu}[(1+x^2a^{-2})^{\frac{1}{2}}]$ $\operatorname{Re}(\nu\pm\mu) > -\frac{1}{2}$	$2^{3/2-\nu}a^{\frac{1}{2}}[\Gamma(\frac{1}{4}+\frac{1}{2}\nu+\frac{1}{2}\mu)\Gamma(\frac{1}{4}+\frac{1}{2}\nu-\frac{1}{2}\mu)]^{-1} \cdot$ $\cdot\, y^{-1}K_\mu(ay)$
9.19	$x^{\frac{1}{2}}\{p_{\mu-\frac{1}{2}}^{-\frac{1}{2}\nu}[(1+x^2a^{-2})^{\frac{1}{2}}]\}^2$ $\operatorname{Re}(\frac{1}{2}\pm\mu) > -\frac{1}{2}$	$2a\pi^{-1}[\Gamma(\frac{1}{2}+\frac{1}{2}\nu+\mu)\Gamma(\frac{1}{2}+\frac{1}{2}\nu-\mu)]^{-1} \cdot$ $\cdot\, y^{-\frac{1}{2}}[K_\mu(\frac{1}{2}ay)]^2$

	$f(x)$	$g(y) = \int\limits_0^\infty f(x)(xy)^{\frac{1}{2}}J_\nu(xy)dx$
9.20	$x^{\frac{1}{2}}(a^2+x^2)^{-\frac{1}{2}} \cdot$ $\cdot\ p_{\mu-\frac{1}{2}}^{-\frac{1}{2}\nu-\frac{1}{2}}[(1+x^2a^{-2})^{\frac{1}{2}}] \cdot$ $\cdot\ p_{\mu-\frac{1}{2}}^{-\frac{1}{2}\nu+\frac{1}{2}}[(1+x^2a^{-2})^{\frac{1}{2}}]$ $\mathrm{Re}(\frac{1}{2}\pm\mu) > -1$	$a\pi^{-1}[\Gamma(1+\frac{1}{2}\nu+\mu)\Gamma(1+\frac{1}{2}\nu-\mu)]^{-1} \cdot$ $\cdot\ y^{\frac{1}{2}}[K_\mu(\frac{1}{2}ay)]^2$
9.21	$x^{\frac{1}{2}}(a^2+x^2)^{-\frac{1}{2}} \cdot$ $\cdot\ p_{-\mu}^{-\frac{1}{2}\nu}[(1+x^2a^{-2})^{\frac{1}{2}}] \cdot$ $\cdot\ p_\mu^{-\frac{1}{2}\nu}[(1+x^2a^{-2})^{\frac{1}{2}}]$ $\mathrm{Re}\ \nu>-1,\ \mathrm{Re}(\frac{1}{2}\nu\pm\mu)>-1$	$a\pi^{-1}[\Gamma(1+\frac{1}{2}\nu+\mu)\Gamma(1+\frac{1}{2}\nu-\mu)]^{-1} \cdot$ $\cdot\ y^{\frac{1}{2}}K_{\mu-\frac{1}{2}}(\frac{1}{2}ay)K_{\mu+\frac{1}{2}}(\frac{1}{2}ay)$
9.22	$q_{\nu-\frac{1}{2}}(\dfrac{a^2+b^2+x^2}{2bx})$ $\mathrm{Re}\ \nu > -\frac{1}{2}$	$\pi b^{\frac{1}{2}}y^{-\frac{1}{2}}e^{-ay}J_\nu(by)$
9.23	$|a^2-x^2|^{\frac{1}{2}\mu}e^{i\frac{\pi}{2}\mu} \cdot$ $\cdot\ q_{\nu-\frac{1}{2}}^{-\frac{1}{2}\mu}(\dfrac{a^2+x^2}{2ay})$ $\mathrm{Re}\ \mu>-1,\ \mathrm{Re}(\mu-2\nu)<1$	$2^\mu(\pi a)^{\frac{1}{2}}\Gamma(\frac{1}{2}+\frac{1}{2}\mu)y^{-\mu-\frac{1}{2}}J_\nu(ay)$

	$f(x)$	$g(y) = \int\limits_0^\infty f(x)(xy)^{\frac{1}{2}}J_\nu(xy)\,dx$
9.24	$x^{-\mu-\frac{1}{2}}(a^2+x^2)^{-\frac{1}{2}\mu-\frac{1}{4}}$ · $\cdot e^{-i\pi(\mu-\frac{1}{2}\nu+\frac{1}{4})}q_{\nu-\frac{1}{2}}^{\mu+\frac{1}{2}}(i\frac{x}{a})$ $\operatorname{Re}\nu > -1,\ \operatorname{Re}\mu < \frac{1}{2}$ $\operatorname{Re}(\nu+\mu) > -1$	$(\tfrac{1}{2}\pi)^{\frac{1}{2}}a^{-\mu+\frac{1}{2}}y^{\mu+\frac{1}{2}}I_\nu(\tfrac{1}{2}ay)K_\mu(\tfrac{1}{2}ay)$
9.25	$x^{\nu-\frac{1}{2}}(a^2+x^2)^{\frac{1}{2}\nu-\frac{1}{4}}$ · $\cdot e^{i\pi(\nu+\frac{1}{2})}q_{-\frac{1}{2}}^{-\nu-\frac{1}{2}}(1+2a^2x^{-2})$ $-1 < \operatorname{Re}\nu < 0$	$\pi^{-\frac{1}{2}}2^\nu a^{\nu+\frac{1}{2}}y^{-\nu-\frac{1}{2}}[K_{\nu+\frac{1}{2}}(\tfrac{1}{2}ay)]^2$
9.26	$x^{-\nu-\frac{1}{2}}(a^2+x^2)^{-\frac{1}{4}}$ · $\cdot e^{i\pi(\frac{1}{2}-\nu)}q_{-\frac{1}{2}}^{\nu-\frac{1}{2}}(1+2a^2x^{-2})$ $0 < \operatorname{Re}\nu < {}^3\!/_2$	$\pi^{\frac{1}{2}}2^{-\nu}a^{\frac{1}{2}-\nu}y^{\nu-\frac{1}{2}}I_{\nu-\frac{1}{2}}(\tfrac{1}{2}ay)K_{\nu-\frac{1}{2}}(\tfrac{1}{2}ay)$
9.27	$(a^2+x^2)^{-\frac{1}{2}\nu-\frac{1}{4}}e^{-i\pi(\frac{1}{2}+\nu)}$ · $\cdot q_{\mu-\frac{1}{2}}^{\frac{1}{2}+\nu}(1+2x^2a^{-2})$ $-1<\operatorname{Re}\nu<\frac{1}{2},\ \operatorname{Re}(\nu+\mu)>-1$	$\pi^{-\frac{1}{2}}2^{-\nu-1}ay^{\frac{1}{2}+\nu}I_\mu(\tfrac{1}{2}ay)K_\mu(\tfrac{1}{2}ay)$
9.28	$x^{-\frac{1}{2}}(a^2+x^2)^{-\frac{1}{2}\mu}e^{-i\pi\mu}$ · $\cdot q_{\frac{1}{2}\nu-\frac{1}{2}}^{\mu}(1+2a^2x^{-2})$ $\operatorname{Re}\nu>-1,\ \operatorname{Re}(\nu+2\mu)>-1$	$2^{-\mu}y^{\mu-\frac{1}{2}}K_\mu(ay)$

	$f(x)$	$g(y) = \int_0^\infty f(x)(xy)^{\frac{1}{2}}J_\nu(xy)dx$
9.29	$x^{\nu+\frac{1}{2}}[(a+2b^2x^2)^2-1]^{-\frac{1}{2}\nu-\frac{1}{4}}\cdot$ $\cdot\ e^{-i\pi(\nu+\frac{1}{2})}\cdot$ $\cdot\ q_{\mu-\frac{1}{2}}^{\nu+\frac{1}{2}}(a+2b^2x^2)\cdot$ $a\geq1,\ \mathrm{Re}\ \nu>-1,\mathrm{Re}(\nu+\mu)>-1$	$\pi^{\frac{1}{2}}2^{-2\nu-3/2}b^{-2\nu-2}y^{\frac{1}{2}+\nu}\cdot$ $\cdot\ I_\mu\{2^{-3/2}b^{-1}y[(a+1)^{\frac{1}{2}}-(a-1)^{\frac{1}{2}}]\}\cdot$ $\cdot\ K_\mu\{2^{-3/2}b^{-1}y[(a+1)^{\frac{1}{2}}+(a-1)^{\frac{1}{2}}]\}$
9.30	$(a^2+x^2)^{-\frac{1}{2}}e^{i\pi\mu}\cdot$ $\cdot\ q_{\nu-\frac{1}{2}}^{-\mu}[(1+a^2x^{-2})^{\frac{1}{2}}]$ $\mathrm{Re}(\nu\pm\mu)>-1$	$\pi^{\frac{1}{2}}2^{-\nu}[\Gamma(^3/_4+\frac{1}{2}\nu+\frac{1}{2}\mu)\Gamma(^3/_4+\frac{1}{2}\nu-\frac{1}{2}\mu)]^{-1}\cdot$ $\cdot\ \Gamma(\frac{1}{2}+\nu-\mu)K_\mu(ay)$
9.31	$e^{i\pi\mu}q_{\nu-\frac{1}{2}}^{-\mu}[(1+a^2x^{-2})^{\frac{1}{2}}]$ $\mathrm{Re}(\nu\pm\mu)>-1$	$2^{1-\nu}\pi^{\frac{1}{2}}[\Gamma(\frac{1}{4}+\frac{1}{2}\nu+\frac{1}{2}\mu)\Gamma(\frac{1}{4}+\frac{1}{2}\nu-\frac{1}{2}\mu)]^{-1}\cdot$ $\cdot\ \Gamma(\frac{1}{2}+\nu-\mu)y^{-1}K_\mu(ay)$
9.32	$(a^2+x^2)^{-\frac{1}{2}}e^{-i\pi\mu}\cdot$ $\cdot\ q_{\nu-\frac{1}{2}}^{\mu}[(1+a^2x^{-2})^{\frac{1}{2}}]$ $\mathrm{Re}(\nu\pm\mu)>-\ ^3/_2$	$2^{\mu-\frac{1}{2}}\dfrac{\Gamma(\frac{1}{4}+\frac{1}{2}\nu+\frac{1}{2}\mu)}{\Gamma(^3/_4+\frac{1}{2}\nu-\frac{1}{2}\mu)}\ K_\mu(ay)$
9.33	$x^{-\frac{1}{2}}(a^2+x^2)^{-\frac{1}{2}}e^{i2\pi\mu}\cdot$ $\cdot\ q_{\frac{1}{2}\nu}^{-\mu}[(1+a^2x^{-2})^{\frac{1}{2}}]\cdot$ $\cdot\ q_{\frac{1}{2}\nu-1}^{-\mu}[(1+a^2x^{-2})^{\frac{1}{2}}]$ $\mathrm{Re}(\frac{1}{2}\pm\mu)>-1$	$\dfrac{\Gamma(\frac{1}{2}\nu-\mu)}{2\Gamma(1+\frac{1}{2}\nu+\mu)}\ y^{\frac{1}{2}}[K_\mu(\frac{1}{2}ay)]^2$

	$f(x)$	$g(y) = \int\limits_{0}^{\infty} f(x)(xy)^{\frac{1}{2}}J_{\nu}(xy)\,dx$
9.34	$x^{-\frac{1}{2}}(a^2+x^2)^{-\frac{1}{2}}e^{i2\pi\mu} \cdot$ $\cdot\, q_{\frac{1}{2}\nu-\frac{1}{2}}^{\frac{1}{2}-\mu}[(1+a^2x^{-2})^{\frac{1}{2}}] \cdot$ $\cdot\, q_{\frac{1}{2}\nu-\frac{1}{2}}^{-\frac{1}{2}-\mu}[(1+a^2x^{-2})^{\frac{1}{2}}]$ $\text{Re } \nu>-1,\ \text{Re}(\tfrac{1}{2}\nu\pm\mu)>-1$	$\dfrac{\Gamma(\frac{1}{2}\nu-\mu)}{2\Gamma(1+\frac{1}{2}\nu+\mu)}\ y^{\frac{1}{2}}K_{\mu-\frac{1}{2}}(\tfrac{1}{2}ay)\,K_{\mu+\frac{1}{2}}(\tfrac{1}{2}ay)$
9.35	$x^{-\frac{1}{2}}e^{i2\pi\mu} \cdot$ $\cdot\, \{q_{\frac{1}{2}\nu-\frac{1}{2}}^{\mu}[(1+a^2x^{-2})^{\frac{1}{2}}]\}$ $\text{Re}(\tfrac{1}{2}\pm\mu) > -\tfrac{1}{2}$	$y^{-\frac{1}{2}}\ \dfrac{\Gamma(\frac{1}{2}+\frac{1}{2}\nu-\mu)}{\Gamma(\frac{1}{2}+\frac{1}{2}\nu+\mu)}\ [K_{\mu}(\tfrac{1}{2}ay)]^2$
9.36	$x^{-\frac{1}{2}}p_{\frac{1}{2}\nu-\frac{1}{2}}^{-\mu}[(1+a^2x^{-2})^{\frac{1}{2}}] \cdot$ $\cdot\, e^{-i\pi\mu}q_{\frac{1}{2}\nu-\frac{1}{2}}^{\mu}[(1+a^2x^{-2})^{\frac{1}{2}}]$ $\text{Re } \nu>-1,\ \text{Re}(\nu+2\mu)>-1$	$y^{-\frac{1}{2}}I_{\mu}(\tfrac{1}{2}ay)K_{\mu}(\tfrac{1}{2}ay)$
9.37	$x^{\frac{1}{2}}p_{\mu-\frac{1}{2}}^{-\frac{1}{2}\nu}[(1+x^2a^{-2})^{\frac{1}{2}}] \cdot$ $\cdot\, e^{-i\frac{\pi}{2}\nu}q_{\mu-\frac{1}{2}}^{\frac{1}{2}\nu}[(1+x^2a^{-2})^{\frac{1}{2}}]$ $\text{Re } \nu>-1,\ \text{Re}(\nu+2\mu)>-1$	$ay^{-\frac{1}{2}}I_{\mu}(\tfrac{1}{2}ay)K_{\mu}(\tfrac{1}{2}ay)$

1.10 Bessel Functions of Argument x

	$f(x)$	$g(y) = \int\limits_0^\infty f(x)\, x^{\frac{1}{2}} J_\nu(xy)\, dx$
10.1	$x^{\frac{1}{2}+\nu+\mu} J_\mu(ax)$ $-1 < \mathrm{Re}(\nu+\mu) < 0$	$-\pi^{-1} 2^{\nu+\mu+1} \sin(\pi\nu)\, \Gamma(\mu+\nu+1)\, a^\mu y^{\nu+\frac{1}{2}} \cdot$ $\cdot\, (a^2-y^2)^{-\mu-\nu-1} \qquad\qquad y < a$ $-\pi^{-1} 2^{\nu+\mu+1} \sin(\pi\mu)\, \Gamma(\mu+\nu+1)\, a^\mu y^{\nu+\frac{1}{2}} \cdot$ $\cdot\, (y^2-a^2)^{-\mu-\nu-1} \qquad\qquad y > a$
10.2	$x^{\frac{1}{2}+\nu-\mu} J_\mu(ax)$ $-1 < \mathrm{Re}\ \nu < \mathrm{Re}\ \mu$	$\dfrac{2^{\nu-\mu+1}}{\Gamma(\mu-\nu)}\, a^{-\mu} y^{\nu+\frac{1}{2}} (a^2-y^2)^{\mu-\nu-1}, \quad y < a$ $0 \qquad\qquad\qquad\qquad , \quad y > a$
10.3	$x^{\frac{1}{2}+\mu-\nu} J_\mu(ax)$ $\mathrm{Re}(\nu+\mu) > -1$	$0 \qquad\qquad\qquad\qquad , \quad y < a$ $\dfrac{2^{\mu-\nu+1}}{\Gamma(\nu-\mu)}\, a^\mu y^{\frac{1}{2}-\nu} (y^2-a^2)^{\nu-\mu-1}, \quad y > a$
10.4	$x^{-\lambda-\frac{1}{2}} J_\nu(ax)$ $\mathrm{Re}\ \lambda > -1,\ \mathrm{Re}(\lambda-2\nu) < 1$	$y^{\frac{1}{2}} \dfrac{\Gamma(\frac{1}{2}+\nu-\frac{1}{2}\lambda)}{\Gamma(\frac{1}{2}\lambda+\frac{1}{2})}\, 2^{-\lambda} \cdot$ $\cdot\, (|a^2-y^2|)^{\frac{1}{2}\lambda-\frac{1}{2}-\nu} P_{\frac{1}{2}\lambda-\frac{1}{2}}^{-\nu}\left(\dfrac{a^2+y^2}{|a^2-y^2|}\right)$ $= 2^{-\lambda}(\pi a)^{-\frac{1}{2}} [\Gamma(\frac{1}{2}+\frac{1}{2}\lambda)]^{-1} |a^2-y^2|^{\frac{1}{2}\lambda} \cdot$ $\cdot\, e^{i\frac{\pi}{2}\lambda}\, q_{\nu-\frac{1}{2}}^{-\frac{1}{2}\lambda}\left(\dfrac{a^2+y^2}{2ay}\right)$

	$f(x)$	$g(y) = \int\limits_0^\infty f(x)(xy)^{\frac{1}{2}} J_\nu(xy)\,dx$
10.5	$x^{-1} J_\mu(ax)$ $\mathrm{Re}(\mu+\nu) > -\tfrac{1}{2}$	$2^{\nu-\frac{1}{2}} \dfrac{\Gamma(\frac{1}{4}+\frac{1}{2}\mu+\frac{1}{2}\nu)}{\Gamma(\,^3/_4+\frac{1}{2}\mu-\frac{1}{2}\nu)} \left(\dfrac{y}{a}\right)^{\frac{1}{2}} P_{\mu-\frac{1}{2}}^{-\nu}\left[(1-\dfrac{y^2}{a^2})^{\frac{1}{2}}\right]$ $y < a$ $2^{\mu-\frac{1}{2}} \dfrac{\Gamma(\frac{1}{4}+\frac{1}{2}\nu+\frac{1}{2}\mu)}{\Gamma(\,^3/_4+\frac{1}{2}\nu-\frac{1}{2}\mu)} P_{\nu-\frac{1}{2}}^{-\mu}\left[(1-\dfrac{a^2}{y^2})^{\frac{1}{2}}\right]$ $y > a$
10.6	$J_\mu(ax)$ $\mathrm{Re}(\nu+\mu) > -\,^3/2$	$2^{\nu+\frac{1}{2}} \dfrac{\Gamma(\frac{1}{2}\mu+\frac{1}{2}\nu+\,^3/_4)}{(\frac{1}{2}\mu-\frac{1}{2}\nu+\frac{1}{4})} \left(\dfrac{y}{a}\right)^{\frac{1}{2}} (a^2-y^2)^{-\frac{1}{2}} \cdot$ $\cdot\, P_{\mu-\frac{1}{2}}^{-\nu}\left[(1-\dfrac{y^2}{a^2})^{\frac{1}{2}}\right],\ y<a$ $= 2^{\mu+\frac{1}{2}} \dfrac{\Gamma(\frac{1}{2}\mu+\frac{1}{2}\nu+\,^3/_4)}{\Gamma(\frac{1}{2}\nu-\frac{1}{2}\mu+\frac{1}{4})} (y^2-a^2)^{-\frac{1}{2}} \cdot$ $\cdot\, P_{\nu-\frac{1}{2}}^{-\mu}\left[(1-\dfrac{a^2}{y^2})^{\frac{1}{2}}\right]\qquad,\quad y > a$
10.7	$x^{-\frac{1}{2}} J_\mu(ax) e^{-bx}$	Watson, G. N., 1934: Journal London Math. Soc., 9, 20.
10.8	$x^{\lambda-\frac{1}{2}} J_\mu(ax) e^{-bx}$	Eason, G. Noble, B. and Sneddon, I. N., 1955: Phil. Trans. Roy. Soc. London (A), 247, 529

	$f(x)$	$g(y) = \int_0^\infty f(x)(xy)^{\frac{1}{2}} J_\nu(xy)\,dx$
10.9	$x^{\mu-\frac{1}{2}} J_\mu(ax)$ $\mathrm{Re}\ \mu < 1$ $\mathrm{Re}(2\mu+\nu) > -1$	$\pi^{-\frac{1}{2}} 2^\mu \cos(\tfrac{1}{2}\pi\nu) a^{-\frac{1}{2}} \Gamma(\tfrac{1}{2}+\tfrac{1}{2}\nu+\mu)\ \cdot$ $\cdot\ y^{\frac{1}{2}}(a^2-y^2)^{-\frac{1}{2}\mu-\frac{1}{4}} P_{\mu-\frac{1}{2}}^{-\frac{1}{2}\nu}(z) =$ $= \pi^{-1}2^{\mu+1}\cos(\tfrac{1}{2}\pi\nu)\ \cdot$ $\cdot\ y^{-\frac{1}{2}}(a^2-y^2)^{-\frac{1}{2}\mu} e^{-i\pi\mu} q_{\frac{1}{2}\nu-\frac{1}{2}}^{\mu}(2a^2y^{-2}-1)$ $y < a,\quad z = (1-\tfrac{1}{2}y^2a^{-2})(1-y^2a^{-2})^{-\frac{1}{2}}$ $2^\mu \Gamma(\tfrac{1}{2}+\tfrac{1}{2}\nu+\mu)[\Gamma(\tfrac{1}{2}+\tfrac{1}{2}\nu-\mu)]^{-1} y^{-\frac{1}{2}}\ \cdot$ $\cdot\ (y^2-a^2)^{-\frac{1}{2}\mu} P_{\frac{1}{2}\nu-\frac{1}{2}}^{-\mu}(1-2a^2y^{-2}) \qquad y > a$
10.10	$x^{-\mu-\frac{1}{2}} J_\mu(ax)$ $\mathrm{Re}(\nu,\mu) > -1$	$2^{1-\mu}[\Gamma(\tfrac{1}{2}+\mu+\tfrac{1}{2}\nu)\Gamma(\tfrac{1}{2}+\mu-\tfrac{1}{2}\nu)]^{-1} y^{-\frac{1}{2}} e^{-i\pi\mu}\ \cdot$ $\cdot\ (a^2-y^2)^{\frac{1}{2}\mu} q_{\frac{1}{2}\nu-\frac{1}{2}}^{\mu}(2a^2y^{-2}-1) \qquad y < a$ $2^{-\mu} y^{-\frac{1}{2}}(y^2-a^2)^{\frac{1}{2}\mu} P_{\frac{1}{2}\nu-\frac{1}{2}}^{-\mu}(1-2a^2y^{-2})\quad y > a$
10.11	$x^{-\frac{1}{2}} e^{-ax} J_\nu(bx)$ $\mathrm{Re}\ \nu > -\tfrac{1}{2}$	$\pi^{-1} b^{-\frac{1}{2}} q_{\frac{1}{2}\nu-\frac{1}{2}}[(2by)^{-1}(a^2+b^2+y^2)]$
10.12	$x^{\frac{1}{2}}(b^2+x^2)^{-1} J_\nu(ax)$ $\mathrm{Re}\ \nu > -1$	$y^{\frac{1}{2}} K_\nu(ab) I_\nu(by) \qquad\qquad y \leq a$ $y^{\frac{1}{2}} I_\nu(ab) K_\nu(by) \qquad\qquad y \geq a$

	$f(x)$	$g(y) = \int\limits_0^\infty f(x)(xy)^{\frac{1}{2}}J_\nu(xy)\,dx$
10.13	$x^{\frac{1}{2}-2n}(b^2+x^2)^{-1}J_\nu(ax)$ $n = 0,\ 1,\ 2,\cdots$ $\text{Re }\nu > n-1$	$(-1)^n b^{-2n}y^{\frac{1}{2}}I_\nu(by)K_\nu(ab)$ $\quad y \le a$ $(-1)^n b^{-2n}y^{\frac{1}{2}}I_\nu(ab)K_\nu(by)$ $\quad y \ge a$
10.14	$x^{\nu-\mu+\frac{1}{2}}(b^2+x^2)^{-1}J_\mu(ax)$ $1+\text{Re }\mu > \text{Re }\nu > -1$	$b^{\nu-\mu}y^{\frac{1}{2}}I_\mu(ab)K_\nu(by)$ $\quad y \ge a$
10.15	$x^{\nu-\mu+2n+\frac{1}{2}} \cdot$ $\cdot\ (b^2+x^2)^{-1}J_\mu(ax)$ $n = 0,\ 1,\ 2,\cdots$ $\text{Re }\mu-2n+1 > \text{Re }\nu >$ $> -n-1$	$(-1)^n b^{\nu-\mu+2n}y^{\frac{1}{2}}I_\mu(ab)K_\nu(by)$ $\quad y \ge a$
10.16	$x^{\mu-\nu+\frac{1}{2}}(b^2+x^2)^{-1}J_\mu(ax)$ $1+\text{Re }\nu > \text{Re }\mu > -1$	$y^{\frac{1}{2}}b^{\mu-\nu}K_\mu(ab)I_\nu(by)$ $\quad y \le a$
10.17	$x^{\mu-\nu+2n+\frac{1}{2}} \cdot$ $\cdot\ (b^2+x^2)^{-1}J_\mu(ax)$ $n = 0,\ 1,\ 2,\cdots$ $\text{Re }\nu+1-2n < \text{Re }\mu < -u-1$	$(-1)^n b^{\mu-\nu+2n}y^{\frac{1}{2}}K_\mu(ab)I_\nu(by)$ $\quad y \le a$

	$f(x)$	$g(y) = \int\limits_0^\infty f(x)(xy)^{1/2}J_\nu(xy)\,dx$	
10.18	$x^{1/2}(b^2+x^2)^{-1}J_{\nu-2n}(ax)$ $n = 0, 1, 2, \cdots$ $\mathrm{Re}\ \nu > n-1$	$(-1)^n y^{1/2}I_\nu(by)K_{\nu-2n}(ab)$ $(-1)^n y^{1/2}I_{\nu-2n}(ab)K_\nu(by)$	$y < a$ $y > a$
10.19	$x^{1/2}(b^2+x^2)^{-1}Y_{\nu-2n-1}(ax)$ $\mathrm{Re}\ \nu > n - \tfrac{1}{2}$ $n = 0, 1, 2 \cdots$	$(-1)^n y^{1/2}K_{\nu-2n-1}(ab)I_\nu(by),$	$y \leq a$
10.20	$x^{1/2}(b^2+x^2)^{-1} \cdot$ $\cdot\{\cos[\tfrac{1}{2}\pi(\nu-\mu)]J_\mu(ax)+$ $+\sin[\tfrac{1}{2}\pi(\nu-\mu)]Y_\mu(ax)\}$ $\mathrm{Re}(\nu\pm\mu) > -2$	$y^{1/2}K_\mu(ab)I_\nu(by)$	$y \leq a$
10.21	$x^{\rho+1/2}(b^2+x^2)^{-1} \cdot$ $\cdot\{\cos[\tfrac{1}{2}\pi(\rho-\mu+\nu)]J_\mu(ax)+$ $+\sin[\tfrac{1}{2}\pi(\rho-\mu+\nu)]Y_\mu(ax)\}$ $\mathrm{Re}\ \rho < 1$ $\mathrm{Re}(\nu\pm\mu+\rho) > -2$	$y^{1/2}K_\mu(ab)I_\nu(by)b^\rho$	$y \leq a$

	$f(x)$	$g(y) = \int\limits_{0}^{\infty} f(x)(xy)^{\frac{1}{2}}J_{\nu}(xy)\,dx$
10.22	$x^{\frac{1}{2}}J_{\frac{1}{2}\nu}(ax)Y_{\frac{1}{2}\nu}(ax)$ $\mathrm{Re}\ \nu > -\frac{1}{2}$	$\begin{array}{ll} 0 & ,\quad y < 2a \\[2mm] -2\pi^{-1}y^{-\frac{1}{2}}(y^2-4a^2)^{-\frac{1}{2}} & ,\quad y > 2a \end{array}$
10.23	$x^{\nu+\frac{1}{2}}J_{\nu}(ax)Y_{\nu}(ax)$ $-\frac{1}{2} < \mathrm{Re}\ \nu < \frac{1}{2}$	$\begin{array}{ll} 0 & y < 2a \\[2mm] -\dfrac{\pi^{-\frac{1}{2}}2^{3\nu+1}a^{2\nu}}{\Gamma(\frac{1}{2}-\nu)}\, y^{-\nu-\frac{1}{2}}(y^2-4a^2)^{-\nu-\frac{1}{2}}, & y>2a \end{array}$
10.24	$x^{\rho+\frac{1}{2}}(x^2+c^2)^{-1}J_{\mu}(bx)\ \cdot$ $\cdot\{\cos[\frac{1}{2}\pi(\rho+\mu)]J_{\nu}(ax)+$ $+\sin[\frac{1}{2}\pi(\rho+\mu)]Y_{\nu}(ax)\}$ $\mathrm{Re}\ \rho < {}^{3}/_{2},\ \mathrm{Re}(\rho+\mu) > -2$ $\mathrm{Re}(\rho+\mu+2\nu) > -2$	$c^{\rho}y^{\frac{1}{2}}I_{\mu}(bc)I_{\nu}(cy)K_{\nu}(ac)$ $y < a - b$
10.25	$x^{\rho+\frac{1}{2}}(x^2+b^2)^{-1}\ \cdot$ $\cdot\prod\limits_{i=1}^{k}[J_{\mu_i}(c_i x)]\ \cdot$ $\cdot\{\cos[\frac{\pi}{2}(\rho+\delta-\mu)]J_{\mu}(ax)+$ $+\sin[\frac{\pi}{2}(\rho+\delta-\mu)]Y_{\mu}(ax)\}$ $\delta=\nu+\sum\limits_{i=1}^{k}\mu_i,$ $k+3 > 2\,\mathrm{Re}\ \rho$ $\mathrm{Re}(\rho+\delta) > |\mathrm{Re}\ \mu|-2$	$b^{\rho}y^{\frac{1}{2}}I_{\nu}(by)K_{\mu}(ab)\prod\limits_{i=1}^{k}I_{\mu_i}(bc_i)$ $y < a - \sum\limits_{i=1}^{k}c_i$

	$f(x)$	$g(y) = \int\limits_0^\infty f(x)(xy)^{\frac{1}{2}}J_\nu(xy)\,dx$
10.26	$x^{-\lambda-\frac{1}{2}}J_\mu(ax)$ $\mathrm{Re}(\mu+\nu)+1 > \mathrm{Re}\ \lambda > -1$	$2^{-\lambda}a^{\lambda-\nu-1}\ \dfrac{\Gamma(\frac{1}{2}\mu+\frac{1}{2}\nu-\frac{1}{2}\lambda+\frac{1}{2})}{\Gamma(1+\nu)\Gamma(\frac{1}{2}\mu+\frac{1}{2}\lambda-\frac{1}{2}\nu+\frac{1}{2})}\ y^{\nu+\frac{1}{2}}\ \cdot$ $\cdot {}_2F_1(\frac{1}{2}\mu+\frac{1}{2}\nu-\frac{1}{2}\lambda+\frac{1}{2},\frac{1}{2}\nu-\frac{1}{2}\lambda-\frac{1}{2}\mu+\frac{1}{2};\nu+1;y^2a^{-2})$ $\hspace{6cm} y < a$ $2^{-\lambda}a^\mu y^{\lambda-\mu-\frac{1}{2}}\ \dfrac{\Gamma(\frac{1}{2}+\frac{1}{2}\mu-\frac{1}{2}\lambda+\frac{1}{2}\nu)}{\Gamma(1+\mu)\Gamma(\frac{1}{2}\lambda+\frac{1}{2}\nu-\frac{1}{2}\mu+\frac{1}{2})}\ \cdot$ $\cdot {}_2F_1(\frac{1}{2}\mu+\frac{1}{2}\nu-\frac{1}{2}\lambda+\frac{1}{2},\frac{1}{2}\mu-\frac{1}{2}\lambda-\frac{1}{2}\nu+\frac{1}{2};\mu+1;a^2y^{-2})$ $\hspace{6cm} y > a$
10.27	$x^{-\lambda-\frac{1}{2}}Y_\mu(ax)$ $\mathrm{Re}(\nu\pm\mu-\lambda+1) > 0$	$\pi^{-1}[\Gamma(1+\nu)]^{-1}\sin[\frac{1}{2}\pi(\nu-\mu-\lambda)]2^{-\lambda}a^{\lambda-\nu-1}\ \cdot$ $\cdot\Gamma(\frac{1}{2}+\frac{1}{2}\nu+\frac{1}{2}\mu-\frac{1}{2}\lambda)\Gamma(\frac{1}{2}+\frac{1}{2}\nu-\frac{1}{2}\mu-\frac{1}{2}\lambda)y^{\frac{1}{2}+\nu}\ \cdot$ $\cdot {}_2F_1(\frac{1}{2}+\frac{1}{2}\nu+\frac{1}{2}\mu-\frac{1}{2}\lambda,\frac{1}{2}+\frac{1}{2}\nu-\frac{1}{2}\mu-\frac{1}{2}\lambda;\nu+1;y^2a^{-2})$ $=\pi^{-1}\sin[\frac{1}{2}\pi(\nu-\mu-\lambda)]\int\limits_0^\infty x^{-\lambda-\frac{1}{2}}I_\nu(xy)(xy)^{\frac{1}{2}}K_\mu(ax)\,dx$ $\hspace{6cm} y < a$ $\cot(\pi\mu)\int\limits_0^\infty x^{-\lambda-\frac{1}{2}}J_\mu(ax)(xy)^{\frac{1}{2}}J_\nu(xy)\,dx$ $-\csc(\pi\mu)\int\limits_0^\infty x^{-\lambda-\frac{1}{2}}J_{-\mu}(ax)(xy)^{\frac{1}{2}}J_\nu(xy)\,dx$ $\hspace{6cm} y > a$ For the integrals see (10.26 and 4.65 Ch. II) for y > a

	$f(x)$	$g(y) = \int\limits_0^\infty f(x)(xy)^{\frac{1}{2}} J_\nu(xy)\,dx$
10.28	$x^{\frac{1}{2}+\nu+\mu} Y_\mu(ax)$ $-1 < \mathrm{Re}\,(\nu+\mu) < 0$ $\mathrm{Re}\,\nu > -1$	$\pi^{-1} 2^{\nu+\mu+1} \cos(\pi\nu)\,\Gamma(\mu+\nu+1) y^{\nu+\frac{1}{2}} a^\mu \cdot$ $\cdot\,(a^2-y^2)^{-\mu-\nu-1} \qquad\qquad , \qquad y < a$ $-\,\pi^{-1} 2^{\nu+\mu+1} \cos(\pi\nu)\,\Gamma(\mu+\nu+1) a^\mu y^{\nu+\frac{1}{2}} \cdot$ $\cdot\,(y^2-a^2)^{-\mu-\nu-1}$ $\qquad\qquad\qquad , \qquad y > a$
10.29	$x^{\frac{1}{2}+\nu-\mu} Y_\mu(ax)$ $\mathrm{Re}\,\nu > -1,$ $-1 < \mathrm{Re}\,(\nu-\mu) < 0$	$-\pi^{-1} 2^{\nu-\mu+1} \Gamma(\nu-\mu+1) a^{-\mu} y^{\nu+\frac{1}{2}} \cdot$ $\cdot\,(y^2-a^2)^{\mu-\nu-1} \qquad\qquad , \qquad y > a$ $\pi^{-1} 2^{\nu-\mu+1} \cos[\pi(\mu-\nu)]\,\Gamma(\nu-\mu+1) a^{-\mu} y^{\nu+\frac{1}{2}} \cdot$ $\cdot\,(a^2-y^2)^{\mu-\nu-1} \qquad\qquad , \qquad y < a$
10.30	$x^{\frac{1}{2}} [J_{\frac{1}{2}\nu}(ax)]^2$ $\mathrm{Re}\,\nu > -1$	$= 2\pi^{-1} y^{-\frac{1}{2}} (4a^2-y^2)^{-\frac{1}{2}} \qquad y < 2a$ $0 \qquad\qquad\qquad\qquad y > 2a$
10.31	$x^{\frac{1}{2}} J_{\frac{1}{2}\nu+\mu}(ax) \cdot$ $\cdot\,J_{\frac{1}{2}\nu-\mu}(ax)$ $\mathrm{Re}\,\nu > -1$	$\pi^{-1}(2a)^{-2\mu} y^{-\frac{1}{2}} (4a^2-y^2)^{-\frac{1}{2}} \cdot$ $\cdot\{[y+i(4a^2-y^2)^{\frac{1}{2}}]^{2\mu} + [y-i(4a^2-y^2)^{\frac{1}{2}}]^{2\mu}\}$ $= 2\pi^{-1} y^{-\frac{1}{2}} (4a^2-y^2)^{-\frac{1}{2}} \cos[2\mu \arccos(\frac{1}{2}ya^{-1})]$ $\qquad\qquad\qquad\qquad y < 2a$ $0 \qquad\qquad\qquad\qquad y > 2a$

	$f(x)$	$g(y) = \int\limits_{0}^{\infty} f(x)\,(xy)^{\frac{1}{2}} J_\nu(xy)\,dx$
10.32	$x^{2n-\mu-3/2}(x^2+c^2)^{-1}$ $J_\nu(ax)J_\mu(bx)$ $n=0,1,2,\cdots\ a>b$ $\mathrm{Re}(\nu+\mu)>-1,$ $\mathrm{Re}\ \mu > 2n-\,{}^{9}/_{2}$	$(-1)^{n+1}c^{2n-\mu-2}y^{\frac{1}{2}}I_\mu(bc)I_\nu(yc)K_\nu(ac)$ $y < a-b$
10.33	$x^{\frac{1}{2}}J_{\frac{1}{2}\nu-\frac{1}{4}}(ax)J_{\frac{1}{2}\nu+\frac{1}{4}}(ax)$ $\mathrm{Re}\ \nu>-1$	$\pi^{-1}(ay)^{-\frac{1}{2}}(2a-y)^{-\frac{1}{2}}\ ,\quad 0<y<2a$ $0\qquad,\qquad y>2a$
10.34	$x^{\rho-\mu-\nu+\frac{1}{2}}J_\mu(ax)J_\rho(bx)$ $\mathrm{Re}\ \rho>-1,\ \mathrm{Re}(\rho-\mu-\nu)<\frac{1}{2}$ $b>a>0$	$0\quad,\quad 0<y<b-a$
10.35	$x^{\rho-\mu-\nu-3/2}J_\mu(ax)J_\rho(bx)$ $b>a>0$ $\mathrm{Re}\ \rho>0,\ \mathrm{Re}(\rho-\mu-\nu)<{}^{5}/_{2}$	$2^{\rho-\mu-\nu-1}a^\mu b^{-\rho}\ \dfrac{\Gamma(\rho)}{\Gamma(1+\mu)\Gamma(1+\nu)}\ y^{\nu+\frac{1}{2}}$
10.36	$x^{\frac{1}{2}-\nu}J_\nu(ax)J_\nu(bx)$ $\mathrm{Re}\ \nu>-\frac{1}{2}$	$\dfrac{\pi^{-\frac{1}{2}}2^{\nu-1}(ab)^{-\nu}}{\Gamma(\frac{1}{2}+\nu)}\ y^{-\nu+\frac{1}{2}}\ \Delta^{2\nu-1}$ Δ is the area of a triangle of sides $a,\ b,\ y$ $\Theta\ ,\qquad$ otherwise

	$f(x)$	$g(y) = \int\limits_0^\infty f(x)(xy)^{\frac12}J_\nu(xy)\,dx$
10.37	$x^{\frac12-\nu}J_\nu(ax)J_\nu(bx)$ $\mathrm{Re}\ \nu > -\tfrac12$	$0\quad,\quad y < \lvert a-b\rvert\ \ \text{and}\ y > a+b$ $\dfrac{\pi^{-\frac12}(ab)^{-\nu}2^{1-3\nu}}{\Gamma(\frac12+\nu)}\ y^{\frac12-\nu}$ $[y^2-(a-b)^2]^{\nu-\frac12}[(a+b)^2-y^2]^{\nu-\frac12}$ $\lvert a-b\rvert < y < a+b$
10.38	$x^{-\frac12}[J_\mu(ax)]$ $\mathrm{Re}(\nu+2) > -1$	$\dfrac{\Gamma(\frac12+\frac12\nu+\mu)}{\Gamma(\frac12+\frac12\nu-\mu)}\ \cdot$ $\cdot\ y^{-\frac12}\{P^{-\mu}_{-\frac12+\frac12\nu}[(1-\tfrac{4a^2}{y^2})^{\frac12}]^2\},\quad y > 2a$
10.39	$x^{-\frac12}J_\mu(ax)J_{-\mu}(ax)$ $\mathrm{Re}\ \nu > -1$	$y^{-\frac12}P^\mu_{\frac12\nu-\frac12}[(1-4\tfrac{a^2}{y^2})^{\frac12}]P^{-\mu}_{\frac12\nu-\frac12}[(1-4\tfrac{a^2}{y^2})^{\frac12}]$ $y > 2a$
10.40	$x^{\frac12}J_{\frac12(\nu+n)}(ax)J_{\frac12(\nu-n)}(ax)$ $\mathrm{Re}\ \nu > -1,\ n=0,1,2,\cdots$	$2\pi^{-1}y^{-\frac12}(4a^2-y^2)^{-\frac12}T_n(\tfrac{y}{2a}),\ 0 < y < 2a$ $0\quad,\qquad y > 2a$
10.41	$x^{\frac12-\nu}[J_\nu(ax)]^2$ $\mathrm{Re}\ \nu > -\tfrac12$	$\dfrac{\pi^{-\frac12}2^{1-3\nu}a^{-2\nu}}{\Gamma(\frac12+\nu)}\ y^{\nu-\frac12}(4a^2-y^2)^{\nu-\frac12},\quad < y < 2a$ $0 < y > 2a$

	$f(x)$	$g(y) = \int\limits_{0}^{\infty} f(x)(xy)^{\frac{1}{2}}J_{\nu}(xy)dx$
10.42	$x^{\frac{1}{2}+\nu}[J_{\nu}(ax)]^{2}$ $-\frac{1}{2} < \mathrm{Re}\ \nu < \frac{1}{2}$	$\pi^{-3/2}2^{1+3\nu}a^{2\nu}\Gamma(\frac{1}{2}+\nu)y^{-\nu-\frac{1}{2}}(4a^2-y^2)^{-\nu-\frac{1}{2}},$ $\qquad\qquad\qquad\qquad 0 < y < 2a$ $-\pi^{-3/2}2^{1+3\nu}a^{2\nu}\Gamma(\frac{1}{2}+\nu)\sin(\pi\nu)\ \cdot$ $\cdot\ y^{-\nu-\frac{1}{2}}(y^2-4a^2)^{-\nu-\frac{1}{2}}, \qquad y > 2a$
10.43	$x^{\nu+\frac{1}{2}}J_{\mu}(ax)J_{-\mu}(ax)$ $-1 < \mathrm{Re}\ \nu < \frac{1}{2}$	$\pi^{-\frac{1}{2}}2^{\nu}a^{-1}(4a^2-y^2)^{-\frac{1}{2}\nu-\frac{1}{4}}P_{\mu-\frac{1}{2}}^{\nu+\frac{1}{2}}\!\left(\dfrac{y^2}{2a^2}-1\right)$ $\qquad\qquad\qquad\qquad\qquad 0 < y < 2a$ $\qquad 0 \qquad\qquad , \quad y > 2a$
10.44	$x^{\frac{1}{2}-\nu}\ \cdot$ $\cdot\{\,[J_{\mu}(ax)]^{2}-[J_{-\mu}(ax)]^{2}\,\}$ $-1 < \mathrm{Re}\ \mu < 1,\ \mathrm{Re}\ \nu > -\frac{1}{2}$	$-\pi^{-\frac{1}{2}}2^{1-\nu}a^{-1}\sin(\pi\mu)(y^2-4a^2)^{\frac{1}{2}\nu-\frac{1}{4}}\ \cdot$ $\cdot\ P_{\mu-\frac{1}{2}}^{\frac{1}{2}-\nu}\!\left(\dfrac{y^2}{2a^2}-1\right) \qquad\qquad , \quad y > 2a$ $\qquad 0 \qquad\qquad , \qquad 0 < y < 2a$
10.45	$x^{\frac{1}{2}-\mu}J_{\mu}(ax)J_{\nu}(ax)$ $\mathrm{Re}\ \mu > -\frac{1}{2},\ \mathrm{Re}(\mu+\nu) > -1$	$\pi^{-\frac{1}{2}}2^{-\mu}a^{-\mu-\frac{1}{2}}y^{\mu-\frac{1}{2}}(4a^2-y^2)^{\frac{1}{2}\mu-\frac{1}{4}}\ \cdot$ $\cdot\ P_{\nu-\frac{1}{2}}^{\frac{1}{2}-\mu}\!\left(\dfrac{y}{2a}\right) \qquad\qquad , \qquad 0 < y < 2a$ $\qquad 0 \qquad\qquad , \qquad y > 2a$
10.46	$x^{\frac{1}{2}+\nu}J_{\nu}(ax)J_{-\nu}(ax)$ $-1 < \mathrm{Re}\ \nu < \frac{1}{2}$	$\pi^{-\frac{1}{2}}2^{3\nu+1}a^{2\nu}[\Gamma(\frac{1}{2}-\nu)]^{-1}\ \cdot$ $\cdot\ y^{-\nu-\frac{1}{2}}(4a^2-y^2)^{-\nu-\frac{1}{2}}, \qquad 0 < y < 2a$ $\qquad 0 \qquad\qquad , \qquad y > 2a$

	$f(x)$	$g(y) = \int_0^\infty f(x)\,(xy)^{\frac{1}{2}} J_\nu(xy)\,dx$
10.47	$x^{\frac{1}{2}-\nu} J_\mu(ax) J_\mu(bx)$ $\mathrm{Re}\ \mu > -1,\ \mathrm{Re}\ \nu > -\tfrac{1}{2}$	$0 \qquad\qquad 0 < y < \|a-b\|$ $(2\pi)^{-\frac{1}{2}}(ab)^{\nu-1} y^{\frac{1}{2}-\nu}(1-z_1^2)^{\frac{1}{2}\nu-\frac{1}{4}} P_{\mu-\frac{1}{2}}^{\frac{1}{2}-\nu}(z_1)$ $\|a-b\| < y < a+b$ $-(\tfrac{1}{2}\pi^3)^{-\frac{1}{2}}(ab)^{\nu-1} y^{\frac{1}{2}-\nu}(z_2^2-1)^{\frac{1}{2}\nu-\frac{1}{4}}\ \cdot$ $\cdot\ \sin[\pi(\mu-\nu)]e^{i\pi(\nu-\frac{1}{2})} q_{\mu-\frac{1}{2}}^{\frac{1}{2}-\nu}(z_2)$ $y > a+b$ $z_1 = \dfrac{a^2+b^2-y^2}{2ab}$ $z_2 = \dfrac{y^2-a^2-b^2}{2ab}$
10.48	$x^{\frac{1}{2}+\nu} J_\mu(ax) J_\mu(bx)$ $\mathrm{Re}(\nu+\mu) > -1,\ \mathrm{Re}\ \nu < \tfrac{1}{2}$	$-\,(\tfrac{1}{2}\pi^3)^{-\frac{1}{2}}\sin(\pi\nu)(ab)^{-\nu-1} y^{\frac{1}{2}+\nu}\ \cdot$ $\cdot\,(z_1^2-1)^{-\frac{1}{2}\nu-\frac{1}{4}} e^{-i\pi(\nu+\frac{1}{2})} q_{\mu-\frac{1}{2}}^{\nu+\frac{1}{2}}(z_1)$ $y < \|a-b\|$ $(2\pi)^{-\frac{1}{2}}(ab)^{-\nu-1} y^{\nu+\frac{1}{2}}(1-z_1^2)^{-\frac{1}{2}\nu-\frac{1}{4}}\ \cdot$ $\cdot\,[P_{\mu-\frac{1}{2}}^{\nu+\frac{1}{2}}(z_1)\cos(\pi\nu) - \tfrac{2}{\pi} Q_{\mu-\frac{1}{2}}^{\nu+\frac{1}{2}}(z_1)\sin(\pi\nu)]$ $\|a-b\| < y < a+b$ $-(\tfrac{1}{2}\pi^3)^{-\frac{1}{2}}\sin(\pi\mu)(ab)^{-\nu-1} y^{\nu+\frac{1}{2}}\ \cdot$ $\cdot\,(z_2^2-1)^{-\frac{1}{2}\nu-\frac{1}{4}} e^{-i\pi(\nu+\frac{1}{2})} q_{\mu-\frac{1}{2}}^{\nu+\frac{1}{2}}(z_2)$ $y > a+b$ $z_1\atop{}_2$ as in 10.47

	$f(x)$	$g(y) = \int\limits_0^\infty f(x)(xy)^{\frac{1}{2}} J_\nu(xy)\,dx$
10.49	$x^{-\frac{1}{2}}[J_\mu^2(ax)+Y_\mu^2(ax)]$ $Re(\nu\pm2\mu) > -1$	$2(\pi y)^{-1}csc(\pi\mu)y^{\frac{1}{2}} \cdot$ $\cdot \{P^\mu_{-\frac{1}{2}+\frac{1}{2}\nu}[(1-\dfrac{4a^2}{y^2})^{\frac{1}{2}}]\ Q^{-\mu}_{-\frac{1}{2}+\frac{1}{2}\nu}[(1-\dfrac{4a^2}{y^2})] -$ $- P^{-\mu}_{-\frac{1}{2}+\frac{1}{2}\nu}[(1-\dfrac{4a^2}{y^2})^{\frac{1}{2}}]\ Q^\mu_{-\frac{1}{2}+\frac{1}{2}\nu}[(1-\dfrac{4a^2}{y^2})^{\frac{1}{2}}]\}$ $y > 2a$
10.50	$x^{\nu+\frac{1}{2}}[J_\mu^2(ax)+Y_\mu^2(ax)]$ $Re\ \nu < \frac{1}{2},\ Re\ \nu > -2$ $Re(\nu\pm2\mu) > -2$	$2^{\nu+1}\pi^{-3/2}a^{-1}\Gamma(1+\nu+\mu)\Gamma(1+\nu-\mu) \cdot$ $\cdot (y^2-4a^2)^{-\frac{1}{4}-\frac{1}{2}\nu}P^{-\nu-\frac{1}{2}}_{\mu-\frac{1}{2}}(\dfrac{y^2}{2a^2}-1)\quad y > 2a$

	$f(x)$	$g(y) = \int\limits_{0}^{\infty} f(x)(xy)^{\frac{1}{2}}J_{\nu}(xy)dx$
10.51	$x^{\frac{1}{2}+\nu}J_{\mu}(bx)Y_{\mu}(ax)$ $-1 < \mathrm{Re}\ \nu < \frac{1}{2}$ $\mathrm{Re}(\nu+\mu) > -1$ $a > b$	$(\frac{1}{2}\pi^3)^{-\frac{1}{2}}\cos(\pi\nu)(ab)^{-\nu-1}y^{\nu+\frac{1}{2}} \cdot$ $\cdot (z_1^2-1)^{-\frac{1}{4}-\frac{1}{2}\nu}e^{-i\pi(\nu+\frac{1}{2})}q_{\mu-\frac{1}{2}}^{\nu+\frac{1}{2}}(z_1)$ $\qquad\qquad y < a-b$ $(2\pi)^{-\frac{1}{2}}(ab)^{-\nu-1}y^{\nu+\frac{1}{2}}(1-z_1^2)^{-\frac{1}{4}-\frac{1}{2}\nu} \cdot$ $\cdot[P_{\mu-\frac{1}{2}}^{\nu+\frac{1}{2}}(z_1)\sin(\pi\nu)+\frac{2}{\pi}Q_{\mu-\frac{1}{2}}^{\nu+\frac{1}{2}}(z_1)\cos(\pi\nu)]$ $\qquad\quad a-b < y < a+b$ $-(\frac{1}{2}\pi^3)^{-\frac{1}{2}}\cos(\pi\mu)(ab)^{-\nu-1}y^{\nu+\frac{1}{2}} \cdot$ $\cdot(z_2^2-1)^{-\frac{1}{4}-\frac{1}{2}\nu}e^{-i\pi(\nu+\frac{1}{2})}q_{\mu-\frac{1}{2}}^{\nu+\frac{1}{2}}(z_2)$ $\qquad\qquad y > a+b$ $z_{\genfrac{}{}{0pt}{}{}{2}}^{\ 1} = \pm\left(\dfrac{a^2+b^2-y^2}{2ab}\right)$
10.52	$x^{\frac{1}{2}+\nu}J_{\mu}(ax)Y_{\mu}(bx)$ $-1 < \mathrm{Re}\ \nu < \frac{1}{2}$ $\mathrm{Re}(\nu+\mu) > -1$ $a > b$	$-(2\pi)^{-\frac{1}{2}}(ab)^{-\nu-1}y^{\nu+\frac{1}{2}}(z_1^2-1)^{-\frac{1}{4}-\frac{1}{2}\nu} \cdot$ $\cdot[2p_{\mu-\frac{1}{2}}^{\nu+\frac{1}{2}}(z_1)-\frac{2}{\pi}\cos(\pi\nu)e^{-i\pi(\nu+\frac{1}{2})}q_{\mu-\frac{1}{2}}^{\nu+\frac{1}{2}}(z_1)]$ $\qquad\qquad y < a-b$ For $a-b < y < a+b$ and $y > a+b$ $\qquad$ as above
10.53	$x^{\frac{1}{2}}J_{\frac{1}{2}\nu}(ax)Y_{\frac{1}{2}\nu}(ax)$	$\qquad\qquad 0 \qquad\qquad\qquad y < 2a$ $-2\pi^{-1}y^{-\frac{1}{2}}(y^2-4a^2)^{-\frac{1}{2}} \qquad y > 2a$

	$f(x)$	$g(y) = \int\limits_0^\infty f(x)(xy)^{\frac{1}{2}}J_\nu(xy)dx$
10.54	$x^\lambda J_\mu(ax)J_\rho(bx)$	Bailey, W. N., 1936: Proc. London Math. Soc. (2), 40, 37-48.
10.55	$x^{-\frac{1}{2}}J_\mu(ax\sin\alpha\cos\beta)\cdot$ $\cdot\,J_\rho^{'}(ax)$ $0 < \alpha,\beta < \frac{1}{2}\pi$ $\mathrm{Re}(\mu+\nu+\rho) > -1$	$\dfrac{a^{-\frac{1}{2}}\Gamma(\frac{1}{2}+\frac{1}{2}\sigma+\frac{1}{2}\rho)}{\Gamma(1+\mu)\Gamma(1+\nu)\Gamma(\frac{1}{2}-\frac{1}{2}\sigma+\frac{1}{2}\rho)}$ $(\sin\alpha\cos\beta)^\mu(\sin\beta\cdot\cos\alpha)^{\nu+\frac{1}{2}}\cdot$ $\cdot\,_2F_1(\frac{1}{2}+\frac{1}{2}\sigma-\frac{1}{2}\rho,\frac{1}{2}+\frac{1}{2}\sigma+\frac{1}{2}\rho;1+\mu;\sin^2\alpha)$ $\cdot\,_2F_1(\frac{1}{2}+\frac{1}{2}\sigma-\frac{1}{2}\rho,\frac{1}{2}+\frac{1}{2}\sigma+\frac{1}{2}\rho;1+\nu;\sin^2\beta)$ $\sigma = \mu+\nu,\ y = a\cos\alpha\sin\beta$
10.56	$x^{\frac{1}{2}}J_\mu(ax\sin\alpha\cos\beta)\cdot$ $\cdot\,J_{\nu-\mu}(ax)$ $0 < \alpha,\beta < \frac{1}{2}\pi$ $\mathrm{Re}\ \nu > -1$	$2\pi^{-1}a^{-3/2}\sin(\pi\mu)(\sin\alpha)^\mu(\sin\beta)^{\nu+\frac{1}{2}}$ $(\cos\alpha)^{\frac{1}{2}-\nu}(\cos\beta)^{-\mu}\cdot$ $\cdot\,[\cos(\alpha+\beta) - \cos(\alpha-\beta)]^{-1}$ $y = a\cos\alpha\sin\beta$
10.57	$x^{\nu-\delta+\frac{1}{2}}\prod\limits_{i=1}^{k}J_{\mu_i}(a_ix)$ $\delta = \sum\limits_{i=1}^{k}\mu_i$ $-1 < \mathrm{Re}\ \nu < \mathrm{Re}\ \delta+\frac{1}{2}k-\frac{1}{2}$	$0 \quad,\ \sum\limits_{i=1}^{k}a_i < y < \infty$

	$f(x)$	$g(y) = \int\limits_{0}^{\infty} f(x)(xy)^{\frac{1}{2}}J_{\nu}(xy)\,dx$
10.58	$x^{\nu-\delta-3/2}\ \prod\limits_{i=1}^{k} J_{\mu_i}(a_i x)$ $\delta = \sum\limits_{i=1}^{k}\mu_i$ $0 < \mathrm{Re}\ \nu < \mathrm{Re}\ \delta+\tfrac{1}{2}k+3/2$	$2^{\nu-\delta-1}y^{\frac{1}{2}-\nu}\Gamma(\nu)\ \prod\limits_{i=1}^{k}\dfrac{a_i^{\mu_i}}{\Gamma(1+\mu_i)}$ $\sum\limits_{i=1}^{k} a_i < y < \infty$

1.11 Bessel Functions of Other Arguments

	$f(x)$	$g(y) = \int\limits_0^\infty f(x)(xy)^{\frac{1}{2}} J_\nu(xy)\,dx$
11.1	$x^{\frac{1}{2}} J_{\frac{1}{2}\nu}(ax^2)$ $\operatorname{Re}\nu > -1$	$\tfrac{1}{2} a^{-1} y^{\frac{1}{2}} J_{\frac{1}{2}\nu}\left(\frac{y^2}{4a}\right)$
11.2	$x^{\frac{1}{2}} e^{-ax^2} J_{\frac{1}{2}\nu}(bx^2)$ $\operatorname{Re}\nu > -1$	$\tfrac{1}{2}(a^2+b^2)^{-\frac{1}{2}} y^{\frac{1}{2}} \exp[-\tfrac{1}{4}ay^2(a^2+b^2)^{-1}] \cdot$ $\cdot\, J_{\frac{1}{2}\nu}[\tfrac{1}{4}by^2(a^2+b^2)^{-1}]$
11.3	$x^{3/2} J_{\frac{1}{2}\nu+\frac{1}{2}}(ax^2)$ $\operatorname{Re}\nu > -2$	$\tfrac{1}{4} a^{-2} y^{3/2} J_{\frac{1}{2}\nu-\frac{1}{2}}\left(\frac{y^2}{4a}\right)$
11.4	$x^{3/2} J_{\frac{1}{2}\nu-\frac{1}{2}}(ax^2)$ $\operatorname{Re}\nu > -1$	$\tfrac{1}{4} a^{-2} y^{3/2} J_{\frac{1}{2}\nu+\frac{1}{2}}\left(\frac{y^2}{4a}\right)$
11.5	$x^{1/6-1/3\nu} \sin(\tfrac{1}{4}ax^2) \cdot$ $\cdot\, J_{1/3\nu-1/6}(\tfrac{1}{4}ax^2)$ $\operatorname{Re}\nu > -\tfrac{5}{2}$	$a^{1/3\nu-2/3} y^{1/6-1/3\nu} \sin\left(\frac{\nu+1}{6}\pi - \frac{y^2}{4a}\right) \cdot$ $J_{1/3\nu-1/6}\left(\frac{y^2}{4a}\right)$
11.6	$x^{1/6-1/3\nu} \cos(\tfrac{1}{4}ax^2) \cdot$ $J_{1/3\nu-1/6}(\tfrac{1}{4}ax^2)$ $\operatorname{Re}\nu > -1$	$a^{1/3\nu-2/3} y^{1/6-1/3\nu} \cos\left(\frac{\nu+1}{6}\pi - \frac{y^2}{4a}\right) \cdot$ $\cdot\, J_{1/3\nu-1/6}\left(\frac{y^2}{4a}\right)$

	$f(x)$	$g(y) = \int\limits_0^\infty f(x)(xy)^{\frac{1}{2}}J_\nu(xy)dx$
11.7	$x^{\frac{1}{2}}[J_{\frac{1}{2}\nu}(\tfrac{1}{2}ax^2)]^2$ $\mathrm{Re}\ \nu > -1$	$-a^{-1}y^{\frac{1}{2}}J_{\frac{1}{2}\nu}(\tfrac{y^2}{4a})\,Y_{\frac{1}{2}\nu}(\tfrac{y^2}{4a})$
11.8	$x^{\frac{1}{2}}J_{\frac{1}{2}\nu}(\tfrac{1}{2}ax^2)Y_{\frac{1}{2}\nu}(\tfrac{1}{2}ax^2)$ $\mathrm{Re}\ \nu > -1$	$-a^{-1}y^{\frac{1}{2}}[J_{\frac{1}{2}\nu}(\tfrac{y^2}{4a})]^2$
11.9	$x^{\frac{1}{2}}J_{\frac{1}{2}\nu-\mu}(ax^2)\cdot$ $\cdot\,J_{\frac{1}{2}\nu+\mu}(ax^2)$ $\mathrm{Re}\ \nu > -\tfrac{1}{2}$	$2\pi^{-1}y^{-3/2}[e^{i\frac{\pi}{4}\nu}W_{\mu,\frac{1}{2}\nu}(z_1)W_{-\mu,\frac{1}{2}\nu}(z_1)\,+$ $+\,e^{-i\frac{\pi}{4}\nu}W_{\mu,\frac{1}{2}\nu}(z_2)W_{-\mu,\frac{1}{2}\nu}(z_2)]$ $z_{\frac{1}{2}} = \tfrac{y^2}{8a}e^{\pm i\frac{\pi}{2}}$
11.10	$x^{-\frac{1}{2}}J_\nu(ax^{-1})$ $\mathrm{Re}\ \nu > -\tfrac{1}{2}$	$y^{-\frac{1}{2}}J_{2\nu}[2(ay)^{\frac{1}{2}}]$
11.11	$x^{-5/2}J_\nu(ax^{-1})$ $\mathrm{Re}\ \nu > \tfrac{1}{2}$	$a^{-1}y^{\frac{1}{2}}J_{2\nu}[2(ay)^{\frac{1}{2}}]$
11.12	$x^{-3/2}J_{\nu-1}(ax^{-1})$ $\mathrm{Re}\ \nu > -\tfrac{1}{2}$	$a^{-\frac{1}{2}}J_{2\nu-1}[2(ay)^{\frac{1}{2}}]$

	$f(x)$	$g(y) = \int_0^\infty f(x)(xy)^{\frac{1}{2}} J_\nu(xy)\,dx$
11.13	$x^{\rho-3/2} J_\mu(ax^{-1})$ $-\mathrm{Re}(\nu+3/2) < \mathrm{Re}\ \rho <$ $< \mathrm{Re}(\mu+3/2)$	$A \cdot {}_0F_3(1+\mu, 1+\tfrac{1}{2}\mu-\tfrac{1}{2}\nu-\tfrac{1}{2}\rho, 1+\tfrac{1}{2}\mu+\tfrac{1}{2}\nu-\tfrac{1}{2}\rho; \tfrac{a^2y^2}{16})\ +$ $+B \cdot {}_0F_3(1+\nu, 1+\tfrac{1}{2}\nu-\tfrac{1}{2}\mu+\tfrac{1}{2}\rho, 1+\tfrac{1}{2}\mu+\tfrac{1}{2}\nu+\tfrac{1}{2}\rho; \tfrac{a^2y^2}{16})$ $A = \dfrac{2^{\rho-1-2\mu} a^\mu y^{\mu-\rho+\frac{1}{2}} \Gamma(\tfrac{1}{2}\nu+\tfrac{1}{2}\rho-\tfrac{1}{2}\mu)}{\Gamma(1+\mu)\Gamma(1+\tfrac{1}{2}\nu+\tfrac{1}{2}\mu-\tfrac{1}{2}\rho)}$ $B = \dfrac{2^{-\rho-1-2\nu} a^\nu y^{\nu+\rho+\frac{1}{2}} \Gamma(\tfrac{1}{2}\mu-\tfrac{1}{2}\nu-\tfrac{1}{2}\rho)}{\Gamma(\nu+1)\Gamma(1+\tfrac{1}{2}\nu+\tfrac{1}{2}\mu+\tfrac{1}{2}\rho)}$
11.14	$x^{-2\nu} J_{\frac{1}{2}-\nu}(ax^{-1})$ $-\tfrac{1}{2} < \mathrm{Re}\ \nu < 3$	$-\tfrac{1}{2}i\ \csc(2\pi\nu) \left(\dfrac{y}{a}\right)^{\nu-\frac{1}{2}} \cdot$ $\cdot\ [e^{i2\pi\nu} J_{1-2\nu}(z_1) J_{2\nu-1}(z_2) -$ $-\ e^{-i2\pi\nu} J_{2\nu-1}(z_1) J_{1-2\nu}(z_2)]$ $z_{\frac{1}{2}} = (\tfrac{1}{2}ay)^{\frac{1}{2}} e^{\pm i\frac{\pi}{4}}$
11.15	$x^{\rho-3/2} J_\mu(x^{-1})$ $\mathrm{Re}(\rho+\nu) > -\tfrac{1}{2}$ $\mathrm{Re}(\rho-\nu) < 3/2$	$\tfrac{1}{2}\pi\ \csc[\tfrac{1}{2}\pi(\mu-\nu-\rho)] y^{\nu+\frac{1}{2}} \cdot$ $\cdot\ [A_0F_3(1+\nu, 1+\tfrac{1}{2}\rho-\tfrac{1}{2}\mu+\tfrac{1}{2}\nu, 1+\tfrac{1}{2}\rho+\tfrac{1}{2}\mu+\tfrac{1}{2}\nu; \tfrac{y^2}{16})$ $-y^{-\mu} B_0F_3(1+\mu, 1+\tfrac{1}{2}\mu+\tfrac{1}{2}\nu-\tfrac{1}{2}\rho, 1+\tfrac{1}{2}\mu-\tfrac{1}{2}\nu-\tfrac{1}{2}\rho; \tfrac{y^2}{16})]$ $A^{-1} = 2^{\nu+\rho} \Gamma(1+\nu)\Gamma(1+\tfrac{1}{2}\rho-\tfrac{1}{2}\mu+\tfrac{1}{2}\nu) \cdot$ $\cdot\ \Gamma(1+\tfrac{1}{2}\rho+\tfrac{1}{2}\mu+\tfrac{1}{2}\nu)$ $B^{-1} = 2^{2\mu-\rho} \Gamma(1+\mu)\Gamma(1+\tfrac{1}{2}\mu+\tfrac{1}{2}\nu-\tfrac{1}{2}\rho) \cdot$ $\cdot\ \Gamma(1+\tfrac{1}{2}\mu-\tfrac{1}{2}\nu-\tfrac{1}{2}\rho)$

	$f(x)$	$g(y) = \int\limits_0^\infty f(x)(xy)^{\frac{1}{2}} J_\nu(xy)\,dx$
11.16	$x^{\frac{1}{2}}(b^2+x^2)^{-\frac{1}{2}}\exp\left(\dfrac{-a^2 b}{b^2+x^2}\right)\cdot$ $\cdot\; J_\nu\left(\dfrac{a^2 x}{b^2+x^2}\right)$ Re $\nu > -\frac{1}{2}$	$y^{-\frac{1}{2}}e^{-by}\; J_{2\nu}\left(2ay^{\frac{1}{2}}\right)$
11.17	$J_{2\nu-1}\left(ax^{\frac{1}{2}}\right)$ Re $\nu > -\frac{1}{2}$	$\frac{1}{2}ay^{-3/2}\,J_{\nu-1}\left(\frac{1}{4}a^2 y^{-1}\right)$
11.18	$x^{-\frac{1}{2}}J_{2\nu}\left(ax^{\frac{1}{2}}\right)$ Re $\nu > -\frac{1}{2}$	$y^{-\frac{1}{2}}J_\nu\left(\frac{1}{4}a^2 y^{-1}\right)$
11.19	$x^{-\frac{1}{2}}e^{-bx}J_{2\nu}\left(2ax^{\frac{1}{2}}\right)$ Re $\nu > -\frac{1}{2}$	$y^{\frac{1}{2}}(b^2+y^2)^{-\frac{1}{2}}\exp\left(-\dfrac{a^2 b}{b^2+y^2}\right)\cdot$ $\cdot\; J_\nu\left(\dfrac{a^2 y}{b^2+y^2}\right)$

	$f(x)$	$g(y) = \int\limits_{0}^{\infty} f(x)(xy)^{\frac{1}{2}}J_{\nu}(xy)\,dx$
11.20	$x^{\nu+\frac{1}{2}}(b^2+x^2)^{-\frac{1}{2}\mu}$. $\cdot J_{\mu}[a(b^2+x^2)^{\frac{1}{2}}]$ $\mathrm{Re}\,\mu > \mathrm{Re}\,\nu > -1$	$a^{-\mu}b^{\nu-\mu+1}(a^2-y^2)^{\frac{1}{2}\mu-\frac{1}{2}\nu-\frac{1}{2}}y^{\nu+\frac{1}{2}}$. $\cdot J_{\mu-\nu-1}[b(a^2-y^2)^{\frac{1}{2}}] \qquad y < a$ $\qquad\qquad\qquad 0 \qquad\qquad y > a$
11.21	$x^{\nu+\frac{1}{2}}(b^2+x^2)^{\frac{1}{2}\mu}$. $\cdot J_{\mu}[a(b^2+x^2)^{\frac{1}{2}}]$ $\mathrm{Re}\,\nu > -1,\ \mathrm{Re}\,(\nu+\mu) < 0$	$a^{\mu}b^{\nu+\mu+1}y^{\nu+\frac{1}{2}}(a^2-y^2)^{-\frac{1}{2}\mu-\frac{1}{2}\nu-\frac{1}{2}}$. $\cdot \{\sin(\pi\nu)Y_{\mu+\nu+1}[b(a^2-y^2)^{\frac{1}{2}}] -$ $-\cos(\pi\nu)J_{\mu+\nu+1}[b(a^2-y^2)^{\frac{1}{2}}]\}\ y \lessgtr a$ $-2\pi^{-1}\sin(\pi\mu)a^{\mu}b^{\nu+\mu+1}y^{\nu+\frac{1}{2}}$. $\cdot (y^2-a^2)^{-\frac{1}{2}\mu-\frac{1}{2}\nu-\frac{1}{2}}K_{\mu+\nu+1}[b(y^2-a^2)^{\frac{1}{2}}]$ $\qquad\qquad\qquad\qquad\qquad y > a$
11.22	$x^{\nu+\frac{1}{2}}(b^2+x^2)^{-\frac{1}{2}\mu-1}$. $\cdot J_{\mu-1}[a(b^2+x^2)^{\frac{1}{2}}]$ $\mathrm{Re}\,(\mu+2) > \mathrm{Re}\,\nu > -1$	$(\tfrac{1}{2}a)^{\mu-1}b^{\nu}[\Gamma(\mu)]^{-1}y^{\frac{1}{2}}K_{\nu}(by)$ $\qquad\qquad\qquad\qquad y > a$
11.23	$x^{\nu-3/2}(b^2+x^2)^{-\frac{1}{2}\mu}$. $\cdot J_{\mu}[a(b^2+x^2)^{\frac{1}{2}}]$ $\mathrm{Re}\,(\mu+2) > \mathrm{Re}\,\nu > 0$	$b^{-\mu}2^{\nu-1}\Gamma(\nu)y^{\frac{1}{2}-\nu}J_{\mu}(ab) \qquad y > a$

	$f(x)$	$g(y) = \int_0^\infty f(x)(xy)^{\frac{1}{2}} J_\nu(xy)\,dx$
11.24	$x^{\nu+\frac{1}{2}}(a^2+x^2)^{-1}(b^2+x^2)^{-\frac{1}{2}\mu}\cdot$ $\cdot\, J_\mu[c(b^2+x^2)^{\frac{1}{2}}]$ $-1 < \mathrm{Re}\ \nu < 2+\mathrm{Re}\ \mu$	$a^\nu y^{\frac{1}{2}}(b^2-a^2)^{-\frac{1}{2}\mu} J_\mu[c(b^2-a^2)^{\frac{1}{2}}]K_\nu(ay)$ $y \geq c$
11.25	$x^{\nu+2n-\frac{3}{2}}(a^2+x^2)^{-1}\cdot$ $\cdot\,(b^2+x^2)^{-\frac{1}{2}\mu}\cdot$ $\cdot\, J_\mu[c(b^2+x^2)^{\frac{1}{2}}]$ $n = 0, 1, 2, \cdots$ $-n < \mathrm{Re}\ \nu < 4-2n+\mathrm{Re}\ \mu$	$(-1)^{n+1}a^{\nu+2n-2}(b^2-a^2)^{-\frac{1}{2}\mu}\cdot$ $\cdot\, J_\mu[c(b^2-a^2)^{\frac{1}{2}}]K_\nu(ay)$ $y \leq c$
11.26	$x^{\nu+\frac{1}{2}}(b^2+x^2)^{-\frac{1}{2}\nu-\frac{1}{4}}\cdot$ $\cdot\, C_{2n+1}^{\nu+\frac{1}{2}}[b(b^2+x^2)^{-\frac{1}{2}}]\cdot$ $\cdot\, J_{\nu+3/2+2n}[a(b^2+x^2)^{\frac{1}{2}}]$ $\mathrm{Re}\ \nu > -1,\ n = 0,1,2,\cdots$	$(-1)^n(\tfrac{1}{2}\pi)^{-\frac{1}{2}}a^{-\frac{1}{2}-\nu}y^{\nu+\frac{1}{2}}(a^2-y^2)^{-\frac{1}{2}}\cdot$ $\cdot\, \sin[b(a^2-y^2)^{\frac{1}{2}}]C_{2n+1}^{\nu+\frac{1}{2}}[(1-\tfrac{y^2}{a^2})^{\frac{1}{2}}],$ $y < a$ $0\qquad ,\ y > a$
11.27	$x^{\nu+\frac{1}{2}}(b^2+x^2)^{-\frac{1}{2}\nu-\frac{1}{4}}\cdot$ $\cdot\, C_{2n}^{\nu+\frac{1}{2}}[b(b^2+x^2)^{-\frac{1}{2}}]\cdot$ $\cdot\, J_{\nu+\frac{1}{2}+2n}[a(b^2+x^2)^{\frac{1}{2}}]$ $n = 0, 1, 2, \cdots$ $\mathrm{Re}\ \nu > -1$	$(-1)^n(\tfrac{1}{2}\pi)^{-\frac{1}{2}}a^{-\frac{1}{2}-\nu}y^{\nu+\frac{1}{2}}(a^2-y^2)^{-\frac{1}{2}}\cdot$ $\cdot\, \cos[b(a^2-y^2)^{\frac{1}{2}}]C_{2n}^{\nu+\frac{1}{2}}[(1-\tfrac{y^2}{a^2})^{\frac{1}{2}}],\ y < a$ $0\qquad ,\ y > a$

	$f(x)$	$g(y) = \int_0^\infty f(x)(xy)^{\frac{1}{2}} J_\nu(xy)\,dx$
11.28	$x^{\nu-3/2}(b^2+x^2)^{-\frac{1}{2}n\mu} \cdot$ $\cdot \prod_{i=1}^{n} J_\nu[a_i(b^2+x^2)^{\frac{1}{2}}]$ $\mathrm{Re}\ \nu > 0,$ $\mathrm{Re}(\frac{1}{2}+\frac{1}{2}n+n\mu-\nu) > 0$	$2^{\nu-1}b^{-n\mu}\Gamma(\nu)y^{\frac{1}{2}-\nu} \prod_{i=1}^{n} J_\mu(a_i b)$ $y > \sum_{i=1}^{n} a_i$
11.29	$x^{\nu+\frac{1}{2}} \prod_{i=1}^{n} z_i^{-\mu_i} J_{\mu_i}(a_i z_i)$ $z_i = (x^2+b_i^2)^{\frac{1}{2}}$ $\frac{1}{2}n + \sum_{i=1}^{n} \mu_i - \frac{1}{2} > \mathrm{Re}\ \nu > -1$	$0 \qquad\qquad y > \sum_{i=1}^{n} a_i$
11.30	$x^{\nu-3/2} \prod_{i=1}^{n} z_i^{-\mu_i} J_{\mu_i}(a_i z_i)$ $z_i = (x^2+b_i^2)^{\frac{1}{2}}$ $\frac{1}{2}n + \sum_{i=1}^{n} \mu_i + 3/2 + \mathrm{Re}\ \nu > 0$	$2^{\nu-1}\Gamma(\nu)y^{\frac{1}{2}-\nu} \cdot$ $\cdot \prod_{i=1}^{n} [b_i^{-\mu_i} J_{\mu_i}(a_i b_i)]$ $y > \sum_{i=1}^{n} a_i$

	$f(x)$	$g(y) = \int_0^\infty f(x)(xy)^{\frac{1}{2}} J_\nu(xy)\,dx$
11.31	$x^{\nu+\frac{1}{2}}(a^2-x^2)^{\frac{1}{2}\mu}\cdot$ $\cdot J_\mu[b(a^2-x^2)^{\frac{1}{2}}]$ $\qquad x < a$ $\qquad 0 \quad x > a$ $\text{Re } \mu>-1,\ \text{Re } \nu>-1$	$a^{\mu+\nu+1}b^\mu y^{\nu+\frac{1}{2}}(b^2+y^2)^{-\frac{1}{2}\mu-\frac{1}{2}\nu-\frac{1}{2}}\cdot$ $\cdot J_{\mu+\nu+1}[a(b^2+y^2)^{\frac{1}{2}}]$
11.32	$\qquad 0 \quad x < b$ $x^{\frac{1}{2}-\nu}(x^2-b^2)^{\frac{1}{2}\mu}\cdot$ $\cdot J_\mu[a(x^2-b^2)^{\frac{1}{2}}]$ $\qquad x > b$ $\text{Re } \mu>-1,\ \text{Re } \mu>\text{Re } \nu$	$\qquad 0 \quad y < a$ $a^\mu b^{\mu-\nu+1} y^{\frac{1}{2}-\nu}(y^2-a^2)^{\frac{1}{2}\nu-\frac{1}{2}\mu-\frac{1}{2}}\cdot$ $\cdot J_{\nu-\mu-1}[b(y^2-a^2)^{\frac{1}{2}}] \qquad y > a$
11.33	$\qquad 0 \quad x < b$ $x^{\frac{1}{2}+\nu}(x^2-b^2)^{\frac{1}{2}\mu}\cdot$ $\cdot J_\mu[a(x^2-b^2)]$ $\qquad x > b$ $\text{Re } \mu > -1$ $\text{Re }(\nu+\mu) < 0$	$-2\pi^{-1}\sin(\pi\nu)a^\mu b^{\mu+\nu+1}y^{\frac{1}{2}+\nu}(a^2-y^2)^{-\frac{1}{2}\nu-\frac{1}{2}\mu-\frac{1}{2}}\cdot$ $\cdot K_{\mu+\nu+1}[b(a^2-y^2)^{\frac{1}{2}}] \qquad y < a$ $a^\mu b^{\nu+\mu+1}y^{\frac{1}{2}+\nu}(y^2-a^2)^{-\frac{1}{2}\mu-\frac{1}{2}\nu-\frac{1}{2}}\cdot$ $\cdot \{\sin(\pi\mu)Y_{\mu+\nu+1}[b(y^2-a^2)^{\frac{1}{2}}] -$ $- \cos(\pi\mu)J_{\mu+\nu+1}[b(y^2-a^2)^{\frac{1}{2}}]\}$ $\qquad y > a$

	$f(x)$	$g(y) = \int\limits_0^\infty f(x)\,(xy)^{\frac{1}{2}} J_\nu(xy)\,dx$
11.34	$\begin{array}{ll} 0 & x < a \\[4pt] x^{\frac{1}{2}-\nu}(x^2+c^2)^{-1}\cdot \\[4pt] \cdot (x^2-a^2)^{\frac{1}{2}\mu}\cdot \\[4pt] \cdot J_\mu[b(x^2-a^2)^{\frac{1}{2}}] \\[4pt] \quad\quad x > a \\[4pt] -1 < \mathrm{Re}\ \mu < 2+\mathrm{Re}\ \nu \end{array}$	$c^{-\nu}(a^2+c^2)^{\frac{1}{2}\mu}y^{\frac{1}{2}}K_\mu[b(a^2+c^2)^{\frac{1}{2}}]\cdot$ $\cdot\, I_\nu(cy) \qquad\qquad y < b$
11.35	$\begin{array}{ll} 0 & x < a \\[4pt] x^{\frac{1}{2}-\nu}(x^2+c^2)^{-1}\cdot \\[4pt] \cdot (x^2-a^2)^{\frac{1}{2}\mu+n-1}\cdot \\[4pt] \cdot J_\mu[b(x^2-a^2)^{\frac{1}{2}}] \\[4pt] \quad\quad x > a \\[4pt] -n < \mathrm{Re}\ \mu < 4-2n+\mathrm{Re}\ \nu \\[4pt] n = 0,\ 1,\ 2,\cdots \end{array}$	$(-1)^{n+1}c^{-\nu}(a^2+c^2)^{\frac{1}{2}\mu+n-1}y^{\frac{1}{2}}\cdot$ $\cdot\, K_\mu[b(a^2+c^2)^{\frac{1}{2}}]\,I_\nu(cy)\ ,\qquad y < b$
11.36	$\begin{array}{l} x^{\nu+2n+\frac{1}{2}}(1-x^2)^{\frac{1}{2}\lambda+m}\cdot \\[4pt] \cdot J_\lambda[b(1-x^2)^{\frac{1}{2}}],\ x<1 \\[4pt] \quad\quad 0\ ,\quad x > 1 \\[4pt] \mathrm{Re}\ \lambda>-1,\ \mathrm{Re}\ \nu>-1 \end{array}$	$b^{-\lambda}y^{\frac{1}{2}-\nu}\left(\dfrac{d}{b\,db}\right)^m\left(\dfrac{d}{y\,dy}\right)^n\cdot$ $\cdot\,(b^2+y^2)^{-\frac{1}{2}(\lambda+\nu+m+n+1)}\cdot$ $\cdot\, J_{\lambda+\nu+m+n+1}[(b^2+y^2)^{\frac{1}{2}}]$

$f(x)$	$g(y) = \int_0^\infty f(x)(xy)^{\frac{1}{2}} J_\nu(xy)\,dx$
11.37 $\quad x^\rho(1-x^2)^\mu J_\lambda[a(1-x^2)^{\frac{1}{2}}]$ $\qquad\quad x < 1$ $\qquad 0 \quad x > 1$	Bailey, W. N. 1938: Quart. J. Math. Oxford Series 9, 141-147.
11.38 $\quad x^{\nu+\frac{1}{2}}(b^2+x^2)^{-\frac{1}{2}\mu}\,\cdot$ $\qquad \cdot\, Y_\mu[a(b^2+x^2)^{\frac{1}{2}}]$ $\qquad\quad \mathrm{Re}\ \mu > \mathrm{Re}\ \nu > -1$	$a^{-\mu}b^{\nu-\mu+1}y^{\nu+\frac{1}{2}}(a^2-y^2)^{\frac{1}{2}\mu-\frac{1}{2}\nu-\frac{1}{2}}\,\cdot$ $\quad \cdot\, Y_{\mu-\nu-1}[b(a^2-y^2)^{\frac{1}{2}}] \qquad\quad y < a$ $-\ \frac{2}{\pi}\, a^{-\mu}b^{\nu-\mu+1}y^{\nu+\frac{1}{2}}(y^2-a^2)^{\frac{1}{2}\mu-\frac{1}{2}\nu-\frac{1}{2}}\,\cdot$ $\quad \cdot\, K_{\mu-\nu-1}[b(y^2-a^2)^{\frac{1}{2}}] \qquad\quad y > a$
11.39 $\quad x^{\nu+\frac{1}{2}}(b^2+x^2)^{\frac{1}{2}\mu}\,\cdot$ $\qquad \cdot\, Y_\mu[a(b^2+x^2)^{\frac{1}{2}}]$ $\qquad \mathrm{Re}\ \nu>-1, \mathrm{Re}(\nu+\mu) < 0$	$-a^\mu b^{\nu+\mu+1}y^{\nu+\frac{1}{2}}(a^2-y^2)^{-\frac{1}{2}\mu-\frac{1}{2}\nu-\frac{1}{2}}\,\cdot$ $\quad \cdot\, \{\cos(\pi\nu)Y_{\mu+\nu+1}[b(a^2-y^2)^{\frac{1}{2}}]\ +$ $\quad +\ \sin(\pi\nu)J_{\mu+\nu+1}[b(a^2-y^2)^{\frac{1}{2}}]\},\quad y < a$ $-\ 2\pi^{-1}\cos(\pi\mu)a^\mu b^{\nu+\mu+1}y^{\nu+\frac{1}{2}}\,\cdot$ $\quad \cdot\, (y^2-a^2)^{-\frac{1}{2}\mu-\frac{1}{2}\nu-\frac{1}{2}}K_{\mu+\nu+1}[b(y^2-a^2)^{\frac{1}{2}}]$ $\qquad\qquad\qquad\qquad\qquad\qquad y > a$
11.40 $\quad x^{\frac{1}{2}}J_{\frac{1}{2}\nu}\{a[(b^2+x^2)^{\frac{1}{2}}-b]\}\cdot$ $\qquad \cdot\, J_{\frac{1}{2}\nu}\{a[(b^2+x^2)^{\frac{1}{2}}+b]\}$ $\qquad\quad \mathrm{Re}\ \nu > -1$	$2\pi^{-1}y^{-\frac{1}{2}}(4a^2-y^2)^{-\frac{1}{2}}\cos[b(4a^2-y^2)^{\frac{1}{2}}]$ $\qquad\qquad\qquad\qquad\qquad\qquad y < 2a$ $\qquad\qquad 0 \qquad\qquad\qquad\qquad\quad y > 2a$

	f(x)	$g(y) = \int\limits_{0}^{\infty} f(x)(xy)^{\frac{1}{2}}J_{\nu}(xy)dx$
11.41	$x^{\nu+\frac{1}{2}}J_{\mu}\{b[(a^2+x^2)^{\frac{1}{2}}+x]\}\cdot$ $\cdot J_{\mu}\{b[(a^2+x^2)^{\frac{1}{2}}-x]\}$ $\mathrm{Re}\ \mu > \mathrm{Re}\ \nu > -1$	$2^{-2\mu}a^{\mu}b^{-\mu}[\Gamma(1+\mu)\Gamma(\mu-\nu)]^{-1}\ \cdot$ $\cdot y^{\nu+\frac{1}{2}}(4b^2-y^2)^{\mu-\nu-1}\ \cdot$ $\cdot {}_1F_2[\tfrac{1}{2}+\mu;2\mu+1,\mu-\nu;-(ab-\tfrac{1}{4}ay^2b^{-1})]$ $\qquad\qquad\qquad\qquad\qquad y < 2b$ $\qquad\qquad\qquad 0 \qquad\qquad y > 2b$
11.42	$x^{\frac{1}{2}}Y_{\frac{1}{2}\nu}(ax^2)$ $\mathrm{Re}\ \nu > -1$	$-\tfrac{1}{2}a^{-1}y^{\frac{1}{2}}\mathbf{H}_{\frac{1}{2}\nu}(\tfrac{y^2}{4a})$
11.43	$x^{\frac{1}{2}}J_{\frac{1}{4}\nu}(ax^2)Y_{\frac{1}{4}\nu}(ax^2)$ $\mathrm{Re}\ \nu > -1$	$-\tfrac{1}{2}a^{-1}y^{\frac{1}{2}}[J_{\frac{1}{4}\nu}(\tfrac{y^2}{16a})]^2$
11.44	$x^{-\frac{1}{2}}Y_{\nu}(ax^{-1})$ $-\tfrac{1}{2} < \mathrm{Re}\ \nu < \tfrac{1}{2}$	$-2\pi^{-1}y^{-\frac{1}{2}}[K_{2\nu}(2a^{\frac{1}{2}}y^{\frac{1}{2}})\ -$ $-\tfrac{1}{2}\pi Y_{2\nu}(2a^{\frac{1}{2}}y^{\frac{1}{2}})]$
11.45	$x^{-5/2}Y_{\nu}(ax^{-1})$ $\tfrac{1}{2} < \mathrm{Re}\ \nu < 5/2$	$2y^{\frac{1}{2}}(a\pi)^{-1}[K_{2\nu}(2a^{\frac{1}{2}}y^{\frac{1}{2}})+\tfrac{1}{2}\pi Y_{2\nu}(2a^{\frac{1}{2}}y^{\frac{1}{2}})]$

	$f(x)$	$g(y) = \int_0^\infty f(x)(xy)^{\frac{1}{2}} J_\nu(xy)\,dx$
11.46	$x^{-\frac{1}{2}} Y_{2\nu}(ax^{\frac{1}{2}})$ $\operatorname{Re}\nu > -\tfrac{1}{2}$	$2\sec(\pi\nu) y^{-\frac{1}{2}} [\tfrac{1}{2}\cos(\pi\nu) Y_\nu(\tfrac{a^2}{4y}) -$ $- Y_{-\nu}(\tfrac{a^2}{4y}) + \mathbf{H}_{-\nu}(\tfrac{a^2}{4y})]$
11.47	$\begin{aligned}& 0 \qquad x < a \\[4pt] & x^{\frac{1}{2}-\nu} \cdot \\[2pt] & \cdot (x^2+c^2)^{-1}(x^2-a^2)^{\frac{1}{2}\mu+n-\frac{1}{2}} \cdot \\[2pt] & \cdot Y_\mu[b(x^2-a^2)^{\frac{1}{2}}],\ x > a \\[2pt] & n = 0,\ 1,\ 2,\ \cdots \\[2pt] & -\tfrac{1}{2}-n < \operatorname{Re}\mu < 3-2n+\operatorname{Re}\nu \end{aligned}$	$(-1)^{n+1} c^{-\nu} y^{\frac{1}{2}}(a^2+c^2)^{\frac{1}{2}\mu+n-\frac{1}{2}} \cdot$ $\cdot K_\mu[b(a^2+c^2)^{\frac{1}{2}}] I_\nu(cy)$
11.48	$x^{\frac{1}{2}} J_{\frac{1}{2}\nu}\{a[(b^2+x^2)^{\frac{1}{2}}-b]\} \cdot$ $\cdot Y_{\frac{1}{2}\nu}\{a[(b^2+x^2)^{\frac{1}{2}}+b]\}$ $\operatorname{Re}\nu > -1$	$2\pi^{-1} y^{-\frac{1}{2}}(4a^2-y^2)^{-\frac{1}{2}} \cdot \sin[b(4a^2-y^2)^{\frac{1}{2}}]$ $\qquad\qquad\qquad\qquad,\quad y < 2a$ $-2\pi^{-1} y^{-\frac{1}{2}}(y^2-4a^2)^{-\frac{1}{2}} e^{-b(y^2-a^2)^{\frac{1}{2}}}$ $\qquad\qquad\qquad\qquad,\quad y > 2a$
11.49	$x^{\frac{1}{2}}[H^{(1)}_{\frac{1}{4}\nu+\mu}(ax^2) H^{(1)}_{\frac{1}{4}\nu-\mu}(ax^2)$ $-H^{(2)}_{\frac{1}{4}\nu+\mu}(ax^2) H^{(2)}_{\frac{1}{4}\nu-\mu}(ax^2)]$ $\operatorname{Re}\nu > -\tfrac{1}{2}$ $\operatorname{Re}(\tfrac{1}{2}\pm\mu+\tfrac{1}{2}\nu) > 0$	$-8i\pi^{-1}[\Gamma(1+\tfrac{1}{2}\nu)]^{-2} y^{-3/2} \cdot$ $\cdot \Gamma(\tfrac{1}{2}-\mu+\tfrac{1}{4}\nu)\Gamma(\tfrac{1}{2}+\mu+\tfrac{1}{4}\nu) \cdot$ $\cdot M_{\mu,\frac{1}{4}\nu}(\tfrac{y^2}{8a} e^{\frac{1}{2}i\pi}) M_{\mu,\frac{1}{4}\nu}(\tfrac{y^2}{8a} e^{-\frac{1}{2}i\pi})$

1.12 Modified Bessel Functions of Argument x

	$f(x)$	$g(y) = \int_0^\infty f(x)(xy)^{\frac{1}{2}}J_\nu(xy)\,dx$
12.1	$x^{\frac{1}{2}}e^{-bx^2}I_\nu(ax)$ Re $\nu > -1$	$\frac{1}{2}b^{-1}y^{\frac{1}{2}}\exp\left(\frac{a^2-y^2}{4b}\right)J_\nu\left(\frac{ay}{2b}\right)$
12.2	$\sinh(ax)K_{\nu-\frac{1}{2}}(ax)$ Re $\nu > -1$	$\frac{1}{2}\pi^{\frac{1}{2}}(2a)^{3/2-\nu}y^{-\frac{1}{2}}(4a^2+y^2)^{-\frac{1}{2}}\,\cdot$ $\cdot\,[(4a^2+y^2)^{\frac{1}{2}}+y]^{\nu-1}$
12.3	$e^{-ax}I_{\nu-\frac{1}{2}}(ax)$ Re $\nu > -\frac{1}{2}$	$(2a)^{\nu-\frac{1}{2}}\pi^{-\frac{1}{2}}y^{-\frac{1}{2}}(4a^2+y^2)^{-\frac{1}{2}}\,\cdot$ $\cdot\,[(y^2+4a^2)^{\frac{1}{2}}+y]^{1-\nu}$
12.4	$K_{\nu+\frac{1}{2}}(ax)$ Re $\nu > -1$	$(\tfrac{1}{2}\pi)^{\frac{1}{2}}a^{-\nu-\frac{1}{2}}y^{\nu+\frac{1}{2}}(a^2+y^2)^{-\frac{1}{2}}$
12.5	$x^{\frac{1}{2}}K_\nu(ax)$ Re $\nu > -1$	$a^{-\nu}y^{\nu+\frac{1}{2}}(a^2+y^2)^{-1}$
12.6	$K_\mu(ax)$ Re $(\nu\pm\mu) > -\frac{3}{2}$	$2^{\nu-\frac{1}{2}}a^{-\frac{1}{2}}\Gamma(3/4+\tfrac{1}{2}\nu+\tfrac{1}{2}\mu)\Gamma(3/4+\tfrac{1}{2}\nu-\tfrac{1}{2}\mu)\,\cdot$ $\cdot\,y^{\frac{1}{2}}(a^2+y^2)^{-\frac{1}{2}}p_{\mu-\frac{1}{2}}^{-\nu}[(1+\tfrac{y^2}{a^2})^{\frac{1}{2}}]$ $= 2^\nu\pi^{-\frac{1}{2}}\dfrac{\Gamma(3/4+\tfrac{1}{2}\nu+\tfrac{1}{2}\mu)\Gamma(3/4+\tfrac{1}{2}\nu-\tfrac{1}{2}\mu)}{\Gamma(\tfrac{1}{2}+\nu-\mu)}\,\cdot$ $\cdot\,(a^2+y^2)^{-\frac{1}{2}}e^{i\pi\mu}q_{\nu-\frac{1}{2}}^{-\mu}[(1+\tfrac{a^2}{y^2})^{\frac{1}{2}}]$

	$f(x)$	$g(y) = \int\limits_0^\infty f(x)(xy)^{\frac{1}{2}} J_\nu(xy)\,dx$
12.7	$x^{-1}K_\mu(ax)$ $\mathrm{Re}(\nu\pm\mu) > -\tfrac{1}{2}$	$2^{\nu-3/2}a^{-\frac{1}{2}}\Gamma(\tfrac{1}{4}+\tfrac{1}{2}\nu+\tfrac{1}{2}\mu)\,\Gamma(\tfrac{1}{4}+\tfrac{1}{2}\nu-\tfrac{1}{2}\mu)\;\cdot$ $\cdot\; y^{\frac{1}{2}}P_{\mu-\frac{1}{2}}^{-\nu}[(1+\tfrac{y^2}{a^2})^{\frac{1}{2}}]$ $= 2^{\nu-1}\pi^{-\frac{1}{2}}\dfrac{\Gamma(\tfrac{1}{4}+\tfrac{1}{2}\nu+\tfrac{1}{2}\mu)\,\Gamma(\tfrac{1}{4}+\tfrac{1}{2}\nu-\tfrac{1}{2}\mu)}{\Gamma(\tfrac{1}{2}+\nu-\mu)}\;\cdot$ $\cdot\; e^{i\pi\mu}q_{\nu-\frac{1}{2}}^{-\mu}[(1+\tfrac{a^2}{y^2})^{\frac{1}{2}}]$
12.8	$x^{\mu-\frac{1}{2}}K_\mu(ax)$ $\mathrm{Re}\,\nu > -1,$ $\mathrm{Re}(\nu+2\mu) > -1$	$2^\mu y^{-\frac{1}{2}}(a^2+y^2)^{-\frac{1}{2}\mu}e^{-i\pi\mu}q_{\frac{1}{2}\nu-\frac{1}{2}}^\mu(1+\tfrac{2a^2}{y^2})$
12.9	$x^{\pm\mu+\nu+\frac{1}{2}}K_\mu(ax)$ $\mathrm{Re}(\nu+1) > \lvert\mathrm{Re}\,\mu\rvert$	$2^{\nu\pm\mu}a^{\pm\mu}\Gamma(\pm\mu+\nu+1)y^{\nu+\frac{1}{2}}(a^2+y^2)^{\mp\mu-\nu-1}$
12.10	$x^{-\nu-\frac{1}{2}}K_\mu(ax)$ $-1 < \mathrm{Re}\,\mu < 1$	$\pi 2^{-\nu-1}\sec(\tfrac{1}{2}\pi\mu)a^{-1}y^{\frac{1}{2}}(a^2+y^2)^{\frac{1}{2}\nu}$ $\cdot\; p_{\frac{1}{2}\mu-\frac{1}{2}}^{-\nu}(1+2y^2a^{-2}) =$ $=\pi^{\frac{1}{2}}2^{-\nu-1}\sec(\tfrac{1}{2}\pi\mu)\,[\Gamma(\tfrac{1}{2}+\nu+\tfrac{1}{2}\mu)]^{-1}(a^2+y^2)^{\frac{1}{2}\nu-\frac{1}{4}}\;\cdot$ $\cdot\; e^{-i\frac{\pi}{2}\mu}q_{\nu-\frac{1}{2}}^{\frac{1}{2}\mu}[(y+\tfrac{1}{2}a^2y^{-1})(a^2+y^2)^{-\frac{1}{2}}]$

	$f(x)$	$g(y) = \int_0^\infty f(x)(xy)^{\frac{1}{2}} J_\nu(xy)\,dx$
12.11	$x^{\nu-\frac{1}{2}} K_\mu(ax)$ $\operatorname{Re}(2\nu\pm\mu) > -1$	$2^{\nu-1}a^{-1}\Gamma(\nu+\tfrac{1}{2}+\tfrac{1}{2}\mu)\Gamma(\nu+\tfrac{1}{2}-\tfrac{1}{2}\mu)y^{\frac{1}{2}}(a^2+y^2)^{-\frac{1}{2}\nu} \cdot$ $\cdot\, P_{\frac{1}{2}\mu-\frac{1}{2}}^{-\nu}(1+2y^2a^{-2}) =$ $= \pi^{-\frac{1}{2}}2^{\nu-1}\Gamma(\tfrac{1}{2}+\nu-\tfrac{1}{2}\mu)(a^2+y^2)^{-\frac{1}{2}\nu-\frac{1}{4}} \cdot$ $\cdot\, e^{-i\frac{\pi}{2}\mu} q_{\nu-\frac{1}{2}}^{\frac{1}{2}\mu}[(y+\tfrac{1}{2}a^2y^{-1})(a^2+y^2)^{-\frac{1}{2}}]$
12.12	$x^{-\lambda-\frac{1}{2}} K_\mu(ax)$ $\operatorname{Re}(\nu\pm\mu)>\operatorname{Re}\ \lambda-1$	$2^{-1-\lambda}a^{\lambda-\nu-1}[\Gamma(1+\nu)]^{-1}\Gamma(\tfrac{1}{2}+\tfrac{1}{2}\nu+\tfrac{1}{2}\mu-\tfrac{1}{2}\lambda) \cdot$ $\cdot\, \Gamma(\tfrac{1}{2}+\tfrac{1}{2}\nu-\tfrac{1}{2}\mu-\tfrac{1}{2}\lambda)y^{\nu+\frac{1}{2}} \cdot$ $\cdot\, {}_2F_1(\tfrac{1}{2}\nu+\tfrac{1}{2}\mu-\tfrac{1}{2}\lambda+\tfrac{1}{2},\tfrac{1}{2}\nu-\tfrac{1}{2}\mu-\tfrac{1}{2}\lambda+\tfrac{1}{2};\nu+1;-b^2a^{-2})$
12.13	$x^{-\lambda-\frac{1}{2}} K_\nu(ax)$ $\operatorname{Re}\ \lambda<1,$ $\operatorname{Re}(2\nu-\lambda)>-1$	$2^{-\lambda-1}\Gamma(\tfrac{1}{2}+\nu-\tfrac{1}{2}\lambda)\Gamma(\tfrac{1}{2}-\tfrac{1}{2}\lambda)y^{\frac{1}{2}}(a^2+y^2)^{\frac{1}{2}\lambda-\frac{1}{2}} \cdot$ $\cdot\, P_{\frac{1}{2}\lambda-\frac{1}{2}}^{-\nu}\left(\dfrac{a^2-y^2}{a^2+y^2}\right)$
12.14	$K_\mu(ax)$ $\operatorname{Re}(\nu\pm\mu)>-\tfrac{3}{2}$	$2^{\nu-\frac{1}{2}}a^{-\frac{1}{2}}\Gamma(\tfrac{3}{4}+\tfrac{1}{2}\nu+\tfrac{1}{2}\mu)\Gamma(\tfrac{3}{4}+\tfrac{1}{2}\nu-\tfrac{1}{2}\mu) \cdot$ $\cdot\, y^{\frac{1}{2}}(a^2+y^2)^{-\frac{1}{4}}p_{\mu-\frac{1}{2}}^{-\nu}[(1+y^2a^{-2})^{\frac{1}{2}}] =$ $= 2^{\frac{1}{2}-\mu}\dfrac{\Gamma(\tfrac{3}{4}+\tfrac{1}{2}\nu-\tfrac{1}{2}\mu)}{\Gamma(\tfrac{1}{4}+\tfrac{1}{2}\nu+\tfrac{1}{2}\mu)}(a^2+y^2)^{-\frac{1}{4}}e^{-i\pi\mu}q_{\nu-\frac{1}{2}}^{\mu}[(1+a^2y^{-2})^{\frac{1}{2}}]$

	$f(x)$	$g(y) = \int\limits_0^\infty f(x)(xy)^{\frac12}J_\nu(xy)\,dx$
12.15	$x^{-\lambda-\frac12}K_\mu(ax)$ $\mathrm{Re}(\nu\pm\mu-\lambda) > -1$	$\dfrac{2^{-\lambda-1}a^{\lambda-1-\nu}y^{\nu+\frac12}}{\Gamma(1+\nu)}\,\Gamma(\tfrac12+\tfrac12\nu+\tfrac12\mu-\tfrac12\lambda)\,\cdot$ $\cdot\;\Gamma(\tfrac12+\tfrac12\nu-\tfrac12\mu-\tfrac12\lambda)\,\cdot$ $\cdot\;_2F_1(\tfrac12+\tfrac12\nu+\tfrac12\mu-\tfrac12\lambda,\tfrac12+\tfrac12\nu-\tfrac12\mu-\tfrac12\lambda;1+\nu;\,-\dfrac{y^2}{a^2})$
12.16	$x^{\frac12}I_{\frac12\nu-\frac12\mu}(ax)\,\cdot$ $\cdot\,K_{\frac12\nu+\frac12\mu}(ax)$ $\mathrm{Re}\,\nu>-1,\ \mathrm{Re}(\mu-\nu)<2$	$(2a)^{-\mu}y^{-\frac12}(4a^2+y^2)^{-\frac12}\,\cdot$ $\cdot\;[(4a^2+y^2)^{\frac12}+y]^{\mu}$
12.17	$x^{\frac12}I_{\frac12\nu}(ax)K_{\frac12\nu}(ax)$ $\mathrm{Re}\,\nu > -1$	$y^{-\frac12}(4a^2+y^2)^{-\frac12}$
12.18	$x^{-\frac12}I_\mu(ax)K_\mu(ax)$ $\mathrm{Re}(2\mu+\nu) > -1$ $\mathrm{Re}\,\nu > -1$	$y^{-\frac12}e^{-i\pi\mu}p^{-\mu}_{\frac12\nu-\frac12}[(1+4a^2y^{-2})^{\frac12}]\,\cdot$ $\cdot\;q^{\mu}_{\frac12\nu-\frac12}[(1+4a^2y^{-2})^{\frac12}]$
12.19	$x^{-\frac12}I_\mu(ax)K_\mu(ax)$ $\mathrm{Re}\,\nu>-1,$ $\mathrm{Re}(\nu+2_\mu)>-1$	$(2a)^{-1}y^{\frac12}e^{-i\frac{\pi}{2}\nu}\,\cdot$ $\cdot\;p^{-\frac12\nu}_{\mu-\frac12}[(1+\dfrac{y^2}{4a^2})^{\frac12}]q^{\frac12\nu}_{\mu-\frac12}[(1+\dfrac{y^2}{4a^2})^{\frac12}]$

	$f(x)$	$g(y) = \int\limits_{0}^{\infty} f(x)(xy)^{\frac{1}{2}} J_{\nu}(xy)\,dx$
12.20	$x^{\frac{1}{2}} K_0(ax) J_{\nu}(bx)$ $\mathrm{Re}\ \nu > -1$	$y^{\frac{1}{2}}(z_1 z_2)^{-1}(z_2 - z_1)^{\nu}(z_2 + z_1)^{-\nu}$ $z_{\frac{1}{2}} = [a^2 + (b\mp y)^2]^{\frac{1}{2}}$
12.21	$x^{\nu+\frac{1}{2}} J_{\nu-1}(ax) K_{\nu-1}(ax)$ $0 < \mathrm{Re}\ \nu < \frac{1}{2}$	$\pi^{-\frac{1}{2}} 2^{3\nu-1} a^{2\nu-2} \Gamma(\tfrac{1}{2}+\nu) y^{\nu+5/2} (y^4+4a^4)^{-\nu-\frac{1}{2}}$
12.22	$x^{\nu+\frac{1}{2}} I_{\nu}(ax) K_{\nu}(bx)$ $\mathrm{Re}\ \nu > -\frac{1}{2}$	$\pi^{-\frac{1}{2}} 2^{3\nu} (ab)^{\nu} \Gamma(\tfrac{1}{2}+\nu) y^{\nu+\frac{1}{2}}\ \cdot$ $\cdot\ [(b^2+y^2-a^2)^2 + 4a^2 y^2]^{-\nu-\frac{1}{2}}$
12.23	$x^{\nu-\frac{1}{2}}\ \cdot$ $\cdot I_{\nu-\frac{1}{2}}(\tfrac{1}{2}ax) K_{\nu-\frac{1}{2}}(\tfrac{1}{2}ax)$ $0 < \mathrm{Re}\ \nu < {}^{3}/_{2}$	$(2a)^{\nu-1} \Gamma(\nu) y^{\frac{1}{2}-\nu} p_{-\nu}\left[\dfrac{y^2+2a^2}{2a(y^2+a^2)^{\frac{1}{2}}}\right]$ $= 2^{\nu} \pi^{-\frac{1}{2}} a^{\nu-\frac{1}{2}} y^{-\nu-\frac{1}{2}}\ \cdot$ $\cdot\ (a^2+y^2)^{-\frac{1}{4}} e^{i\pi(\frac{1}{2}-\nu)} q_{-\frac{1}{2}}^{\nu-\frac{1}{2}}(1+2a^2 y^{-2})$
12.24	$x^{\mu+\frac{1}{2}} I_{\nu}(ax) K_{\mu}(ax)$ $\mathrm{Re}\ \nu>-1,\ \mathrm{Re}\ \mu <\frac{1}{2}$ $\mathrm{Re}(\nu+\mu)>-1$	$(2\pi)^{-\frac{1}{2}} a^{-1} y^{-\mu-\frac{1}{2}} e^{-(\mu-\frac{1}{2}\nu+\frac{1}{4})\pi i}\ \cdot$ $\cdot\ (1+ \dfrac{y^2}{4a^2})^{-\frac{1}{2}\mu-\frac{1}{4}} q_{\nu-\frac{1}{2}}^{\mu+\frac{1}{2}}(i\tfrac{y}{2a})$

	$f(x)$	$g(y) = \int_0^\infty f(x)(xy)^{\frac{1}{2}} J_\nu(xy)\,dx$
12.25	$x^{\frac{1}{2}\pm\mu} J_\nu(bx) K_\mu(ax)$ Re $\nu > -1$, Re$(\nu+\mu) > -1$	$(2\pi)^{-\frac{1}{2}} a^{\pm\mu} b^{\mp\mu-1} y^{\mp\mu-\frac{1}{2}} (z^2-1)^{-\frac{1}{4}\mp\frac{1}{2}\mu}\;\cdot$ $e^{-i\pi(\frac{1}{2}\pm\mu)} Q_{\nu-\frac{1}{2}}^{\frac{1}{2}\pm\mu}(z)$ $z = \dfrac{a^2+b^2+y^2}{2by}$
12.26	$x^{\nu-2\mu+\frac{1}{2}} I_\mu(ax) K_\mu(ax)$ Re $\nu > -1$, Re$(\nu-\mu+1) > 0$ Re$(\nu-2\mu) < \frac{1}{2}$	$2^{\nu-2\mu}\,\dfrac{\Gamma(\nu-\mu+1)}{\Gamma(\mu+1)}\;\cdot$ $\cdot\; y^{2\mu-\nu-\frac{3}{2}}\,{}_2F_1(\nu-\mu+1,\ \tfrac{1}{2};\ \mu+1;\ -\dfrac{4a^2}{y^2})$
12.27	$x^{\frac{1}{2}+\mu} J_\nu(bx) K_\mu(ax)$ Re $\nu > -1$, Re$(\nu+\mu) > -1$	$\tfrac{1}{2} a^\mu b^{-\mu-1} \Gamma(\nu+\mu+1) y^{-\mu-\frac{1}{2}}\;\cdot$ $\cdot\;(z^2-1)^{-\frac{1}{2}\mu-\frac{1}{2}} P_\mu^{-\nu}[(z)(z^2-1)^{-\frac{1}{2}}]$ $z = \dfrac{a^2+b^2+y^2}{2by}$
12.28	$x^{\nu+\frac{1}{2}} J_\nu(ax) K_\nu(bx)$ Re $\nu > -\frac{1}{2}$	$\pi^{-\frac{1}{2}} 2^{3\nu} (ab)^\nu \Gamma(\tfrac{1}{2}+\nu) y^{\nu+\frac{1}{2}}\;\cdot$ $\cdot\;[(a^2+b^2+y^2)^2-4a^2y^2]^{-\nu-\frac{1}{2}}$
12.29	$x^{\nu+\frac{1}{2}} J_{\nu-1}(ax) K_{\nu-1}(ax)$ $0 < $ Re $\nu < \frac{1}{2}$	$\pi^{-\frac{1}{2}} 2^{3\nu-1} a^{2\nu-2} \Gamma(\tfrac{1}{2}+\nu) y^{\nu+\frac{5}{2}}(y^4+4a^4)^{-\nu-\frac{1}{2}}$

	$f(x)$	$g(y) = \int_0^\infty f(x)(xy)^{\frac{1}{2}} J_\nu(xy)\,dx$
12.30	$x^{\frac{1}{2}+\nu} K_\mu(ax) I_\mu(bx)$ $\mathrm{Re}\ \nu > -1,\ \mathrm{Re}(\nu+\mu) > -1$ $a > b$ for $a=b$, $-1 < \mathrm{Re}\ \nu < \frac{1}{2}$	$(2\pi)^{-\frac{1}{2}}(ab)^{-\nu-1} y^{\nu+\frac{1}{2}}(z^2-1)^{-\frac{1}{2}\nu-\frac{1}{4}}\ \cdot$ $\cdot\ e^{-i\pi(\nu+\frac{1}{2})} q_{\mu-\frac{1}{2}}^{\frac{1}{2}+\nu}(z) =$ $= \frac{1}{2}(ab)^{-\nu-1}\Gamma(\nu+\mu+1) y^{\nu+\frac{1}{2}}(z^2-1)^{-\frac{1}{2}\nu-\frac{1}{2}}\ \cdot$ $\cdot\ p_\nu^{-\mu}[z(z^2-1)^{-\frac{1}{2}}]$ $z = (2ab)^{-1}(a^2+b^2+y^2)$
12.31	$x^{\frac{1}{2}+\nu} K_\mu(ax) K_\mu(bx)$ $\mathrm{Re}\ \nu > -1,\ \mathrm{Re}(\nu\pm\mu) > -1$	$\pi^{\frac{1}{2}}2^{-3/2}(ab)^{-\nu-1}\Gamma(1+\nu+\mu)\Gamma(1+\nu-\mu)y^{\nu+\frac{1}{2}}\ \cdot$ $\cdot\ (z^2-1)^{-\frac{1}{2}\nu-\frac{1}{4}} p_{\mu-\frac{1}{2}}^{-\nu-\frac{1}{2}}(z) =$ $= \frac{1}{2}(ab)^{-\nu-1}\Gamma(1+\nu+\mu)y^{\nu+\frac{1}{2}}(z^2-1)^{-\frac{1}{2}\nu-\frac{1}{2}}\ \cdot$ $\cdot\ e^{i\pi\mu} q_\nu^{-\mu}[z(z^2-1)^{-\frac{1}{2}}]$ $z = (2ab)^{-1}(a^2+b^2+y^2)$
12.32	$x^{\rho-\mu+\nu+\frac{1}{2}} K_\rho(ax) I_\mu(bx)$ $a > b,\ \mathrm{Re}\ \rho > -1$ $\mathrm{Re}(\rho+\nu) > -1$	$\int_0^\infty x^{\rho-\mu+\nu+\frac{1}{2}} J_\rho(ax) J_\mu(bx)(xy)^{\frac{1}{2}} K_\nu(xy)\,dx$
12.33	$x^{\rho+\nu-\mu+\frac{1}{2}} J_\mu(ax) K_\rho(bx)$ $\mathrm{Re}(\nu,\mu,\rho,\nu+\rho) > -1$	$2^{\rho+\nu-\mu-1}[\Gamma(1+\mu)]^{-1}\Gamma(\rho+\nu+1)\Gamma(1+\rho)\ \cdot$ $\cdot\ a^{\mu-\rho-\nu-2} y^{\frac{1}{2}}(\cosh\sigma-\cos\delta) P_{\rho+\nu-\mu}(\cos\delta)\cdot$ $\cdot\ p_{\rho+\nu-\mu}^{-\nu}(\cosh\alpha)$ $y+ib = i\,a\,\cot(\frac{1}{2}\delta+i\frac{1}{2}\sigma)$

	$f(x)$	$g(y) = \int_0^\infty f(x)(xy)^{\frac{1}{2}}J_\nu(xy)\,dx$
12.34	$x^{\mu+\frac{1}{2}}I_\nu(ax)K_\mu(bx)$ Re $\nu>-1$, Re$(\nu+\mu)>-1$ $b > a$	$(2\pi)^{-\frac{1}{2}}a^{-\mu-1}b^\mu y^{-\mu-\frac{1}{2}}e^{-i\pi(\mu-\frac{1}{2}\nu+\frac{1}{4})}$. $\cdot\ (z^2+1)^{-\frac{1}{2}\mu-\frac{1}{4}}q_{\nu-\frac{1}{2}}^{\mu+\frac{1}{2}}(iz)$ $z = \dfrac{b^2-a^2+y^2}{2ay}$
12.35	$x^\lambda J_\mu(ax)K_\rho(bx)$ $x^\lambda I_\mu(ax)K_\rho(bx)$	Bailey, W. N., 1936: Proc. London Math. Soc., (2)40, 37-48
12.36	$x^{-\nu-\frac{1}{2}}[K_{\nu+\frac{1}{2}}(\frac{1}{2}ax)]^2$ $-1 < \text{Re } \nu < 0$	$\pi(2a)^{-\nu-1}\Gamma(-\nu)y^{\nu+\frac{1}{2}}(a^2+y^2)^{\frac{1}{2}\nu}$. $\cdot\ P_\nu\Big[\dfrac{2a^2+y^2}{2a(a^2+y^2)^{\frac{1}{2}}}\Big] =$ $\pi^{\frac{1}{2}}2^{-\nu}a^{-\nu-\frac{1}{2}}y^{\nu-\frac{1}{2}}$. $\cdot\ (a^2+y^2)^{\frac{1}{2}\nu-\frac{1}{4}}e^{i\pi(\nu+\frac{1}{2})}q_{-\frac{1}{2}}^{-\nu-\frac{1}{2}}(1+2a^2y^{-2})$
12.37	$x^{\frac{1}{2}}[K_\mu(\frac{1}{2}ax)]^2$ Re$(\frac{1}{2}\nu\pm\mu) > -1$	$2\,\dfrac{\Gamma(1+\frac{1}{2}\nu+\mu)}{\Gamma(\frac{1}{2}\nu-\mu)}\,y^{-\frac{1}{2}}(a^2+y^2)^{-\frac{1}{2}}$. $\cdot e^{2\pi i\mu}q_{\frac{1}{2}\nu}^{-\mu}[(1+a^2y^{-2})^{\frac{1}{2}}]q_{\frac{1}{2}\nu-1}^{-\mu}[(1+a^2y^{-2})^{\frac{1}{2}}]$ $= \pi a^{-1}y^{\frac{1}{2}}\Gamma(1+\frac{1}{2}\nu+\mu)\Gamma(1+\frac{1}{2}\nu-\mu)(a^2+y^2)^{-\frac{1}{2}}$. $\cdot p_{\mu-\frac{1}{2}}^{-\frac{1}{2}\nu-\frac{1}{2}}[(1+y^2a^{-2})^{\frac{1}{2}}]p_{\mu-\frac{1}{2}}^{-\frac{1}{2}\nu+\frac{1}{2}}[(1+y^2a^{-2})^{\frac{1}{2}}]$

	$f(x)$	$g(y) = \int\limits_0^\infty f(x)(xy)^{\frac{1}{2}}J_\nu(xy)dx$
12.38	$x^{-\frac{1}{2}}[K_\mu(\tfrac{1}{2}ax)]^2$ $\mathrm{Re}(\tfrac{1}{2}\nu\pm\mu) > -\tfrac{1}{2}$	$\dfrac{\Gamma(\tfrac{1}{2}+\tfrac{1}{2}\nu+\mu)}{\Gamma(\tfrac{1}{2}+\tfrac{1}{2}\nu-\mu)}\, y^{-\frac{1}{2}}e^{2\pi i\mu}\{q^{-\mu}_{\frac{1}{2}\nu-\frac{1}{2}}[(1+a^2y^{-2})^{\frac{1}{2}}]\}^2$ $= \tfrac{1}{2}\pi a^{-1}y^{\frac{1}{2}}\Gamma(\tfrac{1}{2}+\tfrac{1}{2}\nu+\mu)\Gamma(\tfrac{1}{2}+\tfrac{1}{2}\nu-\mu)\ \cdot$ $\cdot\ \{p^{-\frac{1}{2}\nu}_{\mu-\frac{1}{2}}[(1+y^2a^{-2})^{\frac{1}{2}}]\}^2$
12.39	$x^{\frac{1}{2}}K_{\mu-\frac{1}{2}}(\tfrac{1}{2}ax)K_{\mu+\frac{1}{2}}(\tfrac{1}{2}ax)$ $\mathrm{Re}\ \nu>-1,\mathrm{Re}(\tfrac{1}{2}\nu\pm\mu)>-1$	$\pi a^{-1}\Gamma(1+\tfrac{1}{2}\nu+\mu)\Gamma(1+\tfrac{1}{2}\nu-\mu)(a^2+y^2)^{-\frac{1}{2}}\ \cdot$ $\cdot\ y^{\frac{1}{2}}p^{-\frac{1}{2}\nu}_{-\mu}[(1+y^2a^{-2})^{\frac{1}{2}}]p^{-\frac{1}{2}\nu}_{\mu}[(1+y^2a^{-2})^{\frac{1}{2}}]$ $= 2\,\dfrac{\Gamma(1+\tfrac{1}{2}\nu+\mu)}{\Gamma(\tfrac{1}{2}\nu-\mu)}\, y^{-\frac{1}{2}}(a^2+y^2)^{-\frac{1}{2}}\ \cdot$ $\cdot e^{i2\pi\mu}q^{\frac{1}{2}-\mu}_{\frac{1}{2}\nu-\frac{1}{2}}[(1+a^2y^{-2})^{\frac{1}{2}}]q^{-\frac{1}{2}-\mu}_{\frac{1}{2}\nu-\frac{1}{2}}[(1+a^2y^{-2})^{\frac{1}{2}}]$
12.40	$x^\lambda K_\mu(ax)K_\rho(bx)$	Bailey, W. N., 1936: Proc. London Math. Soc. (2) 41, 215-220
12.41	$x^{-\nu-\frac{1}{2}}[K_{\nu+\frac{1}{2}}(\tfrac{1}{2}ax)]^2$ $-1 < \mathrm{Re}\ \nu < 0$	$(2a)^{-\nu-1}\pi\Gamma(-\nu)y^{\nu+\frac{1}{2}}(a^2+y^2)^{\frac{1}{2}\nu}\ \cdot$ $\cdot\ P_\nu[\dfrac{2a^2+y^2}{2a(a^2+y^2)^{\frac{1}{2}}}]$

1.13 Modified Bessel Functions of Other Arguments

	$f(x)$	$g(y) = \int_0^\infty f(x)(xy)^{\frac{1}{2}} J_\nu(xy)\,dx$
13.1	$x^{\frac{1}{2}-\nu}\exp(-\tfrac{1}{4}a^2x^2) \cdot$ $\cdot\ I_\nu(\tfrac{1}{4}a^2x^2)$ $\operatorname{Re}\nu > -\tfrac{1}{2}$	$(\tfrac{1}{2}\pi)^{-\frac{1}{2}}a^{-1}y^{\nu-\frac{1}{2}}\exp(-\dfrac{y^2}{4a^2})D_{-2\nu}(\tfrac{y}{a})$
13.2	$x^{\nu-3/2}\exp(-\tfrac{1}{4}a^2x^2) \cdot$ $\cdot\ I_{\nu+1}(\tfrac{1}{4}a^2x^2)$ $\operatorname{Re}\nu > -1$	$(\tfrac{1}{2}\pi)^{-\frac{1}{2}}y^{\nu+\frac{1}{2}}\exp(-\tfrac{1}{4}\dfrac{y^2}{a^2})\ D_{-2\nu-3}(\tfrac{y}{a})$
13.3	$x^{\frac{1}{2}}\exp(-\tfrac{1}{4}ax^2)I_{\frac{1}{2}\nu}(\tfrac{1}{4}ax^2)$ $\operatorname{Re}\nu > -1$	$(\tfrac{1}{2}\pi a y)^{-\frac{1}{2}}\exp(-\dfrac{y^2}{2a})$
13.4	$x^{1/3\,(\nu+\frac{1}{2})}\exp(-\tfrac{1}{4}ax^2) \cdot$ $\cdot\ I_{1/3\,(\nu+\frac{1}{2})}(\tfrac{1}{4}ax^2)$ $-1 < \operatorname{Re}\nu < {}^{5}\!/{}_{2}$	$\pi^{-1}a^{-1/3\,(\nu+2)}y^{1/3\,(\nu+\frac{1}{2})}\exp(-\tfrac{1}{4}\dfrac{y^2}{a}) \cdot$ $\cdot\ K_{1/3\,(\nu+\frac{1}{2})}(\tfrac{1}{4}\dfrac{y^2}{a})$
13.5	$x^{-1/3\,(\nu-\frac{1}{2})}\exp(-\tfrac{1}{4}ax^2)\cdot$ $\cdot\ I_{1/3\,(\nu-\frac{1}{2})}(\tfrac{1}{4}ax^2)$ $\operatorname{Re}\nu > -1$	$a^{1/3\,(\nu-2)}y^{1/3\,(\frac{1}{2}-\nu)}\exp(-\tfrac{1}{4}\dfrac{y^2}{a}) \cdot$ $\cdot\ I_{1/3\,(\nu-\frac{1}{2})}(\tfrac{1}{4}\dfrac{y^2}{a})$

	$f(x)$	$g(y) = \int\limits_0^\infty f(x)(xy)^{1/2} J_\nu(xy)\,dx$
13.6	$x^{1/2} e^{-ax^2} I_{1/2\nu}(bx^2)$ Re $\nu > -1$, $a > b$	$\frac{1}{2}(a^2-b^2)^{-1/2} y^{1/2} \exp[-\frac{1}{4}ay^2(a^2-b^2)^{-1}] \cdot$ $\cdot\, I_{1/2\nu}[\frac{1}{4}by^2(a^2-b^2)^{-1}]$
13.7	$x^{1/2+2\mu-\nu} \exp(-\frac{1}{4}ax^2) \cdot$ $\cdot\, I_\mu(\frac{1}{4}ax^2)$ Re $\nu > 2$ Re $\mu+\frac{1}{2} > -\frac{1}{2}$	$\pi^{-1/2}\, \dfrac{\Gamma(\frac{1}{2}+\mu)}{\Gamma(\frac{1}{2}-\mu+\nu)}\, 2^{3/2\mu-1/2\nu+3/4}\, a^{-1/4-1/2\mu+1/2\nu}$ $y^{-\mu-1}\exp(-\frac{y^2}{4a}) M_{1/4+3/2\mu-1/2\nu,\, -1/4-1/2\mu+1/2\nu}\left(\frac{y^2}{2a}\right)$
13.8	$x^{1/2-2\mu+\nu} \exp(-\frac{1}{4}ax^2) \cdot$ $\cdot\, I_\mu(\frac{1}{4}ax^2)$ $-1<$ Re $\nu<\frac{1}{2}+2$ Re μ	$\pi^{-1/2} 2^{3/4+1/2\nu-3/2\mu}\, a^{-1/4-1/2\nu+1/2\mu}\, y^{\mu-1}\exp(-\frac{y^2}{4a}) \cdot$ $\cdot\, W_{1/4+1/2\nu-3/2\mu,\, -1/4+1/2\mu-1/2\nu}\left(\frac{y^2}{2a}\right)$
13.9	$x^{1/2} K_{1/2\nu}(\frac{1}{4}ax^2)$ Re $\nu > -1$	$\pi a^{-1} y^{1/2}\left[I_{1/2\nu}\left(\frac{y^2}{a}\right) - \mathbf{L}_{1/2\nu}\left(\frac{y^2}{a}\right)\right]$
13.10	$x^{3/2} K_{1/2\nu+1/2}(\frac{1}{4}ax^2)$ Re $\nu > -1$	$2\pi a^{-2} y^{3/2}\left[I_{1/2\nu-1/2}\left(\frac{y^2}{a}\right) - \mathbf{L}_{1/2\nu-1/2}\left(\frac{y^2}{a}\right)\right]$
13.11	$x^{1/3(\nu+1/2)} \exp(-\frac{1}{4}ax^2)$ $K_{1/3(\nu+1/2)}(ax^2)$ Re $\nu > -1$	$\pi a^{-1/3(\nu+2)} y^{1/3(\nu+1/2)} \exp(-\frac{y^2}{4a}) I_{1/3(\nu+1/2)}\left(\frac{y^2}{4a}\right)$

	$f(x)$	$g(y) = \int\limits_{0}^{\infty} f(x)(xy)^{\frac{1}{2}} J_{\nu}(xy)\,dx$
13.12	$x^{1/3\,(\nu+\frac{1}{2})} \exp(\frac{1}{4}ax^2)\ \cdot$ $\cdot\ K_{1/3\,(\nu+\frac{1}{2})}(\frac{1}{4}ax^2)$ $-1 < \mathrm{Re}\ \nu < {}^{5}/_{2}$	$a^{-1/3\,(\nu+2)} y^{1/3\,(\frac{1}{2}+\nu)} \exp(\frac{y^2}{4a}) K_{1/3\,(\nu+\frac{1}{2})}(\frac{y^2}{4a})$
13.13	$x^{2\mu+\nu+\frac{1}{2}} \exp(-\frac{1}{4}ax^2)$ $K_{\mu}(\frac{1}{4}ax^2)$ $\mathrm{Re}\ \nu > -1,\ \mathrm{Re}(2\mu+\nu)>-1$	$\pi^{\frac{1}{2}}\,\frac{\Gamma(1+2\mu+\nu)}{\Gamma(\mu+\nu+3/2)}\,2^{3/2\,\mu+\frac{1}{2}\nu+3/4}\,a^{-\frac{1}{4}-\frac{1}{2}\mu-\frac{1}{2}\nu}\ \cdot$ $\cdot\ y^{-1-\mu} \exp(-\frac{y^2}{4a}) M_{\frac{1}{4}+3/2\,\mu+\frac{1}{2}\nu,\ \frac{1}{4}+\frac{1}{2}\mu+\frac{1}{2}\nu}(\frac{y^2}{2a})$
13.14	$x^{2\mu+\nu+\frac{1}{2}} \exp(\frac{1}{4}ax^2)$ $K_{\mu}(\frac{1}{4}ax^2)$ $\mathrm{Re}\ \nu > -1,$ $-1 < \mathrm{Re}(2\mu+\nu) < 1$	$\pi^{\frac{1}{2}}\,\frac{\Gamma(1+\nu+2\mu)}{\Gamma(\frac{1}{2}-\mu)}\,2^{7/4+3/2\,\mu+\frac{1}{2}\nu}\,a^{-\frac{1}{4}-\frac{1}{2}\mu-\frac{1}{2}\nu}\,y^{-\mu-1}\ \cdot$ $\cdot\ \exp(\frac{y^2}{4a}) W_{-\frac{1}{4}-3/2\,\mu-\frac{1}{2}\nu,\ \frac{1}{4}+\frac{1}{2}\mu+\frac{1}{2}\nu}(\frac{y^2}{2a})$
13.15	$x^{\frac{1}{2}} I_{\frac{1}{2}\nu}(\frac{1}{4}ax^2) K_{\frac{1}{2}\nu}(\frac{1}{4}ax^2)$ $\mathrm{Re}\ \nu > -1$	$a^{-1} y^{\frac{1}{2}} I_{\frac{1}{4}\nu}(\frac{y^2}{4a}) K_{\frac{1}{4}\nu}(\frac{y^2}{4a})$
13.16	$x^{\frac{1}{2}} I_{\frac{1}{4}(\nu-\mu)}(\frac{1}{4}ax^2)\ \cdot$ $\cdot\ K_{\frac{1}{4}(\nu+\mu)}(\frac{1}{4}ax^2)$	$2y^{-3/2}\,\frac{\Gamma(\frac{1}{2}+\frac{1}{4}\nu-\frac{1}{4}\mu)}{\Gamma(1+\frac{1}{2}\nu)} W_{\frac{1}{4}\mu,\ \frac{1}{4}\nu}(\frac{y^2}{2a}) M_{-\frac{1}{4}\mu,\ \frac{1}{4}\nu}(\frac{y^2}{2a})$

	$f(x)$	$g(y) = \int_0^\infty f(x)(xy)^{1/2} J_\nu(xy)\,dx$
13.17	$x^{-5/2} K_\nu(ax^{-1})$ $-5/2 < \mathrm{Re}\ \nu < 5/2$	$i\,a^{-1}y^{1/2}[e^{\frac{1}{2}i\pi\nu} K_{2\nu}(2e^{i\frac{\pi}{4}}a^{1/2}y^{1/2}) -$ $- e^{-\frac{1}{2}i\pi\nu} K_{2\nu}(2e^{-i\frac{\pi}{4}}a^{1/2}y^{1/2})$
13.18	$x^{-2\nu} K_{\nu-1/2}(ax^{-1})$ $\mathrm{Re}\ \nu > 1/6$	$(2\pi a)^{1/2} a^{-\nu} y^{\nu-1/2} J_{2\nu-1}[(2ay)^{1/2}] K_{2\nu-1}[(2ay)^{1/2}]$
13.19	$x^{-2\nu-2} K_{\nu-1/2}(ax^{-1})$ $-1/2 < \mathrm{Re}\ \nu < 2$	$(2\pi)^{1/2} a^{-\nu-1/2} y^{\nu+1/2} K_{2\nu}[(2ay)^{1/2}] J_{2\nu}[(2ay)^{1/2}]$
13.20	$K_{2\nu+1}(2ax^{1/2})$ $\mathrm{Re}\ \nu > -1$	$-\tfrac{1}{4}\pi a\,\sec(\pi\nu) y^{-3/2}[\mathbf{H}_{-\nu-1}(\tfrac{a^2}{y}) - Y_{-\nu-1}(\tfrac{a^2}{y})]$
13.21	$x^{-1/2} K_{2\nu}(2ax^{1/2})$ $\mathrm{Re}\ \nu > -1/2$	$\tfrac{1}{4}\pi\sec(\pi\nu) y^{-1/2}[\mathbf{H}_{-\nu}(\tfrac{a^2}{y}) - Y_{-\nu}(\tfrac{a^2}{y})]$
13.22	$x^{1/2} J_\nu(2ax^{1/2}) K_\nu(2ax^{1/2})$ $\mathrm{Re}\ \nu > -1$	$\tfrac{1}{2}y^{-3/2}\,e^{-2\frac{a^2}{y}}$

	$f(x)$	$g(y) = \int_0^\infty f(x)(xy)^{\frac{1}{2}} J_\nu(xy)\,dx$
13.23	$x^{\nu-\frac{1}{2}} J_{2\nu-1}(ax^{\frac{1}{2}}) \cdot$ $\cdot K_{2\nu-1}(ax^{\frac{1}{2}})$ $\mathrm{Re}\ \nu > 0$	$2^{-\nu}\pi^{-\frac{1}{2}} y^{-2\nu} a^{2\nu-1} K_{\nu-\frac{1}{2}}\left(\dfrac{a^2}{2y}\right)$
13.24	$x^{\nu+\frac{1}{2}} J_{2\nu}(2ax^{\frac{1}{2}}) \cdot$ $\cdot K_{2\nu}(2ax^{\frac{1}{2}})$ $\mathrm{Re}\ \nu > -\frac{1}{2}$	$2^{\nu}\pi^{-\frac{1}{2}} a^{2\nu+1} y^{-2\nu-2} K_{\nu-\frac{1}{2}}\left(2\dfrac{a^2}{y}\right)$
13.25	$x^{-\nu-\frac{1}{2}} J_{2\nu+1}(2ax^{\frac{1}{2}}) \cdot$ $\cdot K_{2\nu+1}(2ax^{\frac{1}{2}})$ $\mathrm{Re}\ \nu > -1$	$2^{-\nu-2}\pi^{\frac{1}{2}} a^{-2\nu-1} y^{2\nu} \left[I_{\nu+\frac{1}{2}}\left(2\dfrac{a^2}{y}\right) - \mathbf{L}_{\nu+\frac{1}{2}}\left(2\dfrac{a^2}{y}\right) \right]$
13.26	$x^{-\frac{1}{2}} [K_{2\nu}(2ax^{\frac{1}{2}}) -$ $-\frac{1}{2}\pi Y_{2\nu}(2ax^{\frac{1}{2}})]$ $\mathrm{Re}\ \nu > -\frac{1}{2}$	$-\frac{1}{2}\pi y^{-\frac{1}{2}} Y_\nu\left(\dfrac{a^2}{y}\right)$
13.27	$x^{-\frac{1}{2}} J_\mu(2ax^{\frac{1}{2}}) K_\mu(2ax^{\frac{1}{2}})$ $\mathrm{Re}\ \nu > -1$ $\mathrm{Re}\,(\nu+\mu) > -1$	$\frac{1}{4} a^{-2} \dfrac{\Gamma\left(\frac{1}{2}+\frac{1}{2}\mu+\frac{1}{2}\nu\right)}{\Gamma(1+\mu)} y^{\frac{1}{2}} W_{-\frac{1}{2}\nu,\,\frac{1}{2}\mu}\left(2\dfrac{a^2}{y}\right) \cdot$ $\cdot M_{\frac{1}{2}\nu,\,\frac{1}{2}\mu}\left(2\dfrac{a^2}{y}\right)$

	$f(x)$	$g(y) = \int_0^\infty f(x)(xy)^{\frac{1}{2}}J_\nu(xy)\,dx$
13.28	$x^{-\frac{1}{2}}K_\nu(2ax^{\frac{1}{2}})Y_\nu(2ax^{\frac{1}{2}})$ $\mathrm{Re}\,\nu > -\tfrac{1}{2}$	$-\tfrac{1}{4}a^{-2}y^{\frac{1}{2}}W_{\frac{1}{2}\nu,\frac{1}{2}\nu}(2\tfrac{a^2}{y})W_{-\frac{1}{2}\nu,\frac{1}{2}\nu}(2\tfrac{a^2}{y})$
13.29	$x^{-\frac{1}{2}}K_\mu(2ax^{\frac{1}{2}})\cdot$ $\cdot\{\sin[\tfrac{\pi}{2}(\mu-\nu)]J_\mu(2ax^{\frac{1}{2}})$ $+\cos[\tfrac{\pi}{2}(\mu-\nu)]Y_\mu(2ax^{\frac{1}{2}})\}$ $\mathrm{Re}(\nu\pm\mu) > -1$	$-\tfrac{1}{4}a^{-2}y^{\frac{1}{2}}W_{\frac{1}{2}\nu,\frac{1}{2}\mu}(2\tfrac{a^2}{y})W_{-\frac{1}{2}\nu,\frac{1}{2}\mu}(2\tfrac{a^2}{y})$
13.30	$x^{-\frac{1}{2}}K_\mu[e^{i\frac{\pi}{4}}(2ax)^{\frac{1}{2}}]\cdot$ $\cdot K_\mu[e^{-i\frac{\pi}{4}}(2ax)^{\frac{1}{2}}]$ $\mathrm{Re}(\nu\pm\mu) > -1$	$\tfrac{1}{4}a^{-1}y^{\frac{1}{2}}\Gamma(\tfrac{1}{2}+\tfrac{1}{2}\mu+\tfrac{1}{2}\nu)\Gamma(\tfrac{1}{2}-\tfrac{1}{2}\mu+\tfrac{1}{2}\nu)\cdot$ $\cdot\, W_{-\frac{1}{2}\nu,\frac{1}{2}\mu}(iay^{-1})W_{-\frac{1}{2}\nu,\frac{1}{2}\mu}(-iay^{-1})$
13.31	$x^{-\nu-\frac{1}{2}}K_{2\nu+1}[e^{i\frac{\pi}{4}}(2ax)^{\frac{1}{2}}]\cdot$ $\cdot K_{2\nu+1}[e^{-i\frac{\pi}{4}}(2ax)^{\frac{1}{2}}]$ $-1 < \mathrm{Re}\,\nu < 0$	$-2^{-5/2}\pi^{3/2}\csc(\pi\nu)a^{-\frac{1}{2}-\nu}y^{2\nu}\cdot$ $\cdot\,[\mathbf{H}_{\nu+\frac{1}{2}}(ay^{-1}) - Y_{\nu+\frac{1}{2}}(ay^{-1})]$
13.32	$x^{\nu+\frac{1}{2}}(b^2+x^2)^{-\frac{1}{2}\nu-\frac{1}{4}}\cdot$ $\cdot K_{\nu+\frac{1}{2}}[a(b^2+x^2)^{\frac{1}{2}}]$ $\mathrm{Re}\,\nu > -1$	$(\tfrac{1}{2}\pi)^{\frac{1}{2}}a^{-\nu-\frac{1}{2}}y^{\nu+\frac{1}{2}}(a^2+y^2)^{-\frac{1}{2}}\exp[-b(a^2+y^2)^{\frac{1}{2}}]$

	$f(x)$	$g(y) = \int\limits_{0}^{\infty} f(x)(xy)^{\frac{1}{2}}J_{\nu}(xy)\,dx$
13.33	$x^{\nu+\frac{1}{2}}(b^2+x^2)^{-\frac{1}{2}\mu}\ \cdot$ $\cdot\ K_{\mu}[a(b^2+x^2)^{\frac{1}{2}}]$ $\text{Re } \nu > -1$	$a^{-\mu}b^{\nu+1-\mu}y^{\nu+\frac{1}{2}}(a^2+y^2)^{\frac{1}{2}\mu-\frac{1}{2}\nu-\frac{1}{2}}\ \cdot$ $\cdot\ K_{\mu-\nu-1}[b(a^2+y^2)^{\frac{1}{2}}]$
13.34	$x^{\nu+\frac{1}{2}}(b^2-x^2)^{\frac{1}{2}\mu}\ \cdot$ $\cdot\ Y_{\mu}[a(b^2-x^2)^{\frac{1}{2}}]$ $\qquad x < b$ $-2\pi^{-1}x^{\nu+\frac{1}{2}}(x^2-b^2)^{\frac{1}{2}\mu}\cdot$ $\cdot\ K_{\mu}[a(x^2-b^2)^{\frac{1}{2}}]$ $\qquad x > b$ $\text{Re } \nu > -1,\ \text{Re } \mu > -1$	$a^{\mu}b^{\mu+\nu+1}y^{\nu+\frac{1}{2}}(a^2+y^2)^{-\frac{1}{2}\mu-\frac{1}{2}\nu-\frac{1}{2}}\ \cdot$ $\cdot\ Y_{\mu+\nu+1}[b(a^2+y^2)^{\frac{1}{2}}]$
13.35	$x^{\frac{1}{2}}I_{\frac{1}{2}\nu}\{\frac{1}{2}b[(a^2+x^2)^{\frac{1}{2}}-a]\}\cdot$ $\cdot\,K_{\frac{1}{2}\nu}\{\frac{1}{2}b[(a^2+x^2)^{\frac{1}{2}}+a]\}$ $\text{Re } \nu > -1$	$y^{-\frac{1}{2}}(b^2+y^2)^{-\frac{1}{2}}\exp[-a(b^2+y^2)^{\frac{1}{2}}]$

1.14 Functions Related to Bessel Functions

	$f(x)$	$g(y) = \int\limits_0^\infty f(x)(xy)^{\frac{1}{2}}J_\nu(xy)\,dx$
14.1	$\mathbf{H}_{\nu-\frac{1}{2}}(ax)$ $-3/2 < \operatorname{Re}\nu < 3/2$	$(\tfrac{1}{2}\pi)^{-\frac{1}{2}}a^{\nu-\frac{1}{2}}y^{\frac{1}{2}-\nu}(a^2-y^2)^{-\frac{1}{2}}$ $\quad y < a$ $0 \qquad\qquad y > a$
14.2	$x^{\frac{1}{2}-\nu}[\mathbf{H}_\nu(ax)-Y_\nu(ax)]$ $-2 < \operatorname{Re}\nu < 2$	$\pi^{-\frac{1}{2}}2^{1-2\nu}\Gamma(1-\nu)[\Gamma(\tfrac{1}{2}+\nu)]^{-1}y^{-\frac{1}{2}}\cdot$ $\cdot\,(y^2-a^2)^{\frac{1}{2}\nu-\frac{1}{2}}P_\nu^{\nu-1}(\tfrac{y}{a}) \qquad y > a$ $0 \qquad\qquad y < a$
14.3	$x^{\nu-\mu-\frac{1}{2}}[\mathbf{H}_\mu(ax)-Y_\mu(ax)]$ $-\tfrac{1}{2}\,\operatorname{Re}\nu<3/2,\ \operatorname{Re}(\nu-\mu)>-1$	$\pi^{-\frac{1}{2}}2^{\nu-\mu}a^\mu\dfrac{\Gamma(\tfrac{1}{2}+\nu)\,\Gamma(\tfrac{1}{2}+\nu-\mu)}{\Gamma(\tfrac{1}{2}+\mu)\,\Gamma(1+\nu-\mu)}\,y^{-\nu-\frac{1}{2}}\cdot$ $\cdot\,{}_2F_1(\tfrac{1}{2}+\nu,\tfrac{1}{2};1+\nu-\mu;1-a^2y^{-2})$
14.4	$x^{\nu-\mu+\frac{1}{2}}[\mathbf{H}_\mu(ax)-Y_\mu(ax)]$	$\dfrac{\pi^{-1}2^{1+\nu-\mu}a^{-\mu}}{(\nu-\mu+1)\Gamma(\tfrac{1}{2}+\mu)}\,\Gamma(3/2+\nu)\,y^{2\mu-\nu-3/2}\cdot$ $\cdot\,{}_2F_1(1+\nu-\mu,\tfrac{1}{2}-\mu;2+\nu-\mu;1-a^2y^{-2})$
14.5	$x^{\frac{1}{2}}[\mathbf{H}_{-\nu}(ax)-Y_{-\nu}(ax)]$ $-\tfrac{1}{2} < \operatorname{Re}\nu$	$2a^{-\nu}\pi^{-1}\cos(\pi\nu)\,y^{\nu-\frac{1}{2}}(y+a)^{-1}$
14.6	$x^{\nu+\mu+\frac{1}{2}}[\mathbf{H}_\mu(ax)-Y_\mu(ax)]$ $-1\ \operatorname{Re}(\nu+\mu)<\tfrac{1}{2}-\operatorname{Re}\mu$	$2\pi^{-1}\cos(\pi\mu)\Gamma(2+2\nu+2\mu)a^\mu y^{-\mu-\frac{1}{2}}\cdot$ $\cdot\,(a^2-y^2)^{-\frac{1}{2}\nu-\frac{1}{2}\mu-\frac{1}{2}}P_{-\nu-\mu-1}^{-\nu-\mu-1}(\tfrac{a}{y})$

	$f(x)$	$g(y) = \int\limits_{0}^{\infty} f(x)(xy)^{\frac{1}{2}}J_{\nu}(xy)dx$
14.7	$I_{\nu-\frac{1}{2}}(ax)-\mathbf{L}_{\nu-\frac{1}{2}}(ax)$ $-\frac{1}{2} < \text{Re } \nu < \frac{1}{2}$	$(\frac{1}{2}\pi)^{-\frac{1}{2}}a^{\nu-\frac{1}{2}}y^{\frac{1}{2}-\nu}(a^2+y^2)^{-\frac{1}{2}}$
14.8	$x^{\frac{1}{2}}[I_{\nu}(ax)-\mathbf{L}_{\nu}(ax)]$ $-1 < \text{Re } \nu < -\frac{1}{2}$	$2\pi^{-1}a^{\nu+1}y^{-\nu-\frac{1}{2}}(a^2+y^2)^{-1}$
14.9	$x^{\mu-\nu+\frac{1}{2}}[I_{\mu}(ax)-\mathbf{L}_{\mu}(ax)]$ $-1 < 2\text{Re } \mu+1 < \frac{1}{2}+\text{Re } \nu$	$\pi^{-\frac{1}{2}}2^{\mu-\nu+1}a^{\mu-1}y^{\nu-2\mu-\frac{1}{2}}[\Gamma(\nu-\mu+\frac{1}{2})]^{-1}\cdot$ $\cdot\ _2F_1(1,\frac{1}{2};\nu-\mu+\frac{1}{2};-y^2a^{-2})$
14.10	$x^{\nu-\mu+\frac{1}{2}}[I_{\mu}(ax)-\mathbf{L}_{\mu}(ax)]$ $-1 < \text{Re } \nu < -\frac{1}{2}$	$\pi^{-1}2^{\nu-\mu+1}\Gamma(^3/_2+\nu)[\Gamma(^3/_2+\mu)]^{-1}a^{\mu+1}y^{-\nu-5/2}\cdot$ $\cdot\ _2F_1(1,^3/_2+\nu;\ ^3/_2+\mu;-\dfrac{a^2}{y^2})$
14.11	$x^{\nu-\mu-\frac{1}{2}}[I_{\mu}(ax)-\mathbf{L}_{\mu}(ax)]$ $-\frac{1}{2} < \text{Re } \nu< \frac{1}{2}$	$\pi^{-\frac{1}{2}}2^{\nu-\mu}\Gamma(\frac{1}{2}+\nu)[\Gamma(1+\mu)]^{-1}a^{\mu}y^{-\nu-\frac{1}{2}}\ \cdot$ $\cdot\ _2F_1(\frac{1}{2}+\nu,\frac{1}{2};1+\mu;-a^2y^{-2})$
14.12	$x^{\frac{1}{2}}[I_{\nu}(ax)-\mathbf{L}_{-\nu}(ax)]$ $\text{Re } \nu > \frac{1}{2}$	$2\pi^{-1}a^{1-\nu}\cos(\pi\nu)y^{\nu-\frac{1}{2}}(a^2+y^2)^{-1}$
14.13	$x^{\mu-\nu+\frac{1}{2}}[I_{\mu}(ax)-\mathbf{L}_{-\mu}(ax)]$ $\text{Re } \nu>-\frac{1}{2},\ \text{Re } \mu>-1$	$2^{\mu-\nu+1}a^{-\mu-1}[\Gamma(\frac{1}{2}-\mu)\Gamma(\frac{1}{2}+\nu)]^{-1}y^{\nu-\frac{1}{2}}\ \cdot$ $\cdot\ _2F_1(1,\frac{1}{2}+\mu;\frac{1}{2}+\nu;-y^2a^{-2})$

	$f(x)$	$g(y) = \int_0^\infty f(x)(xy)^{\frac{1}{2}} J_\nu(xy)\, dx$
14.14	$x^{\nu-\mu+\frac{1}{2}}[I_\mu(ax)-\mathbf{L}_{-\mu}(ax)]$ Re $\nu>-\frac{1}{2}$, Re $(\nu-\mu)>-1$ Re $(\nu-2\mu) < \frac{1}{2}$	$\pi^{-3/2}2^{2+\nu-\mu}\cos(\pi\mu)\Gamma(3/2+\nu-\mu)a^{1-\mu}\,\cdot$ $\cdot\, y^{2\mu-\nu-5/2}{}_2F_1(3/2-\mu+\nu,1;3/2;-a^2y^{-2})$
14.15	$x^{\mu+\nu-\frac{1}{2}}[I_\mu(ax)-\mathbf{L}_{-\mu}(ax)]$ -1 Re $\nu<3/2$, Re $(\mu+\nu)>-\frac{1}{2}$	$2^{\mu+\nu}\Gamma(\frac{1}{2}+\mu+\nu)[\Gamma(1+\mu)\Gamma(\frac{1}{2}-\mu)]^{-1}a^\mu\,\cdot$ $\cdot\, y^{-\frac{1}{2}-2\mu-\nu}{}_2F_1(\frac{1}{2}+\mu+\nu,\frac{1}{2}+\mu;1+\mu;-a^2y^{-2})$
14.16	$x^{\frac{1}{2}}\mathbf{H}_{\frac{1}{2}\nu}(ax^2)$ $-2 < $ Re $\nu < 3/2$	$-\frac{1}{2}a^{-1}y^{\frac{1}{2}}Y_{\frac{1}{2}\nu}(\frac{y^2}{4a})$
14.17	$x^{3/2}\mathbf{H}_{\frac{1}{2}\nu-\frac{1}{2}}(ax^2)$ $-2 < $ Re $\nu < \frac{1}{2}$	$-\frac{1}{4}a^{-2}y^{3/2}Y_{\frac{1}{2}\nu+\frac{1}{2}}(\frac{y^2}{4a})$
14.18	$x^{-\frac{1}{2}}[\mathbf{H}_{-\nu}(ax^{-1})-Y_{-\nu}(ax^{-1})]$ $-\frac{1}{2} < $ Re $\nu < \frac{1}{2}$	$4\pi^{-1}\cos(\pi\nu)y^{-\frac{1}{2}}K_{2\nu}[2(ay)^{\frac{1}{2}}]$
14.19	$x^{-3/2}[\mathbf{H}_{-\nu-1}(ax^{-1})\,-$ $-\,Y_{-\nu-1}(ax^{-1})]$ $-\frac{1}{2} < $ Re $\nu < \frac{1}{2}$	$-4\pi^{-1}a^{-\frac{1}{2}}\cos(\pi\nu)K_{-2\nu-1}[2(ay)^{\frac{1}{2}}]$

	$f(x)$	$g(y) = \int_0^\infty f(x)(xy)^{\frac{1}{2}}J_\nu(xy)\,dx$
14.20	$x^{2\nu}[\mathbf{H}_{\nu+\frac{1}{2}}(ax^{-1}) -$ $-Y_{\nu+\frac{1}{2}}(ax^{-1})]$ $-1 < \mathrm{Re}\ \nu < -\,{}^{1}/_{6}$	$-\pi^{-3/2}2^{-5/2}a^{\frac{1}{2}+\nu}\sin(\pi\nu)y^{-\nu-\frac{1}{2}}\ \cdot$ $\cdot\ K_{2\nu+1}[(2aiy)^{\frac{1}{2}}]K[(-2aiy)^{\frac{1}{2}}]$
14.21	$x^{2\nu}[I_{\nu+\frac{1}{2}}(ax^{-1})-\mathbf{L}_{\nu+\frac{1}{2}}(ax^{-1})]$ $-1 < \mathrm{Re}\ \nu < \frac{1}{2}$	$\pi^{-\frac{1}{2}}2^{3/2}a^{\frac{1}{2}+\nu}y^{-\nu-\frac{1}{2}}\ \cdot$ $\cdot\ J_{2\nu+1}[(2ay)^{\frac{1}{2}}]K_{2\nu+1}[(2ay)^{\frac{1}{2}}]$
14.22	$x^{\frac{1}{2}(1-\mu-\nu)}S_{\mu,\nu}(ax)$ $\mathrm{Re}(\mu-\nu) < 1$ $-6<\mathrm{Re}(\mu+\nu) < 4$	$\pi^{\frac{1}{2}}2^{-\nu}\dfrac{\Gamma(1-\frac{1}{2}\mu-\frac{1}{2}\nu)}{\Gamma(\frac{1}{2}-\frac{1}{2}\mu+\frac{1}{2}\nu)}\,a^{\frac{1}{2}(\mu-\nu)}\ \cdot$ $\cdot\ y^{\frac{1}{2}(\nu-\mu-1)}(y^2-a^2)^{\frac{1}{4}(\nu+\mu-2)}\ \cdot$ $\cdot\ P_{\frac{1}{2}(\nu+\mu)}^{\frac{1}{2}(\nu+\mu-2)}(\frac{y}{a})\qquad ,\quad y > a$ $\qquad\qquad\qquad\qquad 0\quad ,\quad y < a$
14.23	$x^{\frac{1}{2}-\nu-\mu}S_{\mu,-2\nu-\mu}(ax)$ $\mathrm{Re}\ \nu>-\frac{1}{2},\mathrm{Re}(\nu+\mu) < 1$	$2^{-2\nu-\mu}\pi^{\frac{1}{2}}\dfrac{\Gamma(1-\mu-\nu)}{\Gamma(\frac{1}{2}+\nu)}\,a^{-\nu}y^{\nu-\frac{1}{2}}\ \cdot$ $\cdot\ (y^2-a^2)^{\frac{1}{2}(\nu+\mu-1)}P_{\nu+\mu}^{\nu+\mu-1}(\frac{y}{a})$
14.24	$x^{\nu+\beta+\frac{1}{2}}S_{\alpha,\beta}(ax)$ $\mathrm{Re}\ \nu>-1,-1<\mathrm{Re}(\nu+\beta)<\frac{1}{2}-\mathrm{Re}\ \beta$ $\mathrm{Re}(2\nu+\alpha+\beta) > -3$	$2^{\alpha+\beta+\nu}a^\beta[(\nu+\beta+1)\Gamma(\dfrac{1-\alpha-\beta}{2})]^{-1}\ \cdot$ $\cdot\ \Gamma(\dfrac{3+\alpha+\beta+2\nu}{2})\,y^{-\nu-2\beta-3/2}\ \cdot$ $\cdot\ {}_2F_1(1+\beta+\nu,\ \dfrac{1-\alpha+\beta}{2};\nu+\beta+2;\ 1-\dfrac{a^2}{y^2})$

	$f(x)$	$g(y) = \int_0^\infty f(x)(xy)^{\frac{1}{2}}J_\nu(xy)\,dx$
14.25	$x^{\nu-\alpha-\frac{1}{2}}S_{\alpha,\beta}(ax)$ $-1 < \mathrm{Re}\ \nu < {}^3/_2$ $\mathrm{Re}(2\nu-\alpha\pm\beta) > -1$	$2^{\nu-1}a^\beta[\Gamma(1+\nu-\alpha)]^{-1}\Gamma(\frac{1-\alpha-\beta+2\nu}{2})\ \cdot$ $\cdot\ \Gamma(\frac{1-\alpha+\beta+2\nu}{2})\ y^{\alpha-\beta-\nu-\frac{1}{2}}\ \cdot$ $\cdot\ {}_2F_1\left(\frac{1-\alpha+\beta+2\nu}{2},\ \frac{1-\alpha+\beta}{2};\ 1-\alpha+\nu;\ 1-\frac{a^2}{y^2}\right)$
14.26	$x^{-\mu-\frac{1}{2}}S_{\nu+\mu,\mu-\nu+1}(ax)$ $\mathrm{Re}\ \mu>-1,\ -1<\mathrm{Re}\ \nu<{}^3/_2$	$2^{\nu-1}a^{\nu-\mu-1}\Gamma(\nu)y^{\frac{1}{2}-\nu}(a^2-y^2)^\mu \quad y < a$ $\qquad\qquad\qquad 0 \qquad\qquad y > a$
14.27	$x^{\frac{1}{2}}[\mathbf{J}_\nu(ax)-J_\nu(ax)]$ $\mathrm{Re}\ \nu > -1$	$\pi^{-1}\sin(\pi\nu)y^{-\frac{1}{2}}(a+y)^{-1}$
14.28	$x^{\frac{1}{2}+\nu}U_{\nu+1}(2a^2b,ax)$ $\mathrm{Re}\ \nu > -1$	$(2b)^{\nu+1}y^{\nu+\frac{1}{2}}\cos[b(a^2-y^2)] \quad y < a$ $\qquad\qquad\qquad 0 \qquad\qquad y > a$
14.29	$x^{\frac{1}{2}+\nu}U_{\nu+2}(a^2b,ax)$ $\mathrm{Re}\ \nu > -1$	$(2b)^{\nu+1}y^{\nu+\frac{1}{2}}\sin[b(a^2-y^2)] \quad y < a$ $\qquad\qquad\qquad 0 \qquad\qquad y > a$

1.15 Parabolic Cylinder Functions

	$f(x)$	$g(y) = \int\limits_0^\infty f(x)(xy)^{\frac{1}{2}}J_\nu(xy)\,dx$
15.1	$x^{\nu-\frac{1}{2}}\exp(-\tfrac{1}{4}a^2x^2)\cdot$ $\cdot D_{2\nu}(ax)$ $\operatorname{Re}\nu > -\tfrac{1}{2}$	$\Gamma(\tfrac{1}{2}+\nu)\,2^{-\frac{1}{2}+\nu}a^{-2\nu-1}y^{\nu+\frac{1}{2}}\exp(-\tfrac{1}{2}y^2a^{-2})\cdot$ $\cdot L_{\nu-\frac{1}{2}}(\tfrac{1}{2}y^2a^{-2})$
15.2	$x^{\nu+\frac{1}{2}}\exp(-\tfrac{1}{4}a^2x^2)\cdot$ $\cdot D_{2\nu+1}(ax)$ $\operatorname{Re}\nu > -1$	$\Gamma(^{3/2}+\nu)\,2^{\nu+\frac{1}{2}}a^{-2\nu-2}y^{\nu+\frac{1}{2}}\exp(-\tfrac{1}{2}y^2a^{-2})\cdot$ $\cdot L_{\nu+\frac{1}{2}}(\tfrac{1}{2}y^2a^{-2})$
15.3	$x^{\nu-\frac{1}{2}}\exp(-\tfrac{1}{4}a^2x^2)$ $D_{2\nu-1}(ax)$ $\operatorname{Re}\nu > -\tfrac{1}{2}$	$-\tfrac{1}{2}\sec(\pi\nu)a^{-2\nu}y^{\nu-\frac{1}{2}}\exp(-\tfrac{1}{4}y^2a^{-2})\cdot$ $\cdot[D_{2\nu-1}(ya^{-1}) - D_{2\nu-1}(-ya^{-1})]$
15.4	$x^{\nu-\frac{1}{2}}\exp(-\tfrac{1}{4}a^2x^2)\cdot$ $\cdot\{[1-2\cos(\pi\nu)]D_{2\nu-1}(ax) -$ $- D_{2\nu-1}(-ax)\}$ $\operatorname{Re}\nu > -\tfrac{1}{2}$	$a^{-2\nu}y^{\nu-\frac{1}{2}}\exp(-\tfrac{1}{4}y^2a^{-2})\cdot$ $\cdot\{[1-2\cos(\pi\nu)]D_{2\nu-1}(ya^{-1}) -$ $- D_{2\nu-1}(-ya^{-1})\}$
15.5	$x^{\nu-\frac{1}{2}}\exp(-\tfrac{1}{4}a^2x^2)\cdot$ $\cdot\{[1+2\cos(\pi\nu)]D_{2\nu-1}(ax) -$ $- D_{2\nu-1}(-ax)\}$ $\operatorname{Re}\nu > -\tfrac{1}{2}$	$- a^{-2\nu}y^{\nu-\frac{1}{2}}\exp(-\tfrac{1}{4}y^2a^{-2})\cdot$ $\cdot\{[1+2\cos(\pi\nu)]D_{2\nu-1}(ya^{-1}) -$ $- D_{2\nu-1}(-ya^{-1})\}$

	$f(x)$	$g(y) = \int\limits_0^\infty f(x)(xy)^{\frac{1}{2}} J_\nu(xy)\,dx$
15.6	$x^{\nu-\frac{1}{2}}\exp(\tfrac{1}{4}a^2x^2)\,\cdot$ $\cdot\, D_{2\nu-1}(ax)$ $-\tfrac{1}{2} < \mathrm{Re}\ \nu < \tfrac{1}{2}$	$\pi\sin(\pi\nu)\,\Gamma(2\nu)\,2^{\frac{1}{2}-\nu}a^{-1}y^{\frac{1}{2}-\nu}\,\cdot$ $\cdot\,\exp(\tfrac{1}{4}y^2a^{-2})K_\nu(\tfrac{1}{4}y^2a^{-2})$
15.7	$x^{\nu-\frac{1}{2}}\exp(-\tfrac{1}{4}a^2x^2)\,\cdot$ $\cdot\, D_{2\nu+1}(ax)$ $\mathrm{Re}\ \nu > -\tfrac{1}{2}$	$\tfrac{1}{2}\sec(\pi\nu)\,a^{-2\nu-1}y^{\nu+\frac{1}{2}}\exp(-\tfrac{1}{4}y^2a^{-2})\,\cdot$ $\cdot\,[D_{2\nu}(ya^{-1}) + D_{2\nu}(-ya^{-1})]$
15.8	$x^{\nu-\frac{1}{2}}\exp(-\tfrac{1}{4}a^2x^2)\,\cdot$ $\cdot\,\{[1+2\cos(\pi\nu)]D_{2\nu+1}(ax)-$ $-\,D_{2\nu+1}(-ax)\}$ $\mathrm{Re}\ \nu > -\tfrac{1}{2}$	$a^{-2\nu-1}y^{\nu+\frac{1}{2}}\exp(-\tfrac{1}{4}y^2a^{-2})\,\cdot$ $\cdot\,\{[1+2\cos(\pi\nu)]D_{2\nu}(ya^{-1})+D_{2\nu}(-ya^{-1})\}$
15.9	$x^{\nu-\frac{1}{2}}\exp(-\tfrac{1}{4}a^2x^2)\,\cdot$ $\cdot\,\{[1-\cos(2\pi\nu)]D_{2\nu+1}(ax)-$ $-\,D_{2\nu+1}(-ax)\}$ $\mathrm{Re}\ \nu > -\tfrac{1}{2}$	$-\,a^{-2\nu-1}y^{\nu+\frac{1}{2}}\exp(-\tfrac{1}{4}y^2a^{-2})\,\cdot$ $\cdot\,\{[1-2\cos(\pi\nu)]D_{2\nu}(ya^{-1})+D_{2\nu}(-ya^{-1})\}$
15.10	$x^{\nu-\frac{1}{2}}\exp(-\tfrac{1}{4}a^2x^2)\,\cdot$ $\cdot\, D_{-2\nu}(ax)$ $\mathrm{Re}\ \nu > -\tfrac{1}{2}$	$(\tfrac{1}{2}\pi)^{\frac{1}{2}}a^{-1}y^{\frac{1}{2}-\nu}\exp(-\tfrac{1}{4}y^2a^{-2})I_\nu(\tfrac{1}{4}y^2a^{-2})$

	$f(x)$	$g(y) = \int\limits_0^\infty f(x)(xy)^{\frac{1}{2}} J_\nu(xy)\,dx$
15.11	$x^{\nu-\frac{1}{2}}\exp(\tfrac{1}{4}a^2x^2) \cdot$ $\cdot\ D_{-2\nu}(ax)$ $\mathrm{Re}\ \nu > -\tfrac{1}{2}$	$a^{-2\nu}y^{\nu-\frac{1}{2}}\exp(\tfrac{1}{4}y^2a^{-2})D_{-2\nu}(ya^{-1})$
15.12	$x^{\nu-\frac{1}{2}}\exp(\tfrac{1}{4}a^2x^2) \cdot$ $\cdot\ D_{-2\nu-2}(ax)$ $\mathrm{Re}\ \nu > -\tfrac{1}{2}$	$(2\nu+1)^{-1}a^{-2\nu-1}y^{\nu+\frac{1}{2}}\exp(\tfrac{1}{4}y^2a^{-2}) \cdot$ $\cdot\ D_{-2\nu-1}(ya^{-1})$
15.13	$x^{\nu-\frac{1}{2}}\exp(-\tfrac{1}{4}a^2x^2) \cdot$ $\cdot\ D_{2\mu}(ax)$ $\mathrm{Re}\ \nu > -\tfrac{1}{2}$	$2^{\frac{1}{2}\nu+\frac{1}{2}\mu}a^{-\nu-\mu}\Gamma(\tfrac{1}{2}+\nu)\,[\Gamma(\nu-\mu+1)]^{-1} \cdot$ $\cdot\ y^{\mu-\frac{1}{2}}\exp(-\tfrac{1}{4}y^2a^{-2}) \cdot$ $\cdot\ M_{\frac{1}{2}\nu+\frac{1}{2}\mu,\ \frac{1}{2}\nu-\frac{1}{2}\mu}(\tfrac{1}{2}y^2a^{-2})$
15.14	$x^{\nu-\frac{1}{2}}\exp(\tfrac{1}{4}a^2x^2) \cdot$ $\cdot\ D_{2\mu}(ax)$ $-\tfrac{1}{2}<\mathrm{Re}\ \nu<\mathrm{Re}(\tfrac{1}{2}-2\mu)$	$2^{\frac{1}{4}+\frac{1}{2}\nu+3/2\,\mu}a^{\frac{1}{2}+\mu-\nu}\Gamma(\tfrac{1}{2}+\nu)\,[\Gamma(\tfrac{1}{2}-\mu)]^{-1} \cdot$ $\cdot\ y^{-\mu-1}\exp(\tfrac{1}{4}y^2a^{-2})W_{\alpha,\beta}(\tfrac{1}{2}y^2a^{-2})$ $2\alpha = \tfrac{1}{2}+\mu-\nu,\ \ 2\beta = \tfrac{1}{2}+\mu+\nu$
15.15	$x^{\nu+\frac{1}{2}}\exp(-\tfrac{1}{4}a^2x^2) \cdot$ $\cdot\ D_{2\nu}(ax)$ $\mathrm{Re}\ \nu > -1$	$\tfrac{1}{2}\sec(\pi\nu)a^{-2\nu-1}\exp(-\tfrac{1}{4}y^2a^{-2}) \cdot$ $\cdot\ [D_{2\nu+1}(ay^{-1}) - D_{2\nu+1}(-ay^{-1})]$

	$f(x)$	$g(y) = \int\limits_{0}^{\infty} f(x)(xy)^{\frac{1}{2}}J_\nu(xy)\,dx$
15.16	$x^{\nu+\frac{1}{2}}\exp(-\frac{1}{4}a^2x^2) \cdot$ $\cdot\{[1+2\cos(\pi\nu)]D_{2\nu}(ax)+$ $+ D_{2\nu}(-ax)\}$ $\mathrm{Re}\ \nu > -1$	$a^{-2\nu-1}y^{\nu-\frac{1}{2}}\exp(-\frac{1}{4}y^2a^{-2}) \cdot$ $\cdot\{[1+2\cos(\pi\nu)]D_{2\nu+1}(ya^{-1}) -$ $- D_{2\nu+1}(-ya^{-1})\}$
15.17	$x^{\nu+\frac{1}{2}}\exp(-\frac{1}{4}a^2x^2) \cdot$ $\cdot\{[1-2\cos(\pi\nu)]D_{2\nu}(ax)+$ $+ D_{2\nu}(-ax)\}$ $\mathrm{Re}\ \nu > -1$	$-a^{-2\nu-1}y^{\nu-\frac{1}{2}}\exp(-\frac{1}{4}y^2a^{-2}) \cdot$ $\cdot\{[1-2\cos(\pi\nu)]D_{2\nu+1}(ya^{-1}) -$ $- D_{2\nu+1}(-ya^{-1})\}$
15.18	$x^{\nu+\frac{1}{2}}\exp(-\frac{1}{4}a^2x^2) \cdot$ $\cdot\ D_{2\nu+2}(ax)$ $\mathrm{Re}\ \nu > -1$	$-\frac{1}{2}\sec(\pi\nu)a^{-2\nu-2}y^{\nu+\frac{1}{2}} \cdot$ $\cdot\exp(-\frac{1}{4}y^2a^{-2})[D_{2\nu+2}(ya^{-1})+D_{2\nu+2}(-ya^{-1})]$
15.19	$x^{\nu+\frac{1}{2}}\exp(-\frac{1}{4}a^2x^2) \cdot$ $\cdot\{[1-2\cos(\pi\nu)]D_{2\nu+2}(ax)+$ $+ D_{2\nu+2}(-ax)\}$ $\mathrm{Re}\ \nu > -1$	$a^{-2\nu-2}y^{\nu+\frac{1}{2}}\exp(-\frac{1}{4}y^2a^{-2}) \cdot$ $\cdot\{[1-2\cos(\pi\nu)]D_{2\nu+2}(ya^{-1}) +$ $+ D_{2\nu+2}(-ya^{-1})\}$

	$f(x)$	$g(y) = \int\limits_0^\infty f(x)(xy)^{\frac{1}{2}} J_\nu(xy)\,dx$
15.20	$x^{\nu+\frac{1}{2}} \exp(-\frac{1}{4}a^2x^2)\cdot$ $\cdot\{[1+2\cos(\pi\nu)]D_{2\nu+2}(ax)+$ $+ D_{2\nu+2}(-ax)\}$ $\mathrm{Re}\ \nu > -1$	$-a^{-2\nu-2}y^{\nu+\frac{1}{2}}\exp(-\frac{1}{4}y^2a^{-2})\cdot$ $\cdot\{[1+2\cos(\pi\nu)]D_{2\nu+2}(ya^{-1})\ +$ $+ D_{2\nu+2}(-ya^{-1})\}$
15.21	$x^{\nu+\frac{1}{2}}\exp(\frac{1}{4}a^2x^2)\cdot$ $\cdot D_{2\nu+2}(ax)$ $-1 < \mathrm{Re}\ \nu < -\,^5/_6$	$\pi^{-1}\sin(\pi\nu)\Gamma(2\nu+3)y^{-\nu-\,^3/_2}$ $\exp(\frac{1}{4}y^2a^{-2})K_{\nu+1}(\frac{1}{4}y^2a^{-2})$
15.22	$x^{\nu+\frac{1}{2}}\exp(\frac{1}{4}a^2x^2)\cdot$ $\cdot D_{-2\nu-1}(ax)$ $\mathrm{Re}\ \nu > -\frac{1}{2}$	$(2\nu+1)a^{-2\nu-1}y^{\nu-\frac{1}{2}}\exp(\frac{1}{4}y^2a^{-2})\cdot$ $\cdot D_{-2\nu-2}(ya^{-1})$
15.23	$x^{\nu+\frac{1}{2}}\exp(-\frac{1}{4}a^2x^2)\cdot$ $\cdot D_{-2\nu-3}(ax)$ $\mathrm{Re}\ \nu > -1$	$(\frac{1}{2}\pi)^{\frac{1}{2}}y^{-\nu-\,^3/_2}\exp(-\frac{1}{4}y^2a^{-2})I_{\nu+1}(\frac{1}{4}y^2a^{-2})$
15.24	$x^{\nu+\frac{1}{2}}\exp(\frac{1}{4}a^2x^2)\cdot$ $\cdot D_{-2\nu-3}(ax)$ $\mathrm{Re}\ \nu > -1$	$a^{-2\nu-2}y^{\nu+\frac{1}{2}}\exp(\frac{1}{4}y^2a^{-2})D_{-2\nu-3}(ya^{-1})$

	$f(x)$	$g(y) = \int\limits_{0}^{\infty} f(x)(xy)^{\frac{1}{2}} J_{\nu}(xy)\,dx$
15.25	$x^{\nu+\frac{1}{2}}\exp(-\frac{1}{4}a^2x^2)\cdot$ $\cdot D_{2\mu}(ax)$ $\mathrm{Re}\ \nu > -1$	$2^{\frac{1}{2}\nu+\frac{1}{2}\mu+3/4}\Gamma(3/2+\nu)[\Gamma(3/2+\nu-\mu)]^{-1}a^{-\nu-\mu-\frac{1}{2}}\cdot$ $\cdot\, y^{\mu-\frac{1}{2}}\exp(-\frac{1}{4}y^2a^{-2})M_{\alpha,\beta}(\frac{1}{2}y^2a^{-2})$ $2\alpha = \nu+\mu+3/2,\quad 2\beta = \nu-\mu+\frac{1}{2}$
15.26	$x^{\nu+\frac{1}{2}}\exp(\frac{1}{4}a^2x^2)\cdot$ $\cdot D_{2\mu}(ax)$ $-1<\mathrm{Re}\ \nu<-\frac{1}{2}-2\ \mathrm{Re}\ \mu$	$\Gamma(3/2+\nu)[\Gamma(-\mu)]^{-1}2^{\frac{1}{2}+\beta+\mu}a^{2\alpha+1}y^{-\mu-3/2}\cdot$ $\cdot\,\exp(\frac{1}{4}y^2a^{-2})W_{\alpha,\beta}(\frac{1}{2}y^2a^{-2})$ $2\alpha = \mu-\nu-1\qquad 2\beta = \mu+\nu+1$
15.27	$x^{\lambda}\exp(-\frac{1}{4}a^2x^2)\cdot$ $\cdot D_{\mu}(ax)$ $\mathrm{Re}(\lambda+\mu) > -3/2$	$\pi^{\frac{1}{2}}\Gamma(2\alpha)[\Gamma(1+\nu)\Gamma(\frac{1}{2}+\beta)]^{-1}2^{-\beta-3/2}\nu a^{-2\alpha}\cdot$ $\cdot\, y^{\nu+\frac{1}{2}}{}_2F_2(\alpha,\alpha+\frac{1}{2};\nu+1,\beta+\frac{1}{2};-\frac{1}{2}y^2a^{-2})$ $2\alpha = \lambda+\nu+3/2,\qquad 2\beta = \lambda-\mu+3/2$
15.28	$D_{-\frac{1}{2}-\nu}[(iax)^{\frac{1}{2}}]\cdot$ $\cdot D_{-\frac{1}{2}-\nu}[(-iax)^{\frac{1}{2}}]$ $\mathrm{Re}\ \nu > -\frac{1}{2}$	$\pi^{\frac{1}{2}}2^{-\nu}[\Gamma(\frac{1}{2}+\nu)]^{-1}y^{-\nu-\frac{1}{2}}\cdot$ $\cdot\,(a+2y)^{-\frac{1}{2}}[(a+2y)^{\frac{1}{2}}-a^{\frac{1}{2}}]^{2\nu}$
15.29	$x^{-\frac{1}{2}}D_{-\frac{1}{2}-\nu}(ae^{i\frac{\pi}{4}}x^{-\frac{1}{2}})\cdot$ $\cdot D_{-\frac{1}{2}-\nu}(ae^{-i\frac{\pi}{4}}x^{-\frac{1}{2}})$ $\mathrm{Re}\ \nu > -\frac{1}{2}$	$(2\pi)^{\frac{1}{2}}[\Gamma(\frac{1}{2}+\nu)]^{-1}y^{-1}\exp[-a(2y)^{\frac{1}{2}}]$

	$f(x)$	$g(y) = \int\limits_{0}^{\infty} f(x)(xy)^{\frac{1}{2}} J_{\nu}(xy)\,dx$
15.30	$D_{\nu-\frac{1}{2}}[(2ax^{-1})^{\frac{1}{2}}] \cdot$ $\cdot\ D_{-\nu-\frac{1}{2}}[(2ax^{-1})^{\frac{1}{2}}]$ $\mathrm{Re}\ \nu > -\frac{1}{2}$	$y^{-1}\exp[-(2ay)^{\frac{1}{2}}]\cos[(2ay)^{\frac{1}{2}}+\tfrac{1}{4}\pi-\dfrac{\pi}{2}\,\nu]$

1.16 Whittaker Functions

$f(x)$	$g(y) = \int\limits_0^\infty f(x)(xy)^{\frac{1}{2}}J_\nu(xy)dx$
16.1	$x^{-1}\exp(-\tfrac{1}{2}a^2x^2)\cdot$ $\cdot M_{\frac{1}{2}\nu-\frac{1}{4},\frac{1}{2}\nu+\frac{1}{4}}(a^2x^2)$ $\mathrm{Re}\ \nu > -\tfrac{1}{2}$	$2^{1-\nu}(\tfrac{1}{2}+\nu)a^{\frac{1}{2}-\nu}y^{\nu-\frac{1}{2}}\mathrm{Erfc}(\tfrac{1}{2}ya^{-1})$
16.2	$x^{-3/2}\exp(-\tfrac{1}{2}a^2x^2)\cdot$ $\cdot M_{\frac{1}{2}\nu+\frac{1}{2},\frac{1}{2}\nu+\frac{1}{2}}(a^2x^2)$ $\mathrm{Re}\ \nu > -1$	$(2a)^{-\nu}\Gamma(\nu+2)[\Gamma(3/2+\nu)]^{-1}y^{\frac{1}{2}+\nu}\mathrm{Erfc}(\tfrac{1}{2}ya^{-1})$
16.3	$x^{2\mu-\nu-\frac{1}{2}}\exp(-\tfrac{1}{2}a^2x^2)\cdot$ $\cdot M_{3\mu-\nu+\frac{1}{2},\mu}(a^2x^2)$ $\mathrm{Re}\ \mu>-\tfrac{1}{2},\ \mathrm{Re}(4\mu-\nu)>-\tfrac{1}{2}$	$2^{\nu-2\mu}a^{2\nu-4\mu}y^{2\mu-\nu-\frac{1}{2}}\cdot$ $\cdot\exp(-\tfrac{1}{8}y^2a^{-2})\,M_{3\mu-\nu+\frac{1}{2},\mu}(\tfrac{1}{4}y^2a^{-2})$
16.4	$x^{\nu-2\mu-\frac{1}{2}}\exp(-\tfrac{1}{2}a^2x^2)\cdot$ $\cdot M_{\nu-\mu,\mu}(a^2x^2)$ $\mathrm{Re}\ \nu > -\tfrac{1}{2}$	$(2a)^{2\mu-2\nu}\Gamma(2\mu+1)[\Gamma(\tfrac{1}{2}+\nu)]^{-1}\cdot$ $\cdot y^{\nu-\frac{1}{2}}\exp(-\tfrac{1}{8}y^2a^{-2})D_{2\nu-4\mu}(2^{-\frac{1}{2}}a^{-1}y)$
16.5	$x^{\nu-2\mu-\frac{1}{2}}\exp(-\tfrac{1}{2}a^2x^2)\cdot$ $\cdot M_{\nu-\mu+1,\mu}(a^2x^2)$ $\mathrm{Re}\ \nu > -1$	$2^{-\frac{1}{2}}(2a)^{2\mu-2\nu-1}\Gamma(2\mu+1)[\Gamma(3/2+\nu)]^{-1}\cdot$ $\cdot y^{\nu+\frac{1}{2}}\exp(-\tfrac{1}{8}y^2a^{-2})D_{2\nu-4\mu+1}(2^{-\frac{1}{2}}a^{-1}y)$

	$f(x)$	$g(y) = \int\limits_0^\infty f(x)(xy)^{\frac{1}{2}} J_\nu(xy)\,dx$
16.6	$x^{-\frac{1}{2}}\exp(-\tfrac{1}{2}a^2 x^2) \cdot$ $\cdot M_{k,\frac{1}{2}\nu}(a^2 x^2)$ $\mathrm{Re}\ \nu > -1,\ \mathrm{Re}\ k < \tfrac{1}{2}$	$(2a)^{-2k}\Gamma(\nu+1)[\Gamma(\tfrac{1}{2}+k+\tfrac{1}{2}\nu)]^{-1} y^{2k-\frac{1}{2}} \cdot$ $\cdot \exp(-\tfrac{1}{4}y^2 a^{-2})$
16.7	$x^{-\frac{2}{3}k-\frac{1}{3}\nu-\frac{5}{6}}\exp(-\tfrac{1}{2}ax^2) \cdot$ $\cdot M_{k,\frac{1}{3}\nu-\frac{1}{3}k-\frac{1}{6}}(ax^2)$ $\mathrm{Re}(\nu-k+1) > 0$ $\mathrm{Re}(\nu+2k+1) > 0$	$\pi^{\frac{1}{2}}\Gamma(\tfrac{2}{3}+\tfrac{2}{3}\nu-\tfrac{2}{3}k)[\Gamma(\tfrac{1}{3}+\tfrac{1}{3}\nu+\tfrac{2}{3}k)]^{-1} \cdot$ $\cdot\, 2^{\frac{1}{3}\nu-\frac{2}{3}-\frac{4}{3}k} a^{\frac{1}{3}\nu-\frac{1}{3}k-\frac{1}{6}} y^{\frac{4}{3}k-\frac{1}{3}\nu+\frac{1}{6}} \cdot$ $\cdot \exp(-\tfrac{1}{8}y^2 a^{-1}) I_{\frac{2}{3}k+\frac{1}{3}\nu-\frac{1}{6}}(\tfrac{1}{8}y^2 a^{-1})$
16.8	$x^{-\frac{2}{3}k+\frac{1}{3}\nu-\frac{5}{6}}\exp(-\tfrac{1}{2}ax^2) \cdot$ $\cdot W_{k,\frac{1}{3}k+\frac{1}{3}\nu+\frac{1}{6}}(ax^2)$ $\mathrm{Re}\ \nu > -1,\ \mathrm{Re}(\nu-2k) > -2$	$\pi^{\frac{1}{2}}2^{-\frac{1}{3}\nu-\frac{2}{3}-\frac{4}{3}k} a^{-\frac{1}{3}\nu-\frac{1}{3}k-\frac{1}{6}} y^{\frac{4}{3}k+\frac{1}{3}\nu+\frac{1}{6}} \cdot$ $\cdot \exp(-\tfrac{1}{8}y^2 a^{-1}) I_{\frac{1}{3}\nu-\frac{2}{3}k+\frac{1}{6}}(\tfrac{1}{8}y^2 a^{-1})$
16.9	$x^{-\frac{2}{3}k+\frac{1}{3}\nu-\frac{5}{6}}\exp(-\tfrac{1}{2}ax^2) \cdot$ $\cdot M_{k,\frac{1}{3}k+\frac{1}{3}\nu+\frac{1}{6}}(ax^2)$ $\mathrm{Re}\ \nu > -1,\ \mathrm{Re}(\nu-8k-\tfrac{5}{2}) < 0$	$\pi^{-\frac{1}{2}}\Gamma(\tfrac{2}{3}k+\tfrac{2}{3}\nu+\tfrac{4}{3})[\Gamma(\tfrac{4}{3}k+\tfrac{1}{3}\nu+\tfrac{2}{3})]^{-1} \cdot$ $\cdot\, 2^{-\frac{1}{3}\nu-\frac{4}{3}k-\frac{2}{3}} a^{-\frac{1}{3}k-\frac{1}{3}\nu-\frac{1}{6}} y^{\frac{4}{3}k+\frac{1}{3}\nu+\frac{1}{6}} \cdot$ $\cdot \exp(-\tfrac{1}{8}y^2 a^{-1}) K_{\frac{2}{3}k-\frac{1}{3}\nu-\frac{1}{6}}(\tfrac{1}{8}y^2 a^{-1})$
16.10	$x^{\frac{2}{3}k+\frac{1}{3}\nu-\frac{5}{6}}\exp(\tfrac{1}{2}ax^2) \cdot$ $\cdot W_{k,\frac{1}{3}\nu-\frac{1}{3}k+\frac{1}{6}}(ax^2)$ $\mathrm{Re}\ \nu > -1,\ \mathrm{Re}(\nu+2k) > -2$ $\mathrm{Re}(\nu+8k-\tfrac{5}{2}) < 0$	$\pi^{-\frac{1}{2}}\Gamma(\tfrac{2}{3}+\tfrac{2}{3}k+\tfrac{1}{3}\nu)[\Gamma(\tfrac{2}{3}-\tfrac{4}{3}k+\tfrac{1}{3}\nu)]^{-1} \cdot$ $\cdot 2^{-\frac{1}{3}\nu+\frac{4}{3}k-\frac{5}{3}} a^{\frac{1}{3}k-\frac{1}{3}\nu-\frac{1}{6}} y^{\frac{1}{3}\nu-\frac{4}{3}k+\frac{1}{6}} \cdot$ $\cdot \exp(\tfrac{1}{8}y^2 a^{-1}) K_{\frac{2}{3}k+\frac{1}{3}\nu+\frac{1}{6}}(\tfrac{1}{8}y^2 a^{-1})$

	$f(x)$	$g(y) = \int_0^\infty f(x)(xy)^{\frac{1}{2}}J_\nu(xy)\,dx$
16.11	$x^{\nu-2\mu-\frac{1}{2}}\exp(\frac{1}{2}a^2x^2)\cdot$ $\cdot W_{3\mu-\nu-\frac{1}{2},\mu}(a^2x^2)$ Re $\nu>-1$, Re$(\nu-2\mu)>-1$ Re$(\nu-4\mu)>-\frac{1}{2}$	$2^{2\mu-\nu}a^{4\mu-2\nu}y^{\nu-2\mu-\frac{1}{2}}\exp(\frac{1}{8}y^2a^{-2})\cdot$ $\cdot W_{3\mu-\nu-\frac{1}{2},\mu}(\frac{1}{4}y^2a^{-2})$
16.12	$x^{2\mu-\nu-\frac{1}{2}}\exp(-\frac{1}{2}ax^2)\cdot$ $\cdot M_{k,\mu}(ax^2)$ Re $\mu > -\frac{1}{2}$ Re$(2\mu-2k-\nu) < \frac{1}{2}$	$\Gamma(2\mu+1)[\Gamma(\nu+k-\mu+\frac{1}{2})]^{-1}\cdot$ $\cdot 2^{\frac{1}{2}+\mu-k}a^{\frac{1}{2}\nu-\frac{1}{2}\mu-\frac{1}{2}k+\frac{1}{4}}y^{k-\mu-1}\cdot$ $\cdot \exp(-\frac{1}{8}y^2a^{-1})M_{\alpha,\beta}(\frac{1}{4}y^2a^{-1})$ $2\alpha = 3\mu+k-\nu+\frac{1}{2},\quad 2\beta = \nu+k-\mu-\frac{1}{2}$
16.13	$x^{\nu-2\mu-\frac{1}{2}}\exp(-\frac{1}{2}ax^2)\cdot$ $\cdot M_{k,\mu}(ax^2)$ Re $\nu > -\frac{1}{2}$ Re$(\nu-2\mu-2k) < \frac{1}{2}$	$\Gamma(2\mu+1)[\Gamma(\mu+k+\frac{1}{2})]^{-1}\cdot$ $\cdot 2^{\frac{1}{2}-\mu-k}a^{\frac{1}{2}\mu-\frac{1}{2}k-\frac{1}{2}\nu+\frac{1}{4}}y^{\mu+k-1}\cdot$ $\cdot \exp(-\frac{1}{8}y^2a^{-1})W_{\alpha,\beta}(\frac{1}{4}y^2a^{-1})$ $2\alpha = k+\nu-3\mu+\frac{1}{2},\quad 2\beta = k+\mu-\nu-\frac{1}{2}$
16.14	$x^{2\mu+\nu-\frac{1}{2}}\exp(-\frac{1}{2}ax^2)\cdot$ $\cdot W_{k,\mu}(ax^2)$ Re $\nu > -1$ Re$(2\mu+\nu) > -1$	$\Gamma(2\mu+\nu+1)[\Gamma(\mu+\nu-k+\frac{3}{2})]^{-1}\cdot$ $\cdot 2^{\mu-k+\frac{1}{2}}a^{-\frac{1}{2}k-\frac{1}{2}\mu-\frac{1}{2}\nu+\frac{1}{4}}y^{k-\mu-1}\cdot$ $\cdot \exp(-\frac{1}{8}y^2a^{-1})M_{\alpha,\beta}(\frac{1}{4}y^2a^{-1})$ $2\alpha = 3\mu+\nu+k+\frac{1}{2},\quad 2\beta = \mu+\nu-k+\frac{1}{2}$

	$f(x)$	$g(y) = \int\limits_{0}^{\infty} f(x)(xy)^{\frac{1}{2}} J_{\nu}(xy)\,dx$
16.15	$x^{\nu+2\mu-\frac{1}{2}} \exp(\frac{1}{2}ax^2) \cdot$ $\cdot\ W_{k,\mu}(ax^2)$ $\operatorname{Re}\nu>-1,\ \operatorname{Re}(2\mu+\nu)>-1$ $\operatorname{Re}(k+\mu+\frac{1}{2}\nu) < \frac{1}{4}$	$\Gamma(2\mu+\nu+1)\,[\Gamma(\frac{1}{2}-\mu-k)]^{-1} \cdot$ $\cdot\ 2^{\frac{1}{2}+k+\mu} a^{\frac{1}{4}+\frac{1}{2}k-\frac{1}{2}\mu-\frac{1}{2}\nu} y^{-\mu-k-1} \cdot$ $\cdot\ \exp({}^1/_8 y^2 a^{-1}) W_{\alpha,\beta}(\frac{1}{4}y^2 a^{-1})$ $2\alpha = k-3\mu-\nu-\frac{1}{2},\quad 2\beta = k+\mu+\nu+\frac{1}{2}$
16.16	$x^{2\rho-\frac{1}{2}}\exp(-\frac{1}{2}ax^2) \cdot$ $\cdot\ W_{k,\mu}(ax^2)$ $\operatorname{Re}(\rho+\frac{1}{2}\nu\pm\mu) > -1$	$\Gamma(1+\mu+\rho+\frac{1}{2}\nu)\,\Gamma(1-\mu+\rho+\frac{1}{2}\nu) \cdot$ $\cdot\ [\Gamma(1+\nu)\,\Gamma({}^3/_2+\frac{1}{2}\nu+\rho-k)]^{-1} \cdot$ $\cdot\ 2^{-\nu-1} a^{-\frac{1}{2}\nu-\rho-\frac{1}{2}} y^{\nu+\frac{1}{2}} \cdot$ $\cdot\ {}_2F_2(\lambda+\mu,\lambda-\mu;\nu+1,\frac{1}{2}+\lambda-k;-\frac{1}{4}y^2 a^{-1})$ $\lambda = 1 + \rho + \frac{1}{2}\nu$
16.17	$x^{-\frac{1}{2}} K_{\frac{1}{2}\nu-\mu}(\frac{1}{2}ax^2) \cdot$ $\cdot\ M_{k,\mu}(ax^2)$ $\operatorname{Re}k > -\frac{1}{4},\ \operatorname{Re}\mu > -\frac{1}{2},$ $\operatorname{Re}\nu > -1$	$\frac{1}{2}\Gamma(1+2\mu)\,[\Gamma(1+k+\frac{1}{2}\nu)]^{-1} \cdot$ $\cdot\ y^{-\frac{1}{2}} W_{\frac{1}{2}k-\frac{1}{2}\mu,\,\frac{1}{2}k-\frac{1}{4}\nu}(\frac{1}{2}y^2 a^{-1}) \cdot$ $\cdot\ M_{\frac{1}{2}k+\frac{1}{2}\mu,\,\frac{1}{2}k+\frac{1}{4}\nu}(\frac{1}{2}y^2 a^{-1})$
16.18	$x^{-\frac{1}{2}} M_{-\frac{1}{2}\mu,\,\frac{1}{2}\nu}(ax) \cdot$ $\cdot\ W_{\frac{1}{2}\mu,\,\frac{1}{2}\nu}(ax)$ $\operatorname{Re}\nu > -1,\ \operatorname{Re}\mu<\frac{1}{2}$	$a\Gamma(1+\nu)\,[\Gamma(\frac{1}{2}+\frac{1}{2}\nu-\frac{1}{2}\mu)]^{-1} y^{-\mu-\frac{1}{2}} \cdot$ $\cdot\ (a^2+y^2)^{-\frac{1}{2}}[a+(a^2+y^2)^{\frac{1}{2}}]^{\mu}$

	$f(x)$	$g(y) = \int\limits_0^\infty f(x)(xy)^{\frac{1}{2}}J_\nu(xy)\,dx$
16.19	$x^{-\frac{1}{2}}M_{k+\mu,k+\frac{1}{2}\mu+\frac{1}{4}\nu}(ax^2)\cdot$ $\cdot W_{k,k+\frac{1}{2}\mu-\frac{1}{4}\nu}(ax^2)$ $\mathrm{Re}(\mu+2k)>-\frac{1}{4},\mathrm{Re}\ \mu>-\frac{1}{2}$ $\mathrm{Re}\ \nu>-1$	$2\Gamma(1+2k+\mu+\frac{1}{2}\nu)[\Gamma(1+2\mu)]^{-1}\cdot$ $\cdot y^{-\frac{1}{2}}K_{\frac{1}{2}\nu-\mu}(\frac{1}{4}y^2a^{-1})\cdot$ $\cdot M_{\mu+2k,\mu}(\frac{1}{2}y^2a^{-1})$
16.20	$x^{\frac{1}{2}}M_{\mu,\frac{1}{2}\nu}(iax)\cdot$ $\cdot M_{\mu,\frac{1}{2}\nu}(-iax)$ $2\ \mathrm{Re}\ \mu<1+\mathrm{Re}\ \nu$	$a\Gamma(1+\nu)[B(\frac{1}{2}+\frac{1}{2}\nu+\mu,\frac{1}{2}+\frac{1}{2}\nu-\mu)]^{-1}\cdot$ $\cdot y^{-2\mu-\frac{1}{2}}(a^2-y^2)^{-\frac{1}{2}}\cdot\{[a+(a^2-y^2)^{\frac{1}{2}}]^{2\mu}+$ $+[a-(a^2-y^2)^{\frac{1}{2}}]^{2\mu}\}\qquad y<a$ $0\qquad y>a$
16.21	$x^{-\nu-3/2}M_{\mu,\frac{1}{2}+\nu}(iax)\cdot$ $\cdot M_{\mu,\frac{1}{2}+\nu}(-iax)$ $-3/2-\mathrm{Re}\ \nu<\mathrm{Re}\ \mu<\frac{1}{2}+\mathrm{Re}\ \nu$	$(2\pi)^{\frac{1}{2}}[\Gamma(3/2+\nu)]^2\Gamma(1+\nu+\mu)\Gamma(1+\nu-\mu)]^{-1}\cdot$ $\cdot 2^{-\frac{1}{2}-\nu}a^{\frac{1}{2}-\nu}x^{-\frac{1}{2}-\nu}P_{-\frac{1}{2}+\mu}^{-\frac{1}{2}-\nu}(2a^2x^{-2}-1)\quad x<a$ $0\qquad x>a$
16.22	$x^{2\mu-\nu-\frac{1}{2}}\cdot$ $\cdot W_{k,\mu}(ax)M_{-k,\mu}(ax)$ $\mathrm{Re}\ \mu>-\frac{1}{2},\mathrm{Re}(2\mu+2k-\nu)<\frac{1}{2}$	$\Gamma(1+2\mu)[\Gamma(\frac{1}{2}+\nu-k-\mu)]^{-1}\cdot$ $\cdot 2^{2\mu-\nu+2k}a^{2k}y^{\nu-2\mu-2k-\frac{1}{2}}\cdot$ $\cdot {}_3F_2(\frac{1}{2}-k,1-k,\frac{1}{2}+\mu-k;1-2k,\frac{1}{2}-k-\mu+\nu;-y^2a^{-2})$

	$f(x)$	$g(y) = \int\limits_0^\infty f(x)(xy)^{\frac{1}{2}}J_\nu(xy)dx$
16.23	$x^{2\rho-\nu-5/2}W_{k,\mu}(ax) \cdot$ $\cdot W_{-k,\mu}(ax)$ $-\mathrm{Re}\,\rho < \mathrm{Re}\,\mu < \mathrm{Re}\,\rho$	$\Gamma(2\rho)\Gamma(\rho+\mu)\Gamma(\rho-\mu) \cdot$ $\cdot [\Gamma(\frac{1}{2}+k+\rho)\Gamma(\frac{1}{2}-k+\rho)\Gamma(1+\nu)]^{-1} \cdot$ $\cdot\, 2^{-\nu-1}a^{1-2\rho}y^{\nu+\frac{1}{2}}{}_4F_3(\rho,\frac{1}{2}+\rho,\rho+\mu,$ $\rho-\mu;\frac{1}{2}+k+\rho,\frac{1}{2}-k+\rho,1+\nu;-y^2a^{-2})$
16.24	$x^{-3/2}M_{-\mu,\frac{1}{4}\nu}(ax^2) \cdot$ $\cdot W_{\mu,\frac{1}{4}\nu}(ax^2)$ $\mathrm{Re}\,\nu > -1$	$\Gamma(1+\frac{1}{2}\nu)[\Gamma(\frac{1}{2}-\mu+\frac{1}{4}\nu)]^{-1}(2ay)^{\frac{1}{2}} \cdot$ $\cdot I_{\frac{1}{4}\nu-\mu}(^1/_8y^2a^{-1})K_{\frac{1}{4}\nu+\mu}(^1/_8y^2a^{-1})$
16.25	$x^{-3/2}M_{\alpha-\beta,\frac{1}{2}\nu-\gamma}(ax^2) \cdot$ $\cdot W_{\alpha+\beta,\frac{1}{4}\nu+\gamma}(ax^2)$ $\mathrm{Re}\,\beta < {}^1/_8,\ \mathrm{Re}\,\nu > -1$ $\mathrm{Re}(\nu-4\gamma) > -2$	$\Gamma(1+\frac{1}{2}\nu-2\gamma)[\Gamma(1+\frac{1}{2}\nu-2\beta)]^{-1}(2a)^{5/2}y^{-3/2} \cdot$ $\cdot M_{\alpha-\gamma,\frac{1}{4}\nu-\beta}(\frac{1}{4}y^2a^{-1})W_{\alpha+\gamma,\frac{1}{4}\gamma+\beta}(\frac{1}{4}y^2a^{-1})$
16.26	$x^{\frac{1}{2}}M_{\frac{1}{2}\nu,\mu}(ax^{-1}) \cdot$ $\cdot W_{-\frac{1}{2}\nu,\mu}(ax^{-1})$ $\mathrm{Re}\,\nu > -1,\ \mathrm{Re}\,\mu > -\frac{1}{4}$	$2a\Gamma(1+2\mu)[\Gamma(\frac{1}{2}+\mu+\frac{1}{2}\nu)]^{-1}y^{-\frac{1}{2}} \cdot$ $\cdot J_{2\mu}[(2ay)^{\frac{1}{2}}]K_{2\mu}[(2ay)^{\frac{1}{2}}]$

	$f(x)$	$f(x) = \int\limits_0^\infty f(x)(xy)^{\frac{1}{2}} J_\nu(xy)\,dx$
16.27	$x^{\frac{1}{2}} W_{\frac{1}{2}\nu,\mu}(ax^{-1}) \cdot$ $\cdot W_{-\frac{1}{2}\nu,\mu}(ax^{-1})$ $\mathrm{Re}(\nu \pm 2\mu) > -1$	$-2ay^{-\frac{1}{2}}\{\sin[\pi(\mu-\frac{1}{2}\nu)]J_{2\mu}[(2ay)^{\frac{1}{2}}] +$ $+ \cos[\pi(\mu-\frac{1}{2}\nu)]Y_{2\mu}[(2ay)^{\frac{1}{2}}]\}K_{2\mu}[(2ay)^{\frac{1}{2}}]$
16.28	$x^{\frac{1}{2}} W_{-\frac{1}{2}\nu,\mu}(iax^{-1}) \cdot$ $\cdot W_{-\frac{1}{2}\nu,\mu}(-iax^{-1})$ $\mathrm{Re}\,\nu > -1,\ -\frac{1}{2} < \mathrm{Re}\,\mu < \frac{1}{2}$	$[\Gamma(\frac{1}{2}+\frac{1}{2}\nu+\mu)\Gamma(\frac{1}{2}-\mu+\frac{1}{2}\nu)]^{-1}4ay^{-\frac{1}{2}} \cdot$ $\cdot K_\mu[(2iay)^{\frac{1}{2}}]K_\mu[(-2iay)^{\frac{1}{2}}]$
16.29	$x^{-\frac{1}{2}} \cdot$ $\cdot M_{-\mu,\frac{1}{2}\nu}\{a[(b^2+x^2)^{\frac{1}{2}}-b]\} \cdot$ $\cdot W_{\mu,\frac{1}{2}\nu}\{a[(b^2+x^2)^{\frac{1}{2}}+b]\}$ $\mathrm{Re}\,\nu > -1,\ \mathrm{Re}\,\mu < \frac{1}{4}$	$a\Gamma(1+\nu)[\Gamma(\frac{1}{2}+\frac{1}{2}\nu-\mu)]^{-1}y^{-2\mu-\frac{1}{2}} \cdot$ $\cdot (a^2+y^2)^{-\frac{1}{2}}[a+(a^2+y^2)^{\frac{1}{2}}]^{2\mu}\exp[-b(a^2+y^2)^{\frac{1}{2}}]$

1.17 Gauss' Hypergeometric Function

	$f(x)$	$g(y) = \int_0^\infty f(x)(xy)^{\frac{1}{2}} J_\nu(xy)\,dx$
17.1	$x^{2\alpha+\nu-\frac{1}{2}} \cdot$ $\cdot\, _2F_1(\alpha-\nu-\tfrac{1}{2},\alpha;2\alpha;-\lambda^2x^2)$ $\mathrm{Re}\ \nu<-\tfrac{1}{2},\ \mathrm{Re}\ \lambda>0$ $\mathrm{Re}(\alpha+\nu) > -\tfrac{1}{2}$	$i\pi^{-1} 2^{\nu+2\alpha-1} \lambda^{1-2\alpha} \Gamma(\tfrac{1}{2}+\alpha)\Gamma(\tfrac{1}{2}+\alpha+\nu)\cdot$ $\cdot\, y^{-\nu-3/2} W_{\frac{1}{2}-\alpha,\,-\frac{1}{2}-\nu}(y\lambda^{-1})\cdot$ $\cdot\,[W_{\frac{1}{2}-\alpha,\,-\frac{1}{2}-\nu}(ye^{-i\pi}\lambda^{-1}) - W_{\frac{1}{2}-\alpha,\,-\frac{1}{2}-\nu}(ye^{i\pi}\lambda^{-1})]$
17.2	$x^{2\alpha-\nu-\frac{1}{2}} \cdot$ $\cdot\, _2F_1(\nu+\alpha-\tfrac{1}{2},\alpha;2\alpha;-\lambda^2x^2)$ $\mathrm{Re}\ \alpha>-\tfrac{1}{2},\ \mathrm{Re}\ \nu>\tfrac{1}{2},\ \mathrm{Re}\ \lambda>0$	$2^{2\alpha-\nu}\lambda^{1-2\alpha}[\Gamma(2\nu)]^{-1}\Gamma(\tfrac{1}{2}+\alpha)y^{\nu-3/2}\cdot$ $\cdot\, M_{\alpha-\frac{1}{2},\,\nu-\frac{1}{2}}(y\lambda^{-1}) W_{\frac{1}{2}-\alpha,\,\nu-\frac{1}{2}}(y\lambda^{-1})$
17.3	$x^{\nu+\frac{1}{2}}{}_2F_1(\alpha,\beta;\nu+1;-\lambda^2x^2)$ $-1 < \mathrm{Re}\ \nu <$ $<2\,\mathrm{Max}(\mathrm{Re}\ \alpha,\mathrm{Re}\ \beta) - 3/2$ $\mathrm{Re}\ \lambda > 0$	$\lambda^{-\alpha-\beta}[\Gamma(\alpha)\Gamma(\beta)]^{-1} 2^{\nu-\alpha-\beta+2}\Gamma(1+\nu)\cdot$ $\cdot\, y^{\alpha+\beta-\nu-3/2} K_{\alpha-\beta}(y\lambda^{-1})$
17.4	$x^{\nu+\frac{1}{2}}{}_2F_1(\alpha,\beta;\tfrac{1}{2}\nu+\tfrac{1}{2}\beta+1;-$ $-\lambda^2x^2)$ $-1<\mathrm{Re}\,\nu<2\,\mathrm{Max}(\mathrm{Re}\,\alpha,\mathrm{Re}\,\beta)-3/2$	$\pi^{-\frac{1}{2}}[\Gamma(\alpha)\Gamma(\beta)]^{-1}2^{1-\beta}\lambda^{-\nu-\beta-1}\Gamma(1+\tfrac{1}{2}\nu+\tfrac{1}{2}\beta)\cdot$ $\cdot\,[K_{\frac{1}{2}(\nu-\beta+1)}(\tfrac{1}{2}y\lambda^{-1})]^2$

	$f(x)$	$g(y) = \int\limits_0^\infty f(x)(xy^{\frac{1}{2}}J_\nu(xy)\,dx$
17.5	$x^{-2\alpha-3/2}{}_2F_1(\tfrac{1}{2}+\alpha,1+\alpha;$ $1+2\alpha;-4\lambda^2x^{-2})$ Re $\nu>-1$, Re $\lambda>0$, Re $\alpha>-\tfrac{1}{2}$	$\lambda^{-2\alpha}y^{\frac{1}{2}}I_{\frac{1}{2}\nu+\alpha}(\lambda y)K_{\frac{1}{2}\nu-\alpha}(\lambda y)$
17.6	$x^{\frac{1}{2}+\nu-4\alpha}$ $\cdot{}_2F_1(\alpha,\alpha+\tfrac{1}{2};\nu+1;-\lambda^2x^{-2})$ $-1+\mathrm{Re}\,\alpha<\mathrm{Re}\,\nu<4\mathrm{Re}\,\alpha-3/2$ Re $\lambda>0$	$2^\nu\lambda^{1-2\alpha}\Gamma(\nu)[\Gamma(2\alpha)]^{-1}y^{2\alpha-\nu-\frac{1}{2}}\cdot$ $\cdot\ I_\nu(\tfrac{1}{2}\lambda y)K_{2\alpha-\nu-1}(\tfrac{1}{2}\lambda y)$
17.7	$x^{\nu+\frac{1}{2}}(1+x)^{-2\alpha}\cdot$ $\cdot{}_2F_1[\alpha,\tfrac{1}{2}+\nu;2\nu+1;4x(1+x)^{-2}]$ $-1<\mathrm{Re}\,\nu<2\,\mathrm{Re}\,\alpha-3/2$	$[\Gamma(\alpha)]^{-1}\Gamma(\nu+1)\Gamma(1+\nu-\alpha)2^{2\nu-2\alpha+1}\cdot$ $\cdot\ y^{2\alpha-2\nu-3/2}J_\nu(y)$

Chapter II. Integral Transforms with Modified Bessel Functions as Kernel

A representation of a given function $f(x)$ by means of a double integral involving modified Bessel functions of order ν is

$$f(x) = (\pi i)^{-1} \int_{c-i\infty}^{c+i\infty} I_\nu(tx)(tx)^{\frac{1}{2}} dt \int_0^\infty K_\nu(ut)(ut)^{\frac{1}{2}} f(u) du$$

or also

$$f(x) = (2\pi i)^{-1} \int_{c-i\infty}^{c+i\infty} [I_\nu(tx)+I_{-\nu}(tx)](tx)^{\frac{1}{2}} dt \int_0^\infty K_\nu(ut)(ut)^{\frac{1}{2}} f(u) du$$

This is equivalent with the pair of inversion formulas

(1) $g(y;\nu) = \int_0^\infty f(x)(xy)^{\frac{1}{2}} K_\nu(xy) dx$

(2) $f(x) = (\pi i)^{-1} \int_{c-i\infty}^{c+i\infty} g(y,\nu)(xy)^{\frac{1}{2}} I_\nu(xy) dy$

or

(3) $g(y;\nu) = \int_0^\infty f(x)(xy)^{\frac{1}{2}} K_\nu(xy) dx$

(4) $f(x) = (2\pi i)^{-1} \int_{c-i\infty}^{c+i\infty} g(y;\nu)(xy)^{\frac{1}{2}} [I_\nu(xy)+I_{-\nu}(xy)] dy$

Since for $\nu = \pm\frac{1}{2}$

$$K_{\pm\frac{1}{2}}(z) = (\tfrac{1}{2}\pi z^{-1})^{\frac{1}{2}} e^{-z}$$

$$I_{\frac{1}{2}}(z) = (\tfrac{1}{2}\pi z)^{-\frac{1}{2}} \sinh z, \quad I_{-\frac{1}{2}}(z) = (\tfrac{1}{2}\pi z)^{-\frac{1}{2}} \cosh z$$

the equations (3) and (4) become

$$(2\pi^{-1})^{\frac{1}{2}} g(y; \pm\frac{1}{2}) = \int_0^\infty f(x) e^{-xy} \, dx$$

$$f(x) = (2\pi i)^{-1} \int_0^\infty (2\pi^{-1}) \, g(y; \pm\frac{1}{2}) e^{xy} \, dy$$

These are the Laplace transform formulas.

Since

$$\pi J_\nu(z) = e^{i\frac{\pi}{2}(\nu+1)} K_\nu(ze^{i\frac{\pi}{2}}) + e^{-i\frac{\pi}{2}(\nu+1)} K_\nu(ze^{-i\frac{\pi}{2}})$$

$$\pi Y_\nu(z) = e^{-i\frac{\pi}{2}(\nu+1)} K_\nu(ze^{-i\frac{\pi}{2}}) - e^{i\frac{\pi}{2}(\nu+1)} K_\nu(ze^{i\frac{\pi}{2}})$$

it is possible to evaluate integral transforms with Bessel function (Chapter I) or Neumann function kernel (Chapter III) by means of the above relations.

REFERENCES

Boas, R. P., 1942: Proc. Nat. Acad. Sci. U.S.A. 28, 21-24.

Boas, R. P., 1942: Bull. Amer. Math. Soc. 48, 286-294.

Meijer, C. S., 1940: Proc. Amsterdam Akad. Wet.   599-608;
                      702-711.

2.1 General Formulas

	$f(x)$	$g(y;\nu) = \int\limits_{0}^{\infty} f(x)(xy)^{\frac{1}{2}}K_{\nu}(xy)\,dx \quad y>0$
1.1	$\dfrac{1}{\pi i}\int\limits_{c-i\infty}^{c+i\infty} g(y)(xy)^{\frac{1}{2}}I_{\nu}(xy)\,dy$	$g(y)$
1.2	$\dfrac{1}{2\pi i}\int\limits_{c-i\infty}^{c+i\infty} g(y)(xy)^{\frac{1}{2}} \cdot$ $\cdot\,[I_{\nu}(xy)+I_{-\nu}(xy)]\,dy$	$g(y)$
1.3	$f(ax) \qquad , \quad a>0$	$a^{-1}g(ya^{-1};\nu)$
1.4	$x^{m}f(x),\ m=0,1,2,\cdots$	$y^{\frac{1}{2}-\nu}\left(-\dfrac{d}{y\,dy}\right)^{m}[y^{m+\nu-\frac{1}{2}}g(y;m+\nu)]$
1.5	$x^{m}f(x),\ m=0,1,2,\cdots$	$y^{\frac{1}{2}+\nu}\left(-\dfrac{d}{y\,dy}\right)^{m}[y^{m-\nu-\frac{1}{2}}g(y;\nu-m)]$
1.6	$x^{-1}f(x)$	$\tfrac{1}{2}\nu^{-1}[yg(y;\nu+1)-yg(y;\nu-1)]$
1.7	$x^{-\mu}f(x)$ $\qquad \mathrm{Re}\,\mu>0$	$2^{1-\mu}[\Gamma(\mu)]^{-1}y^{\nu+\frac{1}{2}}\int\limits_{y}^{\infty}\tau^{\frac{1}{2}-\mu-\nu}(\tau^{2}-y^{2})^{\mu-1}\cdot$ $\cdot\,g(\tau;\nu+\mu)\,d\tau$
1.8	$f'(x)$	$\tfrac{1}{2}\nu^{-1}[(\nu-\tfrac{1}{2})yg(y;\nu+1)+(\nu+\tfrac{1}{2})yg(y;\nu-1)]$

2.2 Transforms of Order Zero

	$f(x)$	$g(y) = \int_0^\infty f(x)(xy)^{\frac{1}{2}}K_0(xy)\,dx$
2.1	$x^{\frac{1}{2}}(a^2+x^2)^{-\frac{1}{2}}$	$2y^{-\frac{1}{2}}[\sin(ay)\mathrm{Ci}(ay)-\cos(ay)\mathrm{si}(ay)]$
2.2	$x^{-\frac{1}{2}}(a^2+x^2)^{-\frac{1}{2}} \cdot$ $\cdot\,[x+(a^2+x^2)^{\frac{1}{2}}]^{-2\mu}$	$\frac{1}{4}\pi a^{-2\mu}y^{\frac{1}{2}}\frac{\partial}{\partial\nu}[J_{\mu+\frac{1}{2}\nu}(\frac{1}{2}ay)Y_{\mu-\frac{1}{2}\nu}(\frac{1}{2}ay) -$ $-\,Y_{\mu+\frac{1}{2}\nu}(\frac{1}{2}ay)J_{\mu-\frac{1}{2}\nu}(\frac{1}{2}ay)]_{\nu=0}$
2.3	$x^{-\frac{1}{2}}(a^2+x^2)^{-\frac{1}{2}} \cdot$ $\cdot\,\{[(a^2+x^2)^{\frac{1}{2}}+x]^{2\mu} +$ $+\,[(a^2+x^2)^{\frac{1}{2}}-x]^{2\mu}\}$	$\frac{1}{4}\pi^2a^{2\mu}y^{\frac{1}{2}}\{[J_\mu(\frac{1}{2}ay)]^2 +[Y_\mu(\frac{1}{2}ay)]^2\}$
2.4	$\quad\ 0\quad\ x<a$ $x^{\frac{1}{2}-\nu}(x^2-a^2)^{\nu-\frac{1}{2}}\quad x>a$ $\mathrm{Re}\ \nu > -\ \frac{1}{2}$	$\pi^{\frac{1}{2}}2^{\nu-1}\ (\frac{1}{2}+\nu)y^{-\nu-\frac{1}{2}}e^{-ay}$
2.5	$\quad\ 0\quad\ x<a$ $x^{\frac{1}{2}-\nu}(x^2-a^2)^{\nu-3/2}\quad x>a$ $\mathrm{Re}\ \nu > \frac{1}{2}$	$\pi^{\frac{1}{2}}2^{\nu-2}a^{-1}\Gamma(\nu-\frac{1}{2})y^{\frac{1}{2}-\nu}e^{-ay}$
2.6	$\quad\ 0\quad\ x<a$ $x^{-\frac{1}{2}}(x^2-a^2)^{-\frac{1}{2}} \cdot$ $\cdot\,\{[2x^2-a^2+2x(x^2-a^2)^{\frac{1}{2}}]^{\mu+\frac{1}{2}}+$ $+\,[2x^2-a^2-2x(x^2-a^2)^{\frac{1}{2}}]^{\mu+\frac{1}{2}}\}$	$a^{2\mu+1}y^{\frac{1}{2}}[K_{\mu+\frac{1}{2}}(\frac{1}{2}ay)]^2$

	$f(x)$	$g(y) = \int\limits_{0}^{\infty} f(x)(xy)^{\frac{1}{2}}K_0(xy)\,dx$
2.7	$x^{\frac{1}{2}}(b^2+x^2)^{-\frac{1}{2}}$	$y^{-\frac{1}{2}}[\sin(by)\text{Ci}(by) -$ $- \cos(by)\text{si}(by)]$
2.8	$x^{-\frac{1}{2}}e^{-ax}$	$y^{\frac{1}{2}}(a^2-y^2)^{-\frac{1}{2}}\log[ay^{-1}+(a^2y^{-2}-1)^{\frac{1}{2}}] \quad y<a$ $y^{\frac{1}{2}}(y^2-a^2)^{-\frac{1}{2}}\arccos(ay^{-1}) \qquad\qquad y>a$
2.9	$x^{\frac{1}{2}}(b^2+x^2)^{-\frac{1}{2}} \cdot$ $\cdot \exp[-a(b^2+x^2)^{\frac{1}{2}}]$	$\frac{1}{2}y^{\frac{1}{2}}(a^2-y^2)^{-\frac{1}{2}} \cdot$ $\cdot \{\exp[b(a^2-y^2)^{\frac{1}{2}}]\text{Ei}(-z_1) -$ $- \exp[-b(a^2-y^2)^{\frac{1}{2}}]\text{Ei}(-z_2)\}$ $z_{\genfrac{}{}{0pt}{}{1}{2}} = b[a\pm(a^2-y^2)^{\frac{1}{2}}]$
2.10	$x^{\frac{1}{2}}e^{-ax^{-1}}$	$2ay^{-\frac{1}{2}}K_1[(2iay)^{\frac{1}{2}}]K_1[(-2iay)^{\frac{1}{2}}]$
2.11	$x^{-3/2}e^{-ax^{-1}}$	$2y^{\frac{1}{2}}K_0[(2iay)^{\frac{1}{2}}]K_0[(-2iay)^{\frac{1}{2}}]$
2.12	$x^{-\frac{1}{2}}\sin(ax)$	$y^{\frac{1}{2}}(a^2+y^2)^{-\frac{1}{2}}\log[ay^{-1}+(1+a^2y^{-2})^{\frac{1}{2}}]$
2.13	$x^{-\frac{1}{2}}\cos(ax)$	$\frac{1}{2}\pi y^{\frac{1}{2}}(a^2+y^2)^{-\frac{1}{2}}$

	$f(x)$	$g(y) = \int_0^\infty f(x)(xy)^{\frac{1}{2}}K_0(xy)\,dx$
2.14	$x^{-\frac{1}{2}}\log(ax)\cos(bx)$	$\frac{1}{2}\pi y^{\frac{1}{2}}(b^2+y^2)^{-\frac{1}{2}} \cdot$ $\cdot\, [\log(\frac{1}{2}ay)-\gamma-\log(b^2+y^2)]$
2.15	$x^{\frac{1}{2}}\sin(ax^{-1})$	$\pi ay^{-\frac{1}{2}}[J_1[(2ay)^{\frac{1}{2}}]K_1[(2ay)^{\frac{1}{2}}]$
2.16	$x^{\frac{1}{2}}\cos(ax^{-1})$	$-\pi ay^{-\frac{1}{2}}[Y_1[(2ay)^{\frac{1}{2}}]K_1[(2ay)^{\frac{1}{2}}]$
2.17	$x^{-3/2}\sin(ax^{-1})$	$\pi y^{\frac{1}{2}}J_0[(2ay)^{\frac{1}{2}}]K_0[(2ay)^{\frac{1}{2}}]$
2.18	$x^{-3/2}\cos(ax^{-1})$	$-\pi y^{\frac{1}{2}}Y_0[(2ay)^{\frac{1}{2}}]K_0[(2ay)^{\frac{1}{2}}]$
2.19	$x^{\frac{1}{2}}(b^2+x^2)^{-\frac{1}{2}} \cdot$ $\cdot\, \cos[a(b^2+x^2)^{\frac{1}{2}}]$	$\frac{1}{2}y^{\frac{1}{2}}(a^2+y^2)^{-\frac{1}{2}}\{\sin u[\mathrm{Ci}(z_1)+\mathrm{Ci}(z_2)] -$ $-\cos u\,[\mathrm{si}(z_1)+\mathrm{si}(z_2)]\}$ $u = b(a^2+y^2)^{\frac{1}{2}}$ $z_{\substack{1\\2}} = b[(a^2+y^2)^{\frac{1}{2}}\pm a]$
2.20	$x^{\frac{1}{2}}(b^2+x^2)^{-\frac{1}{2}} \cdot$ $\cdot\sin[a(b^2+x^2)^{\frac{1}{2}}]$	$\frac{1}{2}y^{\frac{1}{2}}(a^2+y^2)^{-\frac{1}{2}}\{\cos u[\mathrm{Ci}(z_1)-\mathrm{Ci}(z_2)] +$ $+\sin u[\mathrm{si}(z_1)-\mathrm{si}(z_2)]\}$ $u,\ z_1,\ z_2\quad$ as in $\quad 2.19.$

	$f(x)$	$g(y) = \int\limits_0^\infty f(x)(xy)^{\frac{1}{2}}K_0(xy)\,dx$
2.21	$x^{\frac{1}{2}}(b^2-x^2)^{-\frac{1}{2}} \cdot$ $\cdot \cos[a(b^2-x^2)^{\frac{1}{2}}]$ $\qquad y < b$ $\qquad 0 \qquad y > b$	$\frac{1}{2}y^{\frac{1}{2}}(a^2-y^2)^{-\frac{1}{2}} \cdot$ $\cdot \{\cos u[\operatorname{si}(z_1) - \operatorname{si}(z_2)] -$ $- \sin u[\operatorname{Ci}(z_1) + \operatorname{Ci}(z_2)]\} \quad y < a$ $u = b(a^2-y^2)^{\frac{1}{2}}$ $z_1 = b[a \pm (a^2-y^2)^{\frac{1}{2}}]$ _2
2.22	$x^{-\frac{1}{2}}(a^2-x^2)^{-\frac{1}{2}} \cdot$ $\cdot \cos[b(a^2-x^2)^{\frac{1}{2}}] \quad x<a$ $\qquad 0 \qquad\qquad x>a$	$-{}^1\!/\!_8\pi^2\, y^{\frac{1}{2}}[J_0\{\tfrac{1}{2}a[b-(b^2-y^2)^{\frac{1}{2}}]\} \cdot$ $\cdot\, Y_0\{\tfrac{1}{2}a[b+(b^2-y^2)^{\frac{1}{2}}]\} +$ $+ J_0\{\tfrac{1}{2}a[b+(b^2-y^2)^{\frac{1}{2}}]\} \cdot$ $\cdot\, Y_0\{\tfrac{1}{2}a[b-(b^2-y^2)^{\frac{1}{2}}]\}]$
2.23	$\qquad 0 \qquad x < a$ $x^{\frac{1}{2}}(x^2-a^2)^{-\frac{1}{2}} \cdot$ $\cdot \cos[b(x^2-a^2)^{\frac{1}{2}}]$	$\frac{1}{2}\pi y^{\frac{1}{2}} \cdot (b^2+y^2)^{-\frac{1}{2}}\exp[-a(b^2+y^2)^{\frac{1}{2}}]$
2.24	$x^{-\frac{1}{2}}\sin h(ax) \qquad y>a$	$y^{\frac{1}{2}}(y^2-a^2)^{-\frac{1}{2}}\arcsin(ay^{-1})$
2.25	$x^{-\frac{1}{2}}\cos h(ax) \qquad y>a$	$\frac{1}{2}\pi y^{\frac{1}{2}}(y^2-a^2)^{-\frac{1}{2}}$

	$f(x)$	$g(y) = \int_0^\infty f(x)(xy)^{1/2}K_0(xy)dx$
2.26	$x^{-3/2}\sin h(ax)$ $\underline{y\geq a}$	$\frac{1}{2}\pi y^{-1/2}\arcsin(ay^{-1})$
2.27	$x^{-1/2}(a^2-x^2)^{-1/2}$ $\cos h[b(a^2-x^2)^{1/2}]x<a$ 0 $x>a$	$\frac{1}{2}\pi y^{1/2}[I_0\{\frac{1}{2}a[(b^2+y^2)^{1/2}-b]\}K_0\{\frac{1}{2}a[(b^2+y^2)^{1/2}+b]\}+$ $+ I_0\{\frac{1}{2}a[(b^2+y^2)^{1/2}+b]\}K_0\{\frac{1}{2}a[(b^2+y^2)^{1/2}-b]\}]$
2.28	$x^{-1/2}J_0(ax)$	$y^{1/2}(a^2+y^2)^{-1/2}\mathbf{K}[a(a^2+y^2)^{-1/2}]$
2.29	$x^{-1/2}Y_0(ax)$	$-y^{1/2}(a^2+y^2)^{-1/2}\mathbf{K}[y(a^2+y^2)^{-1/2}]$
2.30	$x^{-1/2}I_0(ax)$ $y>a$	$y^{-1/2}\mathbf{K}(ay^{-1})$
2.31	$I_0(ax)$ $y>a$	$2^{1/2}\pi^{-1}\Gamma^2(3/4)(y^2-a^2)^{-1/2}\cdot$ $\cdot\,\mathbf{K}\{[\frac{1}{2}-\frac{1}{2}(1-a^2y^{-2})^{1/2}]^{1/2}\}$
2.32	$x^{-1/2}K_0(ax)$	$\frac{1}{2}a^{-1}y^{1/2}\mathbf{K}[(1-y^2a^{-2})^{1/2}]$ $y<a$ $\frac{1}{2}y^{-1/2}\mathbf{K}[(a^2y^{-2}-1)^{1/2}]$ $y>a$

	$f(x)$	$g(y) = \int_0^\infty f(x)(xy)^{\frac{1}{2}}K_0(xy)dx$
2.33	$x^{-1}K_0(ax)$	$2^{-3/2}\Gamma^2(\tfrac{1}{4})\left(\tfrac{y}{a}\right)^{\frac{1}{2}} \cdot$ $\begin{cases} \mathbf{K}\{[\tfrac{1}{2}-\tfrac{1}{2}(1-y^2a^{-2})^{\frac{1}{2}}]^{\frac{1}{2}}\} + \\ \quad + \mathbf{K}\{[\tfrac{1}{2}+\tfrac{1}{2}(1-y^2a^{-2})^{\frac{1}{2}}]^{\frac{1}{2}}\} \quad y < a \\ (\tfrac{a}{y})^{\frac{1}{2}}\mathbf{K}\{[\tfrac{1}{2}-\tfrac{1}{2}(1-a^2y^{-2})^{\frac{1}{2}}]^{\frac{1}{2}}\} + \\ \quad + (\tfrac{a}{y})^{\frac{1}{2}}\mathbf{K}\{[\tfrac{1}{2}+\tfrac{1}{2}(1-a^2y^{-2})^{\frac{1}{2}}]^{\frac{1}{2}}\} \quad y > a \end{cases}$
2.34	$x^{\frac{1}{2}}K_0(ax)$	$y^{\frac{1}{2}}(a^2-y^2)^{-1}\log(ay^{-1})$
2.35	$x^{-\frac{1}{2}}\cos(bx)I_0(ax)$ $y \geq a$	$y^{\frac{1}{2}}[b^2+(a+y)^2]^{-\frac{1}{2}}\mathbf{K}\{2(ay)^{\frac{1}{2}}[b^2+(a+y)^2]^{-\frac{1}{2}}\}$
2.36	$x^{-\frac{1}{2}}\cos(bx)K_0(ax)$	$\pi y^{\frac{1}{2}}[b^2+(a+y)^2]\mathbf{K}\{[\tfrac{b^2+(a-y)^2}{b^2+(a+y)^2}]^{\frac{1}{2}}\}$
2.37	$x^{-\frac{1}{2}}[J_0(ax)]^2$	$2\pi^{-1}y^{\frac{1}{2}}[2a+(4a^2+y^2)^{\frac{1}{2}}]^{-1} \cdot$ $\cdot \mathbf{K}^2\{2a^{\frac{1}{2}}[2a+(4a^2+y^2)^{\frac{1}{2}}]^{-1}\}$
2.38	$x^{-\frac{1}{2}}[I_0(ax)]^2$ $y > 2a$	$2\pi^{-1}y^{-\frac{1}{2}}\mathbf{K}^2\{[\tfrac{1}{2}-\tfrac{1}{2}(1-4a^2y^{-2})^{\frac{1}{2}}]^{\frac{1}{2}}\}$

	$f(x)$	$g(y) = \int_0^\infty f(x)(xy)^{1/2} K_0(xy)\,dx$
2.39	$x^{-1/2}[K_0(ax)]^2$	$\frac{1}{2}\pi a^{-1} y^{1/2} \mathbf{K}\{[\frac{1}{2}-\frac{1}{2}(1-\frac{1}{4}y^2 a^{-2})^{1/2}]^{1/2}\} \cdot$ $\cdot \mathbf{K}\{[\frac{1}{2}+\frac{1}{2}(1-\frac{1}{4}y^2 a^{-2})^{1/2}]^{1/2}\}$ $\qquad y < 2a$ $\qquad\qquad\qquad\qquad\qquad 0 \quad y > 2a$
2.40	$x^{-1/2} I_0(ax) K_0(ax)$	$y^{-1/2}\mathbf{K}\{[\frac{1}{2}-\frac{1}{2}(1-4a^2 y^{-2})^{1/2}]^{1/2}\} \cdot$ $\cdot \mathbf{K}\{[\frac{1}{2}+\frac{1}{2}(1-4a^2 y^{-2})^{1/2}]^{1/2}\}$ $\qquad y > 2a$
2.41	$x^{3/2} I_0(ax)$ $\qquad y < a$	$2y^{3/2}(y^2-a^2)^{-2}\mathbf{E}(ay^{-1}) - y^{-1/2}(y^2-a^2)^{-1}\mathbf{K}(ay^{-1})$
2.42	$x^{1/2} J_0(ax) J_0(bx)$	$y^{1/2}[(a^2+b^2+y^2)^2-4a^2 b^2]^{-1/2}$
2.43	$x^{1/2}[J_0(\frac{1}{2}ax^2)]^2$	$\frac{1}{16}\pi a^{-1} y^{1/2}\{[J_0(\frac{1}{8}y^2 a^{-1})]^2+[Y_0(\frac{1}{8}y^2 a^{-1})]^2\}$
2.44	$x^{1/2} J_0(ax^{1/2}) I_0(ax^{1/2})$	$y^{-3/2}\cos(\frac{1}{2}a^2 y^{-1})$
2.45	$4J_0(ax^{1/2}) K_0(ax^{1/2}) +$ $+2\pi I_0(ax^{1/2}) Y_0(ax^{1/2})$	$2\pi y^{-3/2}\sin(\frac{1}{2}a^2 y^{-1})$
2.46	$x^{-1/2}[K_\mu(\frac{1}{2}ax^{-1})]^2$	$2\pi y^{-1/2} K_{2\mu}[(2iay)^{1/2}] K_{2\mu}[(-2iay)^{1/2}]$

	$f(x)$	$g(y) = \int\limits_0^\infty f(x)(xy)^{\frac{1}{2}} K_0(xy)\,dx$
2.47	$x^{-\frac{1}{2}} J_\mu(\tfrac{1}{2}ax^{-1}) Y_\mu(\tfrac{1}{2}ax^{-1})$	$-2y^{-\frac{1}{2}} J_{2\mu}[(2ay)^{\frac{1}{2}}] K_{2\mu}[(2ay)^{\frac{1}{2}}]$
2.48	$x^{-\frac{1}{2}}\{[J_\mu(\tfrac{1}{2}ax^{-1})]^2 - [Y_\mu(\tfrac{1}{2}ax^{-1})]^2\}$	$4y^{-\frac{1}{2}} Y_{2\mu}[(2ay)^{\frac{1}{2}}] K_{2\mu}[(2ay)^{\frac{1}{2}}]$
2.49	$x^{-3/2} M_{k,\mu}(iax^2) \cdot$ $\cdot M_{k,\mu}(-iax^2)$ $\text{Re } \mu > -\tfrac{1}{2}$	$a[\Gamma(1+\mu)]^2 y^{-3/2} \cdot$ $\cdot W_{-\mu,k}(\tfrac{1}{4}iy^2 a^{-1}) W_{-\mu,k}[-\tfrac{1}{4}iy^2 a^{-1}]$
2.50	$x^{-3/2} M_{k,0}(iax^2) \cdot$ $\cdot M_{-k,0}(-iax^2)$	$\tfrac{1}{16}\pi y^{\frac{1}{2}}\{[J_k(\tfrac{1}{8}y^2 a^{-1})]^2 + [Y_k(\tfrac{1}{8}y^2 a^{-1})]^2\}$

Transforms of General Order

2.3 Elementary Functions

	$f(x)$	$g(y) = \int_0^\infty f(x)(xy)^{\frac{1}{2}}K_\nu(xy)\,dx$
3.1	$x^{\lambda-1}$ $-\frac{1}{2}-\mathrm{Re}\ \lambda < \mathrm{Re}\ \nu < \frac{1}{2}+\mathrm{Re}\ \lambda$	$2^{\lambda-3/2}y^{-\lambda}\Gamma(\frac{1}{2}\lambda+\frac{1}{4}+\frac{1}{2}\nu)\,\Gamma(\frac{1}{2}\lambda+\frac{1}{4}-\frac{1}{2}\nu)$
3.2	$\begin{cases} 0 & x < a \\ x^{\nu+\frac{1}{2}} & x > a \end{cases}$	$a^{\nu+1}y^{-\frac{1}{2}}K_{\nu+1}(ay)$
3.3	$\begin{cases} 0 & x < a \\ x^{\lambda+\frac{1}{2}} & x > a \end{cases}$	$ay^{-\frac{1}{2}-\lambda}e^{-\frac{1}{2}\pi i\lambda}[K_{\nu-1}(ay)S_{\lambda+1,\nu}(iay)$ $+\,i(\nu+\lambda)K_\nu(ay)S_{\lambda,\nu-1}(iay)]$
3.4	$x^{\mu-\frac{1}{2}}(a+x)^{-1}$ $-1-\mathrm{Re}\ \mu < \mathrm{Re}\ \nu < 1+\mathrm{Re}\ \mu$	$2^{\mu-2}\Gamma(\frac{1}{2}\mu+\frac{1}{2}\nu)\,\Gamma(\frac{1}{2}\mu-\frac{1}{2}\nu)\,y^{\frac{1}{2}-\mu}\ \cdot$ $\cdot\ {}_1F_2(1;1-\frac{1}{2}\mu-\frac{1}{2}\nu,1-\frac{1}{2}\mu+\frac{1}{2}\nu;\frac{1}{4}a^2y^2)\ -$ $-2^{\mu-3}\Gamma(\frac{1}{2}\mu-\frac{1}{2}\nu-\frac{1}{2})\,\Gamma(\frac{1}{2}\mu+\frac{1}{2}\nu-\frac{1}{2})\,ay^{3/2-\mu}\ \cdot$ $\cdot\ {}_1F_2(1;{}^{3}\!/_2-\frac{1}{2}\mu-\frac{1}{2}\nu,{}^{3}\!/_2-\frac{1}{2}\mu+\frac{1}{2}\nu;\frac{1}{4}a^2y^2)\ -$ $-\pi a^\mu y^{\frac{1}{2}}\csc[\pi(\mu-\nu)]\ \cdot$ $\cdot\{K_\nu(ay)+\pi\cos(\pi\mu)\csc[\pi(\nu+\mu)]I_\nu(ay)\}$
3.5	$x^{-\frac{1}{2}}(a+x)^{-1}$ $-1 < \mathrm{Re}\ \nu < 1$	$\frac{1}{2}\pi^2[\csc(\pi\nu)]^2y^{\frac{1}{2}}[I_\nu(ay)+I_{-\nu}(ay)\ -$ $-e^{-\frac{1}{2}i\pi\nu}\mathbf{J}_\nu(iay)-e^{i\frac{1}{2}\pi\nu}\mathbf{J}_{-\nu}(iay)]$

	$f(x)$	$g(y) = \int\limits_0^\infty f(x)(xy)^{\frac{1}{2}}K_\nu(xy)\,dx$
3.6	$x^{-\frac{1}{2}}(a^2+x^2)^{-\frac{1}{2}}$ $-1 < \mathrm{Re}\ \nu < 1$	$\frac{1}{8}\pi^2\sec(\tfrac{1}{2}\pi\nu)y^{\frac{1}{2}}\,\cdot$ $\cdot\,\{[J_{\frac{1}{2}\nu}(\tfrac{1}{2}ay)]^2 + [Y_{\frac{1}{2}\nu}(\tfrac{1}{2}ay)]^2\}$
3.7	$x^{-\frac{1}{2}}(a^2+x^2)^{-1}$ $-1 < \mathrm{Re}\ \nu < 1$	$\frac{1}{4}\pi^2 a^{-1}\sec(\tfrac{1}{2}\pi\nu)y^{\frac{1}{2}}\,\cdot$ $\cdot\,\{\tan(\tfrac{1}{2}\pi\nu)[\mathbf{J}_\nu(ay)-J_\nu(ay)] -$ $-\,\mathbf{E}_\nu(ay) - Y_\nu(ay)\}$
3.8	$x^{-\nu-\frac{1}{2}}(a^2+x^2)^{-1}$ $\mathrm{Re}\ \nu < \tfrac{1}{2}$	$\frac{1}{4}\pi^2\sec(\pi\nu)a^{-\nu-1}y^{\frac{1}{2}}\,\cdot$ $\cdot\,[\mathbf{H}_\nu(ay) - Y_\nu(ay)]$
3.9	$x^{-\mu-\frac{1}{2}}(a^2+x^2)^{-1}$ $\mathrm{Re}(\mu\pm\nu) < \tfrac{1}{2}$	$(2a)^{-1-\mu}\Gamma(\tfrac{1}{2}-\tfrac{1}{2}\mu-\tfrac{1}{2}\nu)\,\Gamma(\tfrac{1}{2}-\tfrac{1}{2}\mu+\tfrac{1}{2}\nu)\,\cdot$ $\cdot\,y^{\frac{1}{2}}S_{\mu,\nu}(ay)$
3.10	$x^{\frac{1}{2}+\nu}(a^2+x^2)^{\mu}$ $\mathrm{Re}\ \nu > -1$	$2^\nu\Gamma(1+\nu)a^{\nu+\mu+1}y^{-\mu-\frac{1}{2}}\,\cdot$ $\cdot\,S_{\mu-\nu,\mu+\nu+1}(ay)$

	$f(x)$	$g(y) = \int\limits_{0}^{\infty} f(x)(xy)^{\frac{1}{2}} K_{\nu}(xy)\,dx$
3.11	$x^{\lambda-3/2}(a^2+x^2)^{\mu}$ $-\operatorname{Re}\lambda < \operatorname{Re}\nu < \operatorname{Re}\lambda$	$a^{\lambda+2\mu}[4\Gamma(-\mu)]^{-1} y^{\frac{1}{2}}[f(\nu)+f(-\nu)] +$ $+ 2^{2\mu+\lambda-2}\Gamma(\tfrac{1}{2}\lambda+\mu-\tfrac{1}{2}\nu)\Gamma(\tfrac{1}{2}\lambda+\mu+\tfrac{1}{2}\nu) \cdot$ $\cdot {}_1F_2(-\mu;1-\mu-\tfrac{1}{2}\lambda-\tfrac{1}{2}\nu,1-\mu-\tfrac{1}{2}\lambda+\tfrac{1}{2}\nu;-\tfrac{1}{4}a^2y^2)$ $f(\nu) = (\tfrac{1}{2}a)^{\nu}\Gamma(-\nu)\Gamma(\tfrac{1}{2}\nu+\tfrac{1}{2}\lambda)\Gamma(-\tfrac{1}{2}\nu-\tfrac{1}{2}\lambda-\mu)$ $\cdot y^{\nu}{}_1F_2(\tfrac{1}{2}\lambda+\tfrac{1}{2}\nu;\tfrac{1}{2}\lambda+\mu+1+\tfrac{1}{2}\nu,1+\nu;-\tfrac{1}{4}a^2y^2)$
3.12.	$x^{-\frac{1}{2}}(a^2+x^2)^{-\frac{1}{2}} \cdot$ $\cdot [(a^2+x^2)^{\frac{1}{2}}+a]^{k} \cdot$ $\cdot [(a^2+x^2)^{\frac{1}{2}}-a]^{-k} \cdot$ $\cdot \exp[-z(a^2+x^2)^{\frac{1}{2}}]$ $\operatorname{Re}(-2k\pm\nu) > -1$	$\tfrac{1}{2}a^{-1}\Gamma(\tfrac{1}{2}+\tfrac{1}{2}\nu-k)\Gamma(\tfrac{1}{2}-\tfrac{1}{2}\nu-k)y^{-\frac{1}{2}} \cdot$ $\cdot W_{k,\frac{1}{2}\nu}\{a[z+(z^2-y^2)^{\frac{1}{2}}]\} \cdot$ $\cdot W_{k,\frac{1}{2}\nu}\{a[z-(z^2-y^2)^{\frac{1}{2}}]\}$
3.13	$x^{-\frac{1}{2}}(a^2-x^2)^{-\frac{1}{2}} \quad x<a$ $\qquad 0 \qquad\quad x>a$ $-1 < \operatorname{Re}\nu < 1$	$\tfrac{1}{2}\pi\sec(\tfrac{1}{2}\pi\nu)K_{\frac{1}{2}\nu}(\tfrac{1}{2}ay)y^{\frac{1}{2}} \cdot$ $\cdot [I_{\frac{1}{2}\nu}(\tfrac{1}{2}ay) + I_{-\frac{1}{2}\nu}(\tfrac{1}{2}ay)]$
3.14	$[x(a^2-x^2)]^{\nu-\frac{1}{2}} \quad x<a$ $\qquad 0 \qquad\qquad x>a$ $\operatorname{Re}\nu > -\tfrac{1}{2}$	$\pi^{\frac{1}{2}}2^{\nu-1}a^{2\nu}\Gamma(\tfrac{1}{2}+\nu)y^{\frac{1}{2}-\nu} \cdot$ $\cdot I_{\nu}(\tfrac{1}{2}ay)K_{\nu}(\tfrac{1}{2}ay)$

	$f(x)$	$g(y) = \int\limits_0^\infty f(x)(xy)^{\frac{1}{2}}K_\nu(xy)\,dx$
3.15	$x^{\frac{1}{2}-\nu}(a^2-x^2)^\mu \quad x<a$ $0 \qquad\qquad x>a$ $\mathrm{Re}\ \mu>-1,\ \mathrm{Re}\ \nu<1$	$2^{-\nu-2}a^{2\mu+2}y^{\nu+\frac{1}{2}}(\mu+1)^{-1}\Gamma(-\nu)\ \cdot$ $\cdot\ {}_1F_2(1;\nu+1,\mu+2;\tfrac{1}{4}a^2y^2)\ +$ $+\pi 2^{\mu-1}a^{\mu-\nu+1}y^{-\mu-\frac{1}{2}}\csc(\pi\nu)\ \cdot$ $\cdot\ \Gamma(\mu+1)I_{\mu-\nu+1}(ay)$
3.16	$x^{\mu-\frac{1}{2}}(a^2-x^2)^\lambda \quad x<a$ $0 \qquad\qquad x>a$ $\mathrm{Re}\ \lambda > -1$ $\mathrm{Re}(\mu\pm\nu) > -1$	$a^{\mu+2\lambda+\frac{1}{2}}[f(\nu) + f(-\nu)]$ $f(\nu) = \dfrac{2^{\nu-2}\Gamma(\nu)\Gamma(1+\lambda)\Gamma(\frac{1}{2}+\frac{1}{2}\mu-\frac{1}{2}\nu)}{\Gamma(\frac{3}{2}+\frac{1}{2}\mu-\frac{1}{2}\nu+\lambda)}\ \cdot$ $\cdot\,(ay)^{\frac{1}{2}-\nu}{}_1F_2(\tfrac{1}{2}+\tfrac{1}{2}\mu-\tfrac{1}{2}\nu;1-\nu,\tfrac{3}{2}+\tfrac{3}{2}\mu-\tfrac{3}{2}\nu+\lambda;\tfrac{1}{4}a^2y^2)$
3.17	$0 \qquad\qquad x<a$ $x^{-\frac{1}{2}}(x^2-a^2)^{-\frac{1}{2}} \quad x>a$	$\tfrac{1}{2}y^{\frac{1}{2}}[K_{\frac{1}{2}\nu}(\tfrac{1}{2}ay)]^2$
3.18	$0 \qquad\qquad x<a$ $[x(x^2-a^2)]^{\nu-\frac{1}{2}} \quad x>a$ $\mathrm{Re}\ \nu > -\tfrac{1}{2}$	$\pi^{-\frac{1}{2}}2^{\nu-1}a^{2\nu}\Gamma(\tfrac{1}{2}+\nu)\ \cdot$ $\cdot\ y^{\frac{1}{2}-\nu}[K_\nu(\tfrac{1}{2}ay)]^2$
3.19	$0 \qquad\qquad x<a$ $x^{\frac{1}{2}-\nu}(x^2-a^2)^\mu \quad x>a$ $\mathrm{Re}\ \mu > -1$	$2^\mu a^{\mu-\nu+1}\Gamma(1+\mu)y^{-\mu-\frac{1}{2}}K_{\mu-\nu+1}(ay)$

	$f(x)$	$g(y) = \int\limits_0^\infty f(x)(xy)^{1/2}K_\nu(xy)dx$
3.20	$x^{-1/2}(a^2+x^2)^{-1/2} \cdot$ $\cdot [(a^2+x^2)^{1/2}+x]^{-2\mu}$	$\tfrac{1}{4}\pi^2 a^{-2\mu}y^{1/2}\csc(\pi\nu) \cdot$ $\cdot [J_{\mu+\frac{1}{2}\nu}(\tfrac{1}{2}ay)Y_{\mu-\frac{1}{2}\nu}(\tfrac{1}{2}ay) -$ $- Y_{\mu+\frac{1}{2}\nu}(\tfrac{1}{2}ay)J_{\mu-\frac{1}{2}\nu}(\tfrac{1}{2}ay)]$
3.21	$x^{-1/2}(a^2+x^2)^{-1/2} \cdot$ $\cdot \{[(a^2+x^2)^{1/2}+x]^{2\mu}+$ $[(a^2+x^2)^{1/2}-x]^{2\mu}\}$ $\nu = 0$	$\tfrac{1}{4}\pi^2 a^{2\mu}y^{1/2}\{[J_\mu(\tfrac{1}{2}ay)]^2+[Y_\mu(\tfrac{1}{2}ay)]^2\}$
3.22	$x^{-1/2}(a^2+x^2)^{-1/2} \cdot$ $\{\cos[\tfrac{1}{2}\pi(\nu-\mu)][(a^2+x^2)^{1/2}+x]^{2\mu}+$ $+\cos[\tfrac{1}{2}\pi(\nu+\mu)][(a^2+x^2)^{1/2}-x]^{2\mu}\}$	$\tfrac{1}{4}\pi^2 a^{2\mu}y^{1/2}[J_{\frac{1}{2}\nu+\mu}(\tfrac{1}{2}ay)J_{\frac{1}{2}\nu-\mu}(\tfrac{1}{2}ay) +$ $+Y_{\frac{1}{2}\nu+\mu}(\tfrac{1}{2}ay)Y_{\frac{1}{2}\nu-\mu}(\tfrac{1}{2}ay)]$
3.23	$x^{-1/2-2\mu}(a^2+x^2)^{-1/2} \cdot$ $\cdot [(a^2+x^2)^{1/2}+a^{2\mu}$ $\mathrm{Re}(2\mu\pm\nu) < 1$	$\tfrac{1}{2}a^{-1}y^{-1/2}\Gamma(\tfrac{1}{2}-\mu+\tfrac{1}{2}\nu)\Gamma(\tfrac{1}{2}-\mu-\tfrac{1}{2}\nu) \cdot$ $\cdot W_{\mu,\frac{1}{2}\nu}(iay)W_{\mu,\frac{1}{2}\nu}(-iay)$
3.24	$\begin{array}{ll} 0 & x < a \\ x^{-1/2}(x^2-a^2)^{-1/2} \cdot \\ \cdot \{[x+(x^2-a^2)^{1/2}]^{2\mu}+ \\ +[x-(x^2-a^2)^{1/2}]^{2\mu}\} \\ & x > a \end{array}$	$a^{2\mu}y^{1/2}K_{\frac{1}{2}\nu+\mu}(\tfrac{1}{2}ay)K_{\frac{1}{2}\nu-\mu}(\tfrac{1}{2}ay)$

	$f(x)$	$g(y) = \int_0^\infty f(x)(xy)^{\frac{1}{2}}K_\nu(xy)\,dx$
3.25	$\begin{aligned}&0 \quad x < a\\ &x^{-\frac{1}{2}-2\mu}(x^2-a^2)^{-\frac{1}{2}}\cdot\\ &\cdot\{[a+i(x^2-a^2)^{\frac{1}{2}}]^{2\mu}+\\ &+[a-i(x^2-a^2)^{\frac{1}{2}}]^{2\mu}\}\\ &\qquad x > a\end{aligned}$	$\pi a^{-1}y^{-\frac{1}{2}}W_{\mu,\frac{1}{2}\nu}(ay)W_{-\mu,\frac{1}{2}\nu}(ay)$
3.26	$x^{-\frac{1}{2}}(x^2-a^2)^{-1}$ $-1 < \text{Re }\nu < 1$ principal value	$\sec(\tfrac{1}{2}\pi\nu)y^{\frac{1}{2}}\{\tfrac{1}{2}\pi(1-\nu^2)^{-1}y\cdot$ $\cdot\,{}_1F_2(1;\,{}^{3}/_{2}-\tfrac{1}{2}\nu,\,{}^{3}/_{2}+\tfrac{1}{2}\nu;\tfrac{1}{4}a^2y^2)\,-$ $-\,{}^{1}/_{8}\pi^2a^{-1}\sec(\tfrac{1}{2}\pi\nu)[I_\nu(ay)+I_{-\nu}(ay)]\}$
3.27	$x^{\nu-\frac{1}{2}}(x^2-a^2)^{-1}$ $\text{Re }\nu > -\tfrac{1}{2}$ principal value	$\tfrac{1}{4}\pi^2a^{\nu-1}\sec(\pi\nu)[\mathbf{L}_{-\nu}(ay)-I_\nu(ay)]$
3.28	$x^{\mu-3/2}(x^2-a^2)^{-1}$ $\text{Re}(\mu\pm\nu) > 0$ principal value	$2^{\mu-4}y^{5/2-\mu}\Gamma(\tfrac{1}{2}\mu-\tfrac{1}{2}\nu-1)\Gamma(\tfrac{1}{2}\mu+\tfrac{1}{2}\nu-1)\cdot$ $\cdot\,{}_1F_2(1;2-\tfrac{1}{2}\mu-\tfrac{1}{2}\nu,2-\tfrac{1}{2}\mu+\tfrac{1}{2}\nu;\tfrac{1}{4}a^2y^2)\,-$ $-\tfrac{1}{4}\pi^2a^{\mu-2}\csc(\pi\nu)y^{\frac{1}{2}}\cdot$ $\cdot\{\cot[\tfrac{1}{2}\pi(\mu-\nu)]I_{-\nu}(ay)\,+$ $+\cot[\tfrac{1}{2}\pi(\mu+\nu)]I_\nu(ay)\}$

	$f(x)$	$g(y) = \int\limits_0^\infty f(x)(xy)^{\frac{1}{2}}K_\nu(xy)\,dx$
3.29	$x^{-\frac{1}{2}}e^{-ax}$ $-1 < \mathrm{Re}\ \nu < 1$	$\frac{1}{2}\pi\csc(\pi\nu)y^{\frac{1}{2}-\nu}(a^2-y^2)^{-\frac{1}{2}}\ \cdot$ $\cdot\{[a+(a^2-y^2)^{\frac{1}{2}}]^\nu-[a-(a^2-y^2)^{\frac{1}{2}}]^\nu\}\quad y < a$ $\pi\csc(\pi\nu)y^{\frac{1}{2}}(y^2-a^2)^{-\frac{1}{2}}\ \cdot$ $\cdot\ \sin[\nu\arccos(ay^{-1})]\qquad\qquad y > a$
3.30	$x^{-\frac{1}{2}}e^{ax}$ $-1 < \mathrm{Re}\ \nu < 1,\quad y > a$	$\pi\csc(\pi\nu)y^{\frac{1}{2}}(y^2-a^2)^{-\frac{\nu}{2}}\ \cdot$ $\cdot\ \sin\{\nu[\tfrac{1}{2}\pi+\arcsin(ay^{-1})]\}$
3.31	$x^{\mu-\frac{1}{2}}e^{-ax}$ $\mathrm{Re}(\mu\pm\nu) > -1$ $\quad a > -y$	$(\tfrac{1}{2}\pi)^{\frac{1}{2}}\Gamma(\mu+1-\nu)\Gamma(\mu+1+\nu)\ \cdot$ $\cdot\ (y^2-a^2)^{-\frac{1}{2}\mu-\frac{1}{4}}P_{\nu-\frac{1}{2}}^{-\mu-\frac{1}{2}}(ay^{-1}) =$ $= (\tfrac{1}{2}\pi)^{\frac{1}{2}}\Gamma(\mu+1-\nu)\Gamma(\mu+1+\nu)\ \cdot$ $\cdot\ (a^2-y^2)^{-\frac{1}{2}\mu-\frac{1}{4}}P_{\nu-\frac{1}{2}}^{-\mu-\frac{1}{2}}(ay^{-1})$ $= \Gamma(\mu-\nu+1)y^{\frac{1}{2}}(a^2-y^2)^{-\frac{1}{2}\mu-\frac{1}{4}}e^{-i\pi\nu}\ \cdot$ $\cdot\ q_\mu^\nu[a(a^2-y^2)^{-\frac{1}{2}}]$
3.32	$x^{-3/2}\exp(-ax^{-1}-bx)$	$2y^{\frac{1}{2}}K_\nu\{a^{\frac{1}{2}}[(b+y)^{\frac{1}{2}}+(b-y)^{\frac{1}{2}}]\}\ \cdot$ $\cdot\ K_\nu\{a^{\frac{1}{2}}[(b+y)^{\frac{1}{2}}-(b-y)^{\frac{1}{2}}]\}$

	$f(x)$	$g(y) = \int_0^\infty f(x)(xy)^{\frac{1}{2}} K_\nu(xy)\,dx$
3.33	$x^{-\frac{1}{2}} \exp(-ax^2)$ $-1 < \mathrm{Re}\,\nu < 1$	$\tfrac{1}{4} \sec(\tfrac{1}{2}\pi\nu) \left(\tfrac{\pi y}{a}\right)^{\frac{1}{2}} \cdot$ $\cdot \exp\left(\tfrac{y^2}{8a}\right) K_{\frac{1}{2}\nu}\left(\tfrac{y^2}{8a}\right)$
3.34	$x^{-\frac{1}{2}-2\mu} \exp(-ax^2)$ $2\,\mathrm{Re}\,\mu < 1-\mathrm{Re}(\pm\nu)$	$\tfrac{1}{2} a^\mu y^{-\frac{1}{2}} \Gamma(\tfrac{1}{2}-\mu+\tfrac{1}{2}\nu)\,\Gamma(\tfrac{1}{2}-\mu-\tfrac{1}{2}\nu) \cdot$ $\cdot \exp\left(\tfrac{y^2}{8a}\right) W_{\mu,\frac{1}{2}\nu}\left(\tfrac{y^2}{4a}\right)$
3.35	$x^{-\frac{1}{2}}(a^2+x^2)^{-\frac{1}{2}} \cdot$ $\cdot \exp[-b(x^2+a^2)^{\frac{1}{2}}]$ $-1 < \mathrm{Re}\,\nu < 1$ $b+y > 0$	$\tfrac{1}{2} y^{\frac{1}{2}} \sec(\tfrac{1}{2}\pi\nu) \cdot$ $\cdot K_{\frac{1}{2}\nu}\{\tfrac{1}{2}a[b+(b^2-y^2)^{\frac{1}{2}}]\} \cdot$ $\cdot K_{\frac{1}{2}\nu}\{\tfrac{1}{2}a[b-(b^2-y^2)^{\frac{1}{2}}]\}$
3.36	$x^{-1} \cos(ax^{\frac{1}{2}})$ $-\tfrac{1}{2} < \mathrm{Re}\,\nu < \tfrac{1}{2}$	$\tfrac{1}{2}\pi \sec(\pi\nu)\, \{D_{\nu-\frac{1}{2}}[a(2y)^{-\frac{1}{2}}] \cdot$ $\cdot D_{-\nu-\frac{1}{2}}[-a(2y)^{-\frac{1}{2}}] +$ $+ D_{\nu-\frac{1}{2}}[-a(2y)^{-\frac{1}{2}}]D_{-\nu-\frac{1}{2}}[a(2y)^{-\frac{1}{2}}]\}$
3.37	$x^{-\frac{1}{2}} \cos(ax^2)$ $-1 < \mathrm{Re}\,\nu < 1$	$\tfrac{\pi}{8} \sec(\tfrac{1}{2}\pi\nu)\left(\tfrac{\pi y}{a}\right)^{\frac{1}{2}} \cdot$ $\cdot [J_{\frac{1}{2}\nu}\left(\tfrac{y^2}{8a}\right) \cos\left(\tfrac{y^2}{8a} - \tfrac{\pi}{4} - \tfrac{\pi}{4}\nu\right) +$ $+ Y_{\frac{1}{2}\nu}\left(\tfrac{y^2}{8a}\right) \sin\left(\tfrac{y^2}{8a} - \tfrac{\pi}{4} - \tfrac{\pi}{4}\nu\right)]$

	$f(x)$	$\bar{f}(x) = \int_0^\infty f(x)(xy)^{\frac{1}{2}}K_\nu(xy)\,dx$
3.38	$x^{-\frac{1}{2}}\sin(ax^2)$ $-3 < \text{Re }\nu < 3$	$\frac{\pi}{8}\sec(\frac{1}{2}\pi\nu)(\pi\frac{y}{a})^{\frac{1}{2}}\cdot$ $\cdot[J_{\frac{1}{2}\nu}(\frac{y^2}{8a})\sin(\frac{y^2}{8a}-\frac{\pi}{4}-\frac{\pi}{4}\nu) -$ $- Y_{\frac{1}{2}\nu}(\frac{y^2}{8a})\cos(\frac{y^2}{8a}-\frac{\pi}{4}-\frac{\pi}{4}\nu)]$
3.39	$x^{-1}\exp(-ax^{\frac{1}{2}})\cdot$ $\cdot\cos(ax^{\frac{1}{2}}+\frac{1}{4}\pi-\frac{1}{2}\nu\pi)$ $-\frac{1}{2} < \text{Re }\nu < \frac{1}{2}$	$(\frac{1}{2}\pi)^{\frac{1}{2}}\Gamma(\frac{1}{2}-\nu)D_{\nu-\frac{1}{2}}[a(-iy)^{-\frac{1}{2}}]D_{\nu-\frac{1}{2}}[a(iy)^{-\frac{1}{2}}]$
3.40	$x^{-\frac{1}{2}}(a^2-x^2)^{-\frac{1}{2}}\cdot$ $\cdot\cos[b(a^2-x^2)^{\frac{1}{2}}]$ $\qquad\qquad\qquad x < a$ $\qquad 0 \qquad x > a$ $-1 < \text{Re }\nu < 1$	$-\frac{1}{4}\pi^2 y^{\frac{1}{2}}\csc(\frac{1}{2}\pi\nu)[J_{\frac{1}{2}\nu}(u)J_{\frac{1}{2}\nu}(v) -$ $- J_{-\frac{1}{2}\nu}(u)J_{-\frac{1}{2}\nu}(v)]$ $u = \frac{1}{2}a[b+(b^2-y^2)^{\frac{1}{2}}]$ $v = \frac{1}{2}a[b-(b^2-y^2)^{\frac{1}{2}}]$
3.41	$\qquad 0 \qquad x < a$ $x^{-\frac{1}{2}}(x^2-a^2)^{-\frac{1}{2}}\cdot$ $\cdot\cos[b(x^2-a^2)^{\frac{1}{2}}]$ $\qquad\qquad\qquad x > a$	$\frac{1}{2}y^{\frac{1}{2}}K_{\frac{1}{2}\nu}\{\frac{1}{2}a[(b^2+y^2)^{\frac{1}{2}}-b]\}\cdot$ $\cdot K_{\frac{1}{2}\nu}\{\frac{1}{2}a[(b^2+y^2)^{\frac{1}{2}}+b]\}$

	$f(x)$	$g(y) = \int\limits_0^\infty f(x)(xy)^{\frac{1}{2}} K_\nu(xy)\,dx$
3.42	$x^{-\frac{1}{2}}(a^2+x^2)^{-\frac{1}{2}} \cdot$ $\cdot \cos[b(a^2+x^2)^{\frac{1}{2}}]$ $-1 < \mathrm{Re}\,\nu < 1$	$\dfrac{\pi^2}{8}\sec(\tfrac{1}{2}\pi\nu)y^{\frac{1}{2}}[J_{\frac{1}{2}\nu}(u)J_{\frac{1}{2}\nu}(v)+Y_{\frac{1}{2}\nu}(u)Y_{\frac{1}{2}\nu}(v)]$ $u = \tfrac{1}{2}a\,[(b^2+y^2)^{\frac{1}{2}}+b]$ $v = \tfrac{1}{2}a[(b^2+y^2)^{\frac{1}{2}}-b]$
3.43	$x^{-\frac{1}{2}}(a^2+x^2)^{-\frac{1}{2}} \cdot$ $\cdot \sin[b(a^2+x^2)^{\frac{1}{2}}]$	$\dfrac{\pi^2}{8}\sec(\tfrac{1}{2}\pi\nu)y^{\frac{1}{2}}[Y_{\frac{1}{2}\nu}(u)J_{\frac{1}{2}\nu}(v)-J_{\frac{1}{2}\nu}(u)Y_{\frac{1}{2}\nu}(v)]$ $u = \tfrac{1}{2}a\,[(b^2+y^2)^{\frac{1}{2}}+b]$ $v = \tfrac{1}{2}a[(b^2+y^2)^{\frac{1}{2}}-b]$
3.44	$\begin{array}{ll} 0 & x < a \\ x^{\frac{1}{2}-\nu}(x^2-a^2)^{-\frac{1}{2}} \cdot \\ \cdot\cos[b(x^2-a^2)^{\frac{1}{2}}] \\ & x > a \end{array}$	$(\tfrac{1}{2}\pi)^{\frac{1}{2}}(ay)^{\frac{1}{2}-\nu}(b^2+y^2)^{\frac{1}{2}\nu-\frac{1}{4}} \cdot$ $\cdot\, K_{\nu-\frac{1}{2}}[a(b^2+y^2)^{\frac{1}{2}}]$
3.45	$x^{-\frac{1}{2}}\sin h(ax)$ $-2 < \mathrm{Re}\,\nu < 2$	$\tfrac{1}{2}\pi y^{\frac{1}{2}}(y^2-a^2)^{-\frac{1}{2}}\csc(\tfrac{1}{2}\pi\nu)$ $\sin[\nu\arcsin(\tfrac{a}{y})]$
3.46	$x^{-\frac{1}{2}}\cos h(ax)$ $-1 < \mathrm{Re}\,\nu < 1$	$\tfrac{1}{2}\pi y^{\frac{1}{2}}(y^2-a^2)^{-\frac{1}{2}}\sec(\tfrac{1}{2}\pi\nu) \cdot$ $\cdot\,\cos[\nu\arcsin(\tfrac{a}{y})]$

	$f(x)$	$g(y) = \int\limits_0^\infty f(x)(xy)^{\frac{1}{2}}K_\nu(xy)\,dx$
3.47	$x^{-3/2}\sin h(ax)$ $-1 < \mathrm{Re}\ \nu < 1$	$\frac{1}{2}\pi y^{-\frac{1}{2}}\nu^{-1}\sec(\frac{1}{2}\pi\nu)\ \cdot$ $\cdot\ \sin[\nu\arcsin(\frac{a}{y})]$
3.48	$x^{-\frac{1}{2}}(a^2-x^2)^{-\frac{1}{2}}\ \cdot$ $\cdot\cos h[b(a^2-x^2)^{\frac{1}{2}}]$ $\qquad\qquad x < a$ $\qquad 0 \qquad x > a$ $-1 < \mathrm{Re}\ \nu < 1$	$\frac{1}{4}\pi^2 y^{\frac{1}{2}}\csc(\frac{1}{2}\pi\nu)\,[I_{-\frac{1}{2}\nu}(u)I_{-\frac{1}{2}\nu}(v)\ -$ $-\ I_{\frac{1}{2}\nu}(u)I_{\frac{1}{2}\nu}(v)]$ $u = \frac{1}{2}a[(b^2+y^2)^{\frac{1}{2}}+b]$ $v = \frac{1}{2}a[(b^2+y^2)^{\frac{1}{2}}-b)]$
3.49	$\qquad 0 \qquad x < a$ $x^{-\frac{1}{2}}(x^2-a^2)^{-\frac{1}{2}}$ $\cos[\mu\arccos(ax^{-1})]$ $\qquad\qquad x > a$	$\frac{1}{2}\pi a^{-1}y^{-\frac{1}{2}}W_{\frac{1}{2}\mu,\frac{1}{2}\nu}(ay)W_{-\frac{1}{2}\mu,\frac{1}{2}\nu}(ay)$

2.4 Higher Transcendental Functions

	$f(x)$	$g(y) = \int\limits_0^\infty f(x)(xy)^{\frac{1}{2}}K_\nu(xy)\,dx$
4.1	$0 \quad x < a$ $x^{-\frac{1}{2}}(x^2-a^2)^{-\frac{1}{2}}T_n\left(\frac{a}{x}\right)$ $x > a$ $n = 0, 1, 2, \cdots$	$\frac{1}{2}\pi a^{-1}y^{-\frac{1}{2}}W_{\frac{1}{2}n,\frac{1}{2}\nu}(ay)W_{-\frac{1}{2}n,\frac{1}{2}\nu}(ay)$
4.2	$0 \quad x < a$ $P_{\nu-\frac{1}{2}}\left(\frac{x}{a}\right) \quad x > a$	$\left(2\frac{y}{a}\right)^{-\frac{1}{2}}e^{-ay}K_\nu\left(\frac{1}{2}ay\right)$
4.3	$0 \quad x < a$ $x^\mu(x^2-a^2)^{-\frac{1}{2}\mu}P_{\nu-\frac{1}{2}}^\mu\left(\frac{x}{a}\right)$ $x > a$ $\text{Re }\mu < 1$	$\left(\frac{1}{2}\pi\right)^{\frac{1}{2}}y^{-1}e^{-\frac{1}{2}ay}W_{\mu,\nu}(ay)$
4.4	$0 \quad x < a$ $x^{\mu-2}(x^2-a^2)^{-\frac{1}{2}\mu}P_{\nu-\frac{1}{2}}^\mu\left(\frac{x}{a}\right)$ $x > a$ $\text{Re }\mu < 1$	$\left(\frac{1}{2}\pi\right)^{\frac{1}{2}}a^{-1}e^{-\frac{1}{2}ay}W_{\mu-1,\nu}(ay)$
4.5	$0 \quad x < a$ $x^{-\mu}(x^2-a^2)^{-\frac{1}{2}\mu}P_{\nu-\frac{1}{2}}^\mu\left(\frac{x}{a}\right)$ $x > a$	$(2\pi)^{-\frac{1}{2}}a^{1-\mu}y^\mu K_\nu\left(\frac{1}{2}ay\right)K_{\mu-\frac{1}{2}}\left(\frac{1}{2}ay\right)$

	$f(x)$	$g(y) = \int\limits_{0}^{\infty} f(x)(xy)^{\frac{1}{2}} K_{\nu}(xy)\,dx$
4.6	$\begin{array}{c} 0 \qquad x < a \\[4pt] x^{\mu-1}(x^2-a^2)^{-\frac{1}{2}\mu} p^{\mu}_{\nu-3/2}\left(\dfrac{x}{a}\right) \\[4pt] x > a \\[4pt] \mathrm{Re}\ \mu < 1 \end{array}$	$(2\dfrac{ay}{\pi})^{-\frac{1}{2}} e^{-\frac{1}{2}ay} W_{\mu-\frac{1}{2},\,\nu-\frac{1}{2}}(ay)$
4.7	$\begin{array}{c} x^{\frac{1}{2}}(a^2+x^2)^{\frac{1}{2}\nu} p^{\nu}_{\mu}(1+2x^2a^{-2}) \\[4pt] \mathrm{Re}\ \nu < 1 \end{array}$	$2^{-\nu} a y^{-\nu-\frac{1}{2}} S_{2\nu,\,2\mu+1}(ay)$
4.8	$\begin{array}{c} x^{\frac{1}{2}} p^{\nu}_{\mu}[(1+x^2)^{\frac{1}{2}}] \\[4pt] \mathrm{Re}\ \nu < 1 \end{array}$	$y^{-1} S_{\nu+\frac{1}{2},\,\mu+\frac{1}{2}}(y)$
4.9	$\begin{array}{c} x^{\frac{1}{2}}(1+x^2)^{-\frac{1}{2}} p^{\nu}_{\mu}[(1+x^2)^{\frac{1}{2}}] \\[4pt] \mathrm{Re}\ \nu < 1 \end{array}$	$S_{\nu-\frac{1}{2},\,\mu+\frac{1}{2}}(y)$
4.10	$\begin{array}{l} x^{\frac{1}{2}}(a^2+x^2)^{\frac{1}{2}\nu}\cdot \\[4pt] \cdot[(\mu-\nu)p^{\nu}_{\mu}(1+2x^2a^{-2})+ \\[4pt] +(\mu+\nu)p^{\nu}_{-\mu}(1+2x^2a^{-2})] \end{array}$	$2^{1-\nu}\mu y^{-\nu-3/2} S_{2\nu+1,\,2\mu}(ay)$

	$f(x)$	$g(y) = \int_0^\infty f(x)(xy)^{\frac{1}{2}} K_\nu(xy)\,dx$
4.11	$x^{\frac{1}{2}}(a^2+x^2)^{\frac{1}{2}\nu-1} \cdot$ $\cdot\,[p_\mu^\nu(1+2x^2a^{-2}) +$ $+\,p_{-\mu}^\nu(1+2x^2a^{-2})]$ $\text{Re } \nu < 1$	$2^{1-\nu}y^{\frac{1}{2}-\nu}S_{2\nu-1,2\mu}(ay)$
4.12	$\quad 0 \qquad x < a$ $x^{\frac{1}{2}}(x^2-a^2)^{-\frac{1}{2}\nu} \cdot$ $\cdot\,p_\mu^\nu(2x^2a^{-2}-1)$ $\qquad\qquad x > a$ $\text{Re } \nu < 1$	$2^{-\nu}ay^{\nu-\frac{1}{2}}K_{\mu+1}(ay)$
4.13	$\quad 0 \qquad x < a$ $(x^2-a^2)^{\frac{1}{2}\nu-\frac{1}{4}} \cdot$ $\cdot\,p_\mu^{\frac{1}{2}-\nu}(2x^2a^{-2}-1)$ $\qquad\qquad x > a$ $\text{Re } \nu > -\tfrac{1}{2}$	$\pi^{-\frac{1}{2}}2^{\nu-1}ay^{\frac{1}{2}-\nu}[K_{\mu+\frac{1}{2}}(\tfrac{1}{2}ay)]^2$
4.14	$x^{-\nu-\frac{1}{2}}(a^2+x^2)^{\frac{1}{4}-\frac{1}{2}\nu} \cdot$ $\cdot\,q_{-\frac{1}{2}}^{\frac{1}{2}-\nu}(1+2a^2x^{-2})$ $\text{Re } \nu < 1$	$ie^{-i\pi\nu}\pi^{3/2}2^{-\nu-3}a^{\frac{1}{2}-\nu}y^{\nu-\frac{1}{2}}[\Gamma(1-\nu)]^2 \cdot$ $\cdot\,\{[J_{\nu-\frac{1}{2}}(\tfrac{1}{2}ay)]^2 + [Y_{\nu-\frac{1}{2}}(\tfrac{1}{2}ay)]^2\}$

	$f(x)$	$g(y) = \int\limits_{0}^{\infty} f(x)(xy)^{\frac{1}{2}}K_{\nu}(xy)\,dx$
4.15	$x^{-\nu-\frac{1}{2}}(a^2+x^2)^{\frac{1}{4}-\frac{1}{2}\nu}\cdot$ $\cdot q_{\mu}^{\frac{1}{2}-\nu}(1+2a^2x^{-2})$ $\mathrm{Re}\ \mu > -\,^{3}/_{2}$ $\mathrm{Re}\,(\mu-\nu) > -\,^{3}/_{2}$	$i\,e^{-i\pi\nu}\pi^{\frac{1}{2}}2^{-\nu-1}a^{-\nu-\frac{1}{2}}y^{\nu-3/2}[\Gamma(^{3}/_{2}+\mu-\nu)]^2\cdot$ $\cdot W_{-\mu-\frac{1}{2},\,\nu-\frac{1}{2}}(iay)W_{-\mu-\frac{1}{2},\,\nu-\frac{1}{2}}(-iay)$
4.16	$x^{\nu+\mu+\frac{1}{2}}J_{\mu}(ax)$ $\mathrm{Re}\ \mu>-1,\ \mathrm{Re}\,(\nu+\mu)>-1$	$2^{\nu+\mu}\Gamma(\nu+\mu+1)a^{\mu}y^{\nu+\frac{1}{2}}(a^2+y^2)^{-\mu-\nu-1}$
4.17	$J_{\mu}(ax)$ $\mathrm{Re}\,(\mu\pm\nu) > -\,^{3}/_{2}$	$2^{\mu-\frac{1}{2}}\Gamma(^{3}/_{4}+\frac{1}{2}\mu+\frac{1}{2}\nu)\Gamma(^{3}/_{4}+\frac{1}{2}\mu-\frac{1}{2}\nu)\cdot$ $\cdot(a^2+y^2)^{-\frac{1}{2}}p_{\nu-\frac{1}{2}}^{-\mu}[(1+a^2y^{-2})^{\frac{1}{2}}]$
4.18	$x^{\mu-\frac{1}{2}}J_{\mu}(ax)$ $\mathrm{Re}\,(2\mu\pm\nu) > -1$	$2^{\mu-1}\Gamma(\frac{1}{2}+\mu+\frac{1}{2}\nu)\Gamma(\frac{1}{2}+\mu-\frac{1}{2}\nu)\cdot$ $\cdot y^{-\frac{1}{2}}(a^2+y^2)^{-\frac{1}{2}\mu}p_{\frac{1}{2}\nu-\frac{1}{2}}^{-\mu}(1+2a^2y^{-2})$
4.19	$x^{-\mu-\frac{1}{2}}J_{\mu}(ax)$ $-1 < \mathrm{Re}\ \nu < 1$	$\pi 2^{-\mu-1}\sec(\frac{1}{2}\pi\nu)y^{-\frac{1}{2}}\cdot$ $\cdot(a^2+y^2)^{\frac{1}{2}\mu}p_{\frac{1}{2}\nu-\frac{1}{2}}^{-\mu}(1+2a^2y^{-2})$
4.20	$x^{-\lambda-\frac{1}{2}}J_{\nu}(ax)$ $\mathrm{Re}\ \lambda<1,\ \mathrm{Re}\,(2\nu-\lambda)>-1$	$2^{-\lambda-1}\Gamma(\frac{1}{2}+\nu-\frac{1}{2}\lambda)\Gamma(\frac{1}{2}-\frac{1}{2}\lambda)y^{\frac{1}{2}}\cdot$ $\cdot(a^2+y^2)^{-\frac{1}{2}+\frac{1}{2}\lambda}p_{\frac{1}{2}\lambda-\frac{1}{2}}^{-\nu}\left(\dfrac{y^2-a^2}{y^2+a^2}\right)$

	$f(x)$	$g(y) = \int_0^\infty f(x)(xy)^{\frac{1}{2}}K_\nu(xy)dx$
4.21	$x^{-\lambda-\frac{1}{2}}J_\mu(ax)$ $\text{Re}(\mu\pm\nu-\lambda) > -1$	$2^{-1-\lambda}a^\mu[\Gamma(1+\mu)]^{-1}\Gamma(\frac{1}{2}+\frac{1}{2}\mu-\frac{1}{2}\lambda+\frac{1}{2}\nu)\;\cdot$ $\cdot\,\Gamma(\frac{1}{2}+\frac{1}{2}\mu-\frac{1}{2}\lambda-\frac{1}{2}\nu)y^{\lambda-\mu-\frac{1}{2}}\;\cdot$ $\cdot\;{}_2F_1(\frac{1}{2}+\frac{1}{2}\mu-\frac{1}{2}\lambda+\frac{1}{2}\nu,\frac{1}{2}+\frac{1}{2}\mu-\frac{1}{2}\lambda-\frac{1}{2}\nu;\mu+1;-a^2y^{-2})$
4.22	$x^{-\lambda-\frac{1}{2}}Y_\mu(ax)$ $\text{Re}(\pm\mu\pm\nu-\lambda) > -1$	$\int_0^\infty x^{-\lambda-\frac{1}{2}}[\text{ctn}(\pi\mu)J_\mu(ax)-\csc(\pi\mu)J_{-\mu}(ax)]\;\cdot$ $\cdot\;(xy)^{\frac{1}{2}}K_\nu(xy)dx$ (For the integral see formula before.)
4.23	$x^{-\frac{1}{2}}Y_0(ax)$ $-1 < \text{Re }\nu < 1$	$-\frac{1}{2}y^{-\frac{1}{2}}\sec(\frac{1}{2}\pi\nu)\;\cdot$ $\cdot\;[q_{\frac{1}{2}\nu-\frac{1}{2}}(1+2a^2y^{-2}) + q_{-\frac{1}{2}\nu-\frac{1}{2}}(1+2a^2y^{-2})$
4.24	$x^{-\frac{1}{2}}[J_\mu(ax)]^2$ $\text{Re}(2\mu\pm\nu) > -1$	$\frac{1}{2}\Gamma(\frac{1}{2}+\mu+\frac{1}{2}\nu)\Gamma(\frac{1}{2}+\mu-\frac{1}{2}\nu)y^{-\frac{1}{2}}$ $\cdot\,\{p_{\frac{1}{2}\nu-\frac{1}{2}}^{-\mu}[(1+4a^2y^{-2})^{\frac{1}{2}}]\}^2\;=$ $=\frac{1}{2}(\pi a)^{-1}\Gamma(\frac{1}{2}+\mu+\frac{1}{2}\nu)[\Gamma(\frac{1}{2}+\mu-\frac{1}{2}\nu)]^{-1}\;\cdot$ $\cdot\;y^{\frac{1}{2}}e^{i\pi\nu}\{q_{\mu-\frac{1}{2}}^{-\frac{1}{2}\nu}[(1+\frac{1}{4}y^2a^{-2})^{\frac{1}{2}}]\}^2$

	$f(x)$	$g(y) = \int_0^\infty f(x)(xy)^{\frac{1}{2}}K_\nu(xy)dx$
4.25	$x^{-\frac{1}{2}}J_\mu(ax)J_{-\mu}(ax)$ $-1 < \mathrm{Re}\ \nu < 1$	$\frac{1}{2}\pi y^{-\frac{1}{2}}\sec(\frac{1}{2}\pi\nu)\ \cdot$ $\cdot\ p_{\frac{1}{2}\nu-\frac{1}{2}}^{\mu}[(1+4a^2y^{-2})^{\frac{1}{2}}]p_{\frac{1}{2}\nu-\frac{1}{2}}^{-\mu}[(1+4a^2y^{-2})^{\frac{1}{2}}] =$ $= \frac{1}{2}a^{-1}\sec(\frac{1}{2}\pi\nu)[\Gamma(\frac{1}{2}-\frac{1}{2}\nu-\mu)\Gamma(\frac{1}{2}-\frac{1}{2}\nu+\mu)]^{-1}\ \cdot$ $\cdot\ q_{-\mu-\frac{1}{2}}^{-\frac{1}{2}\nu}[(1+\frac{1}{4}y^2a^{-2})^{\frac{1}{2}}]q_{\mu-\frac{1}{2}}^{-\frac{1}{2}\nu}[(1+\frac{1}{4}y^2a^{-2})^{\frac{1}{2}}]$
4.26	$x^{\frac{1}{2}}[J_\mu(ax)]^2$ $\mathrm{Re}(2\mu\pm\nu) > -2$	$\Gamma(1+\mu+\frac{1}{2}\nu)\Gamma(1+\mu-\frac{1}{2}\nu)y^{-3/2}(1+4a^2y^{-2})^{-\frac{1}{2}}\ \cdot$ $\cdot\ p_{\frac{1}{2}\nu}^{-\mu}[(1+4a^2y^{-2})^{\frac{1}{2}}]p_{-\frac{1}{2}\nu}^{-\mu}[(1+4a^2y^{-2})^{\frac{1}{2}}] =$ $= -(\pi a)^{-1}(\mu^2-\frac{1}{4}\nu^2)y^{-\frac{1}{2}}(1+4a^2y^{-2})^{-\frac{1}{2}}\ \cdot$ $\cdot\ q_{\mu-\frac{1}{2}}^{-\frac{1}{2}\nu-\frac{1}{2}}[(1+\frac{1}{4}y^2a^{-2})^{\frac{1}{2}}]q_{\mu-\frac{1}{2}}^{\frac{1}{2}\nu-\frac{1}{2}}[(1+\frac{1}{4}y^2a^{-2})^{\frac{1}{2}}]$
4.27	$x^{\frac{1}{2}}J_\mu(ax)J_{-\mu}(ax)$ $-2 < \mathrm{Re}\ \nu < 2$	$-\frac{1}{2}\pi\csc(\frac{1}{2}\pi\nu)y^{-3/2}z^{-1}\ \cdot$ $\cdot\ [(\mu-\frac{1}{2}\nu)p_{\frac{1}{2}\nu}^{\mu}(z)p_{-\frac{1}{2}\nu}^{-\mu}(z)\ -$ $-\ (\mu+\frac{1}{2}\nu)p_{\frac{1}{2}\nu}^{-\mu}(z)p_{-\frac{1}{2}\nu}^{\mu}(z)]$ $z = (1+4a^2y^{-2})^{\frac{1}{2}}$

	$f(x)$	$g(y) = \int\limits_0^\infty f(x)(xy)^{\frac{1}{2}}K_\nu(xy)\,dx$
4.28	$x^{\frac{1}{2}}J_\mu(ax)J_{1+\mu}(ax)$ $\mathrm{Re}(2\mu\pm\nu) > -3$	$\Gamma(^3\!/_2+\mu+\tfrac12\nu)\Gamma(^3\!/_2+\mu-\tfrac12\nu)y^{-3/2}(1+4a^2y^{-2})^{\frac12}\;\cdot$ $\cdot\; p_{\frac12\nu-\frac12}^{-\mu}[(1+4a^2y^{-2})^{\frac12}]p_{\frac12\nu-\frac12}^{-\mu-1}[(1+4a^2y^{-2})^{\frac12}] =$ $= (\pi a)^{-1}(\tfrac12+\mu+\tfrac12\nu)y^{-\frac12}(1+4a^2y^{-2})^{\frac12}\;\cdot$ $\cdot\; q_{\mu-\frac12}^{\frac12\nu}[(1+\tfrac14y^2a^{-2})^{\frac12}]q_{\mu+\frac12}^{-\frac12\nu}[(1+\tfrac14y^2a^{-2})^{\frac12}]$
4.29	$x^{\frac{1}{2}}J_\mu(ax)J_{1-\mu}(ax)$ $-3 < \mathrm{Re}\ \nu < 3$	$\tfrac14 a\,[\mu(1-\mu)]^{-1}(1-\nu^2)\sec(\tfrac12\pi\nu)y^{-5/2}\;\cdot$ $\cdot\; {}_4F_3(^3\!/_2+\tfrac12\nu,\,^3\!/_2-\tfrac12\nu,\,1,\,^3\!/_2;\,2-\mu,\,1+\mu,\,2;\,-4a^2y^{-2})$
4.30	$x^{\frac{1}{2}}J_\mu(ax)J_{-\mu-1}(ax)$ $-1 < \mathrm{Re}\ \nu < 1$	$\tfrac12\pi\sec(\tfrac12\pi\nu)z^{-1}y^{-3/2}[p_{\frac12\nu-\frac12}^{-\mu}(z)p_{\frac12\nu-\frac12}^{\mu+1}(z) +$ $+(\tfrac12\nu-\tfrac12-\mu)(\tfrac12\nu+\tfrac12+\mu)p_{\frac12\nu-\frac12}^{-\mu-1}(z)p_{\frac12\nu-\frac12}^{\mu}(z)]$ $-\tfrac14\pi^{-1}y^{\frac12}\sin(\pi\mu)\sec(\tfrac12\nu\pi)$ $z = (1+4a^2y^{-2})^{\frac12}$
4.31	$x^{\frac12-\nu}[J_\mu(ax)]^2$ $\mathrm{Re}\ \nu>-\tfrac12,\ \mathrm{Re}\ \mu>-1$ $\mathrm{Re}(\mu-\nu) > -1$	$2^{-\nu}\Gamma(1+\mu-\nu)y^{-\frac12}(y^2+4a^2)^{\frac12\nu-\frac12}\;\cdot$ $\cdot\; p_{\nu-1}^{-\mu}[(y+2a^2y^{-1})(y^2+4a^2)^{-\frac12}] =$ $= \pi^{-\frac12}2^{-\nu}a^{-1}(y^2+4a^2)^{\frac12\nu-\frac14}e^{i\pi(\nu-\frac12)}\;\cdot$ $\cdot\; q_{\mu-\frac12}^{\frac12-\nu}(\tfrac12y^2a^{-2}+1)$

	$f(x)$	$g(y) = \int\limits_0^\infty f(x)(xy)^{\frac{1}{2}}K_\nu(xy)dx$
4.32	$x^{\frac{1}{2}+\nu}[J_\nu(ax)]^2$ $\mathrm{Re}\ \nu > -\frac{1}{2}$	$\pi^{-\frac{1}{2}}2^{3\nu}\Gamma(\frac{1}{2}+\nu)a^{2\nu}y^{-\nu-\frac{1}{2}}(y^2+4a^2)^{-\nu-\frac{1}{2}}$
4.33	$x^{\frac{1}{2}+\mu}J_\nu(ax)J_\mu(bx)$ $\mathrm{Re}\ \mu>-1,\ \mathrm{Re}(\mu+\nu)>-1$	$(2\pi)^{-\frac{1}{2}}a^{-\mu-1}b^\mu y^{-\mu-\frac{1}{2}}[(\frac{b^2-a^2+y^2}{2ay})^2 - 1]^{-\frac{1}{2}\mu-\frac{1}{4}}\cdot$ $\cdot q_{\nu-\frac{1}{2}}^{\mu+\frac{1}{2}}[i(2ay)^{-1}(b^2-a^2+y^2)]e^{-i\pi(\mu+\frac{1}{4}-\frac{1}{2}\nu)}$
4.34	$x^{\nu+\frac{1}{2}}[J_\mu(bx)Y_{-\mu}(ax)+$ $+J_{-\mu}(ax)Y_\mu(bx)]$ $\mathrm{Re}\ \nu>-1,\ \mathrm{Re}\ \mu>-1$ $\mathrm{Re}(\nu+\mu)>-1$	$-(2\pi)^{-\frac{1}{2}}\Gamma(\nu+\mu+1)\Gamma(\nu-\mu+1)(ab)^{-\nu-1}$ $y^{\nu+\frac{1}{2}}[(\frac{a^2+b^2+y^2}{2ab})^2 - 1]^{-\frac{1}{2}\nu-\frac{1}{4}}\cdot$ $\cdot p_{\mu-\frac{1}{2}}^{-\nu-\frac{1}{2}}[(2ab)^{-1}(a^2+b^2+y^2)]$
4.35	$x^{\rho-\mu+\nu+\frac{1}{2}}J_\rho(ax)J_\mu(bx)$ $a>b,\ \mathrm{Re}\ \rho>-1$ $\mathrm{Re}(\rho+\nu)>-1$	$\int\limits_0^\infty x^{\rho-\mu+\nu+\frac{1}{2}}K_\rho(ax)I_\mu(bx)\ \cdot$ $\cdot\ (xy)^{\frac{1}{2}}J_\nu(xy)dx$
4.36	$x^{\frac{1}{2}\pm\nu}J_\mu(ax)J_\mu(bx)$ $\mathrm{Re}\ \mu>-1,\ \mathrm{Re}(\nu+\mu)>-1$	$(2\pi)^{-\frac{1}{2}}(ab)^{\mp\nu-1}y^{\frac{1}{2}\pm\nu}(z^2-1)^{\pm\frac{1}{2}\nu-\frac{1}{4}}\cdot$ $\cdot\ e^{-i\pi(\frac{1}{2}\pm\nu)}q_{\mu-\frac{1}{2}}^{\frac{1}{2}\pm\nu}(z)$ $z = \dfrac{a^2+b^2+y^2}{2ab}$

	$f(x)$	$g(y) = \int\limits_{0}^{\infty} f(x)(xy)^{\frac{1}{2}}K_{\nu}(xy)dx$
4.37	$x^{\sigma+\frac{1}{2}}J_{\mu}(ax)J_{\lambda}(ax)$ $Re(\sigma+\mu+\lambda\pm\nu) > -2$	$2^{\sigma}a^{\mu+\lambda}[\Gamma(1+\mu)\Gamma(1+\lambda)]^{-1} \cdot$ $\cdot \Gamma(1+\frac{1}{2}\mu+\frac{1}{2}\lambda+\frac{1}{2}\nu+\frac{1}{2}\sigma) \cdot$ $\cdot \Gamma(1+\frac{1}{2}\mu+\frac{1}{2}\lambda+\frac{1}{2}\sigma-\frac{1}{2}\nu)y^{-\mu-\lambda-\sigma-3/2} \cdot$ $\cdot {}_{4}F_{3}(\frac{1}{2}+\frac{1}{2}\mu+\frac{1}{2}\lambda,1+\frac{1}{2}\mu+\frac{1}{2}\lambda,1+\frac{1}{2}\mu+\frac{1}{2}\lambda+\frac{1}{2}\sigma+\frac{1}{2}\nu,$ $1+\frac{1}{2}\mu+\frac{1}{2}\lambda+\frac{1}{2}\sigma-\frac{1}{2}\nu;1+\mu,1+\lambda,1+\mu+\lambda;-4a^2y^{-2})$
4.38	$x^{\sigma+\frac{1}{2}}J_{\mu}(ax)J_{\lambda}(bx)$	Bailey, W. N., 1936: Proc. London Math. Soc. 40, 37-48; J. London Math. Soc. 11, 16-20.
4.39	$x^{\frac{1}{2}}J_{\frac{1}{2}\nu}(ax^2)$ $Re\ \nu > -1$	$^{1}/_{8}\pi a^{-1}\sec(\frac{1}{2}\pi\nu) \cdot$ $\cdot y^{\frac{1}{2}}[\mathbf{H}_{-\frac{1}{2}\nu}(\frac{1}{4}y^2a^{-1}) - Y_{-\frac{1}{2}\nu}(\frac{1}{4}y^2a^{-1})]$
4.40	$x^{\frac{1}{2}}Y_{\frac{1}{2}\nu}(ax^2)$ $-1 < Re\ \nu < 1$	$\frac{1}{4}\pi a^{-1}\csc(\pi\nu)y^{\frac{1}{2}} \cdot$ $\cdot [\cos(\frac{1}{2}\pi\nu)\mathbf{H}_{-\frac{1}{2}\nu}(\frac{1}{4}y^2a^{-1}) -$ $- \sin(\frac{1}{2}\pi\nu)J_{-\frac{1}{2}\nu}(\frac{1}{4}y^2a^{-1}) - \mathbf{H}_{\frac{1}{2}\nu}(\frac{1}{4}y^2a^{-1})]$
4.41	$x^{\frac{1}{2}}J_{\frac{1}{4}\nu}(ax^2)J_{-\frac{1}{4}\nu}(ax^2)$ $-2 < Re\ \nu < 2$	$^{1}/_{32}\pi a^{-1}\sec(\frac{1}{4}\pi\nu)y^{\frac{1}{2}} \cdot$ $\cdot \{[J_{\frac{1}{2}\nu}(^{1}/_{16}y^2a^{-1})]^2 + [Y_{\frac{1}{2}\nu}(^{1}/_{16}y^2a^{-1})]^2\}$

	$f(x)$	$g(y) = \int\limits_0^\infty f(x)(xy)^{\frac{1}{2}}K_\nu(xy)\,dx$
4.42	$x^{\frac{1}{2}}J_{\mu+\frac{1}{4}\nu}(ax^2)J_{\mu-\frac{1}{4}\nu}(ax^2)$ $\mathrm{Re}(\mu\pm\frac{1}{4}\nu) > -\frac{1}{2}$	$\pi^{-1}\Gamma(\frac{1}{2}+\mu+\frac{1}{4}\nu)\Gamma(\frac{1}{2}+\mu-\frac{1}{4}\nu)y^{-3/2}\;\cdot$ $\cdot\; W_{-\mu,\frac{1}{4}\nu}(\frac{iy^2}{8a})W_{-\mu,\frac{1}{4}\nu}(-\frac{iy^2}{8a})$
4.43	$x^{-\frac{1}{2}}J_\nu(ax^{-1})$ $-\frac{5}{2} < \mathrm{Re}\ \nu < \frac{5}{2}$	$e^{i\frac{\pi}{2}(\nu+1)}y^{-\frac{1}{2}}K_{2\nu}[2(iay)^{\frac{1}{2}}] +$ $+\,e^{-i\frac{\pi}{2}(\nu+1)}y^{-\frac{1}{2}}K_{2\nu}[2(-iay)^{\frac{1}{2}}]\}$
4.44	$x^{-\frac{5}{2}}J_\nu(ax^{-1})$ $-\frac{1}{2} < \mathrm{Re}\ \nu < \frac{1}{2}$	$a^{-1}y^{\frac{1}{2}}\{e^{i\frac{\pi}{2}\nu}K_{2\nu}[2(iay)^{\frac{1}{2}}] +$ $+\,e^{-i\frac{\pi}{2}\nu}K_{2\nu}[2(-iay)^{\frac{1}{2}}]$
4.45	$x^{2\nu-2}J_{\nu+\frac{1}{2}}(ax^{-1})$ $\mathrm{Re}\ \nu > -\frac{1}{3}$	$(2\pi)^{\frac{1}{2}}(\frac{y}{a})^{\frac{1}{2}-\nu}J_{2\nu}[(2ay)^{\frac{1}{2}}]K_{2\nu}[(2ay)^{\frac{1}{2}}]$
4.46	$x^{-2\nu}J_{\nu-\frac{1}{2}}(ax^{-1})$ $\mathrm{Re}\ \nu < 1$	$(2\pi)^{\frac{1}{2}}(\frac{y}{a})^{\nu-\frac{1}{2}}K_{2\nu-1}[(2ay)^{\frac{1}{2}}]\;\cdot$ $\cdot\; \{\sin(\pi\nu)J_{2\nu-1}[(2ay)^{\frac{1}{2}}] +$ $+\,\cos(\pi\nu)Y_{2\nu-1}[(2ay)^{\frac{1}{2}}]\}$
4.47	$x^{2\nu}J_{\nu+\frac{1}{2}}(ax^{-1})$ $\mathrm{Re}\ \nu > -1$	$(2\pi)^{\frac{1}{2}}(\frac{y}{a})^{-\nu-\frac{1}{2}}J_{2\nu+1}[(2ay)^{\frac{1}{2}}]K_{2\nu+1}[(2ay)^{\frac{1}{2}}]$

	$f(x)$	$g(y) = \int_0^\infty f(x)(xy)^{\frac{1}{2}}K_\nu(xy)dx$
4.48	$x^{-\frac{1}{2}}Y_\nu(ax^{-1})$ $-{}^5/_2 < \mathrm{Re}\ \nu < {}^5/_2$	$-y^{-\frac{1}{2}}\{e^{i\frac{\pi}{2}\nu}K_{2\nu}[2(iay)^{\frac{1}{2}}]+e^{-i\frac{\pi}{2}\nu}K_{2\nu}[2(-iay)^{\frac{1}{2}}]$
4.49	$x^{-{}^5/_2}Y_\nu(ax^{-1})$ $-\frac{1}{2} < \mathrm{Re}\ \nu < \frac{1}{2}$	$a^{-1}y^{\frac{1}{2}}\{e^{i\frac{\pi}{2}(\nu+1)}K_{2\nu}[2(iay)^{\frac{1}{2}}] +$ $+\ e^{-i\frac{\pi}{2}(\nu+1)}K_{2\nu}[2(-iay)^{\frac{1}{2}}]\}$
4.50	$x^{2\nu-2}Y_{\nu+\frac{1}{2}}(ax^{-1})$ $\mathrm{Re}\ \nu > -\ ^1/_3$	$(2\pi)^{\frac{1}{2}}(\frac{y}{a})^{\frac{1}{2}-\nu}Y_{2\nu}[(2ay)^{\frac{1}{2}}]K_{2\nu}[(2ay)^{\frac{1}{2}}]$
4.51	$x^{-2\nu}Y_{\nu-\frac{1}{2}}(ax^{-1})$ $\mathrm{Re}\ \nu < 1$	$-(\frac{1}{2}\pi)^{\frac{1}{2}}(\frac{y}{a})^{\frac{1}{2}-\nu}\sec(\pi\nu)K_{2\nu-1}[(2ay)^{\frac{1}{2}}]\ \cdot$ $\cdot\{J_{2\nu-1}[(2ay)^{\frac{1}{2}}]-J_{1-2\nu}[(2ay)^{\frac{1}{2}}]\}$
4.52	$x^{2\nu}Y_{\nu+\frac{1}{2}}(ax^{-1})$ $\mathrm{Re}\ \nu < -1$	$(2\pi)^{\frac{1}{2}}(\frac{y}{a})^{-\nu-\frac{1}{2}}Y_{2\nu+1}[(2ay)^{\frac{1}{2}}]K_{2\nu+1}[(2ay)^{\frac{1}{2}}]$
4.53	$x^{-\frac{1}{2}}J_\mu(ax^{-1})Y_\mu(ax^{-1})$ $\nu = 0$	$-2y^{-\frac{1}{2}}J_{2\mu}[2(ay)^{\frac{1}{2}}]K_{2\mu}[2(ay^{\frac{1}{2}}]$

	$f(x)$	$g(y) = \int\limits_0^\infty f(x)(xy)^{\frac{1}{2}}K_\nu(xy)\,dx$
4.54	$x^{-\frac{1}{2}}\{[J_\mu(ax^{-1})]^2 + [Y_\mu(ax^{-1})]^2\}$ $\nu = 0$	$4y^{-\frac{1}{2}}Y_{2\mu}[2(ay)^{\frac{1}{2}}]K_{2\mu}[2(ay)^{\frac{1}{2}}]$
4.55	$J_{2\nu-1}(ax^{\frac{1}{2}})$ $\mathrm{Re}\ \nu > -\frac{1}{2}$	$\frac{1}{4}\pi a y^{-3/2}[I_{\nu-1}(\frac{1}{4}a^2y^{-1}) - \mathbf{L}_{\nu-1}(\frac{1}{4}a^2y^{-1})]$
4.56	$x^{-\frac{1}{2}}J_{2\nu}(ax^{\frac{1}{2}})$ $\mathrm{Re}\ \nu > -\frac{1}{2}$	$\frac{1}{2}\pi y^{-\frac{1}{2}}[I_\nu(\frac{1}{4}a^2y^{-1}) - \mathbf{L}_\nu(\frac{1}{4}a^2y^{-1})]$
4.57	$x^{-\frac{1}{2}}Y_{2\nu}(ax^{\frac{1}{2}})$ $-\frac{1}{2} < \mathrm{Re}\ \nu < \frac{1}{2}$	$\frac{1}{2}\pi y^{-\frac{1}{2}}\cdot$ $\cdot[\csc(2\pi\nu)\mathbf{L}_{-\nu}(\frac{1}{4}a^2y^{-1}) - \operatorname{ctn}(2\pi\nu)\mathbf{L}_\nu(\frac{1}{4}a^2y^{-1})$ $- \tan(\pi\nu)I_\nu(\frac{1}{4}a^2y^{-1}) - \pi^{-1}\sec(\pi\nu)K_\nu(\frac{1}{4}a^2y^{-1})]$
4.58	$\begin{array}{l} 0 \qquad x < a \\ x^{\frac{1}{2}-\nu}(x^2-a^2)^{\frac{1}{2}\mu}\cdot \\ \cdot J_\mu[b(x^2-a^2)^{\frac{1}{2}}] \\ \qquad\qquad x > a \\ \mathrm{Re}\ \mu > -1 \end{array}$	$a^{\mu-\nu+1}b^\mu y^{\frac{1}{2}-\nu}(b^2+y^2)^{\frac{1}{2}\nu-\frac{1}{2}\mu-\frac{1}{2}}$ $K_{\nu-\mu-1}[a(b^2+y^2)^{\frac{1}{2}}]$

	$f(x)$	$g(y) = \int_0^\infty f(x)(xy)^{\frac{1}{2}}K_\nu(xy)\,dx$
4.59	$x^{\nu+\mu+\frac{1}{2}}I_\mu(ax)$ Re $\mu>-1$, Re$(\nu+\mu)>-1$ $y > a$	$2^{\nu+\mu}\Gamma(\nu+\mu+1)a^\mu y^{\nu+\frac{1}{2}}(y^2-a^2)^{-\mu-\nu-1}$
4.60	$x^{\mu-\frac{1}{2}}I_\mu(ax)$ Re$(2\mu\pm\nu) > -1$ $y > a$	$2^{\mu-1}\Gamma(\tfrac{1}{2}+\mu-\tfrac{1}{2}\nu)\Gamma(\tfrac{1}{2}+\mu+\tfrac{1}{2}\nu)y^{-\frac{1}{2}} \cdot$ $\cdot\ (y^2-a^2)^{-\frac{1}{2}\mu}P_{\frac{1}{2}\nu-\frac{1}{2}}^{-\mu}(1-2a^2y^{-2})$
4.61	$x^{-\mu-\frac{1}{2}}I_\mu(ax)$ $-1 < $ Re $\nu < 1$ $y > a$	$2^{-\mu-1}\pi\sec(\tfrac{1}{2}\pi\nu)y^{-\frac{1}{2}} \cdot$ $\cdot\ (y^2-a^2)^{\frac{1}{2}\mu}P_{\frac{1}{2}\nu-\frac{1}{2}}^{-\mu}(1-2a^2y^{-2})$
4.62	$I_\mu(ax)$ Re$(\mu\pm\nu)>-\tfrac{3}{2}$ $y > a$	$2^{\mu-\frac{1}{2}}\Gamma(\tfrac{3}{4}+\tfrac{1}{2}\mu+\tfrac{1}{2}\nu)\Gamma(\tfrac{3}{4}+\tfrac{1}{2}\mu-\tfrac{1}{2}\nu) \cdot$ $\cdot\ (y^2-a^2)^{-\frac{1}{2}}P_{\nu-\frac{1}{2}}^{-\mu}[(1-a^2y^{-2})^{\frac{1}{2}}]$
4.63	$x^{-\lambda-\frac{1}{2}}I_\nu(ax)$ Re $\lambda<1$, Re$(2\nu-\lambda)>-1$ $y > a$	$2^{-\lambda-1}\Gamma(\tfrac{1}{2}+\nu-\tfrac{1}{2}\lambda)\Gamma(\tfrac{1}{2}-\tfrac{1}{2}\lambda)y^{\frac{1}{2}} \cdot$ $\cdot\ (y^2-a^2)^{-\frac{1}{2}+\frac{1}{2}\lambda}P_{\frac{1}{2}\lambda-\frac{1}{2}}^{-\nu}\left(\dfrac{a^2+y^2}{a^2-y^2}\right)$

	$f(x)$	$g(y) = \int\limits_0^\infty f(x)(xy)^{\frac{1}{2}}K_\nu(xy)dx$
4.64	$x^{-\lambda-\frac{1}{2}}I_\mu(ax)$ $\mathrm{Re}(\mu\pm\nu-\lambda)>-1$ $y>a$	$2^{1-\lambda}a^\mu[\Gamma(1+\mu)]^{-1}\Gamma(\frac{1}{2}+\frac{1}{2}\mu-\frac{1}{2}\lambda+\frac{1}{2}\nu)\ \cdot$ $\cdot\ \Gamma(\frac{1}{2}+\frac{1}{2}\mu-\frac{1}{2}\lambda-\frac{1}{2}\nu)y^{\lambda-\mu-\frac{1}{2}}\ \cdot$ $\cdot\ _2F_1(\frac{1}{2}+\frac{1}{2}\mu+\frac{1}{2}\nu-\frac{1}{2}\lambda,\frac{1}{2}+\frac{1}{2}\mu-\frac{1}{2}\nu-\frac{1}{2}\lambda;\mu+1;a^2y^{-2})$
4.65	$x^{\frac{1}{2}}K_\nu(ax)$ $-1<\mathrm{Re}\ \nu<1$	$\frac{1}{2}\pi a^{-\nu}y^{\frac{1}{2}-\nu}\csc(\pi\nu)(a^{2\nu}-y^{2\nu})(a^2-y^2)^{-1}$
4.66	$x^{\nu-\frac{1}{2}}K_\mu(ax)$ $-1<\mathrm{Re}\ \mu<1$ $\mathrm{Re}(\frac{1}{2}+\nu\pm\frac{1}{2}\mu)>0$	$2^{\nu-2}\pi\sec(\frac{1}{2}\pi\mu)a^{-1}y^{\frac{1}{2}}\ \cdot$ $\cdot\ \begin{cases} (a^2-y^2)^{-\frac{1}{2}\nu}P_{-\frac{1}{2}-\frac{1}{2}\mu}^{-\nu}(2y^2a^{-2}-1) & y<a \\[2mm] (y^2-a^2)^{-\frac{1}{2}\nu}P_{-\frac{1}{2}-\frac{1}{2}\mu}^{-\nu}(2y^2a^{-2}-1) & y>a \end{cases}$
4.67	$x^{\mu-\frac{1}{2}}K_\mu(ax)$ $-1<\mathrm{Re}\ \nu<1$ $\mathrm{Re}(2\mu\pm\nu)>-1$	$\pi2^{\mu-2}\sec(\frac{1}{2}\pi\nu)\Gamma(\frac{1}{2}+\mu+\frac{1}{2}\nu)\Gamma(\frac{1}{2}+\mu-\frac{1}{2}\nu)\ \cdot$ $\cdot y^{-\frac{1}{2}}\begin{cases} (a^2-y^2)^{-\frac{1}{2}\mu}P_{-\frac{1}{2}-\frac{1}{2}\nu}^{-\mu}(2a^2y^{-2}-1) & y<a \\[2mm] (y^2-a^2)^{-\frac{1}{2}\mu}P_{-\frac{1}{2}-\frac{1}{2}\nu}^{-\mu}(2a^2y^{-2}-1) & y>a \end{cases}$

	$f(x)$	$g(y) = \int\limits_0^\infty f(x)(xy)^{\frac{1}{2}}K_\nu(xy)\,dx$
4.68	$x^{-\lambda-\frac{1}{2}}K_\nu(ax)$ Re $\lambda < 1$ Re$(\lambda\pm2\nu) < 1$	$\pi^{\frac{1}{2}}2^{-\lambda-2}y^{-\lambda-\frac{1}{2}}\Gamma(\tfrac{1}{2}+\nu-\tfrac{1}{2}\lambda)\,\Gamma(\tfrac{1}{2}-\nu-\tfrac{1}{2}\lambda)\;\cdot$ $\cdot\;\Gamma(\tfrac{1}{2}-\tfrac{1}{2}\lambda)(y^2-a^2)^{\frac{1}{2}\lambda}P_{\nu-\frac{1}{2}}^{\frac{1}{2}\lambda}(\dfrac{a^2+y^2}{2ay})$, $y > a$ $\pi^{\frac{1}{2}}2^{-\lambda-2}y^{\frac{1}{2}}a^{-\lambda-1}\Gamma(\tfrac{1}{2}+\nu-\tfrac{1}{2}\lambda)\,\Gamma(\tfrac{1}{2}-\nu-\tfrac{1}{2}\lambda)\;\cdot$ $\cdot\;\Gamma(\tfrac{1}{2}-\tfrac{1}{2}\lambda)(a^2-y^2)^{\frac{1}{2}\lambda}P_{\nu-\frac{1}{2}}^{\frac{1}{2}\lambda}(\dfrac{a^2+y^2}{2ay})$, $y < a$
4.69	$x^{-1}K_\mu(ax)$ Re$(\pm\mu\pm\nu) > -\,^3/_2$	$\pi2^{\mu-5/2}\dfrac{\Gamma(\tfrac{1}{4}+\tfrac{1}{2}\nu+\tfrac{1}{2}\mu)\,\Gamma(\tfrac{1}{4}+\tfrac{1}{2}\nu-\tfrac{1}{2}\mu)}{\cos(\pi\mu)-\sin(\pi\nu)}\;\cdot$ $\cdot\;(\dfrac{y}{a})^{\frac{1}{2}}\{P_{\mu-\frac{1}{2}}^{-\nu}[(1-\dfrac{y^2}{a^2})^{\frac{1}{2}}]+P_{\mu-\frac{1}{2}}^{-\nu}[-(1-\dfrac{y^2}{a^2})^{\frac{1}{2}}]\}$ $y < a$ $\pi2^{\mu-5/2}\dfrac{\Gamma(\tfrac{1}{4}+\tfrac{1}{2}\nu+\tfrac{1}{2}\mu)\,\Gamma(\tfrac{1}{4}-\tfrac{1}{2}\nu+\tfrac{1}{2}\mu)}{\cos(\pi\nu)-\sin(\pi\mu)}\;\cdot$ $\{P_{\nu-\frac{1}{2}}^{-\mu}[(1-\dfrac{a^2}{y^2})^{\frac{1}{2}}]+P_{\nu-\frac{1}{2}}^{-\mu}[-(1-\dfrac{a^2}{y^2})^{\frac{1}{2}}]\}$ $y > a$
4.70	$x^{-\lambda-\frac{1}{2}}K_\mu(ax)$ Re$(\lambda\pm\nu\pm\mu) < 1$	$\dfrac{a^{-\nu-1+\lambda}}{\Gamma(1-\lambda)}\,2^{-2-\lambda}\Gamma(\tfrac{1}{2}+\tfrac{1}{2}\mu+\tfrac{1}{2}\nu-\tfrac{1}{2}\lambda)\;\cdot$ $\cdot\;\Gamma(\tfrac{1}{2}-\tfrac{1}{2}\mu+\tfrac{1}{2}\nu-\tfrac{1}{2}\lambda)\,\Gamma(\tfrac{1}{2}+\tfrac{1}{2}\mu-\tfrac{1}{2}\nu-\tfrac{1}{2}\lambda)\;\cdot$ $\cdot\;\Gamma(\tfrac{1}{2}-\tfrac{1}{2}\mu-\tfrac{1}{2}\nu-\tfrac{1}{2}\lambda)\,y^{\nu+\frac{1}{2}}\;\cdot$ $\cdot\;{}_2F_1(\tfrac{1}{2}+\tfrac{1}{2}\nu+\tfrac{1}{2}\mu-\tfrac{1}{2}\lambda,\tfrac{1}{2}+\tfrac{1}{2}\nu-\tfrac{1}{2}\mu-\tfrac{1}{2}\lambda;1-\lambda;1-\dfrac{y^2}{a^2})$

	$f(x)$	$g(y) = \int\limits_0^\infty f(x)(xy)^{\frac{1}{2}}K_\nu(xy)\,dx$
4.71	$x^{-\frac{1}{2}}I_\mu(ax)I_{-\mu}(ax)$	$\frac{1}{2}\pi y^{-\frac{1}{2}}\sec(\frac{1}{2}\pi\nu)P^\mu_{\frac{1}{2}\nu-\frac{1}{2}}(z)P^{-\mu}_{\frac{1}{2}\nu-\frac{1}{2}}(z)$ $-1 < \mathrm{Re}\ \nu < 1,\ y > 2a,\ z = (1-4\frac{a^2}{y^2})^{\frac{1}{2}}$
4.72	$x^{-\frac{1}{2}}[I_\mu(ax)]^2$	$\frac{1}{2}y^{-\frac{1}{2}}\Gamma(\mu+\frac{1}{2}\nu+\frac{1}{2})\,\Gamma(\mu-\frac{1}{2}\nu+\frac{1}{2})$ $\cdot\ [P^{-\mu}_{\frac{1}{2}\nu-\frac{1}{2}}(z)]^2,\qquad y > 2a$ $\mathrm{Re}(\mu\pm\frac{1}{2}\nu) > -\frac{1}{2}$ (z as above)
4.73	$x^{-\frac{1}{2}}I_\mu(ax)K_\mu(ax)$	$-\frac{\pi y^{\frac{1}{2}}}{4a}\csc(\pi\nu)\,[P^{-\frac{1}{2}\nu}_{\mu-\frac{1}{2}}(z)Q^{\frac{1}{2}\nu}_{\mu-\frac{1}{2}}(z) -$ $-\ P^{\frac{1}{2}\nu}_{\mu-\frac{1}{2}}(z)Q^{-\frac{1}{2}\nu}_{\mu-\frac{1}{2}}(z)],\qquad\qquad y < 2a$ $z = (1-\frac{y^2}{4a^2})^{\frac{1}{2}}$ $\frac{\pi}{4}y^{-\frac{1}{2}}\sec(\frac{1}{2}\pi\nu)\sec(\pi\mu)\ \cdot$ $\cdot\ P^{-\mu}_{-\frac{1}{2}\nu-\frac{1}{2}}(z)[Q^\mu_{-\frac{1}{2}-\frac{1}{2}\nu}(z)+Q^\mu_{-\frac{1}{2}+\frac{1}{2}\nu}(z)]$ $y > 2a,\quad z = (1-\frac{4a^2}{y^2})^{\frac{1}{2}}$ For both $-1 < \mathrm{Re}\ \nu < 1$ $\mathrm{Re}(\mu+\frac{1}{2}\pm\frac{1}{2}\nu) > 0$

	$f(x)$	$g(y) = \int\limits_{0}^{\infty} f(x)(xy)^{\frac{1}{2}}K_{\nu}(xy)\,dx$
4.74	$x^{\frac{1}{2}+\mu}I_{\mu}(ax)I_{\nu}(bx)$ Re $\mu>-1$, Re $(\nu+\mu)>-1$ $y > a + b$	$(2\pi)^{-\frac{1}{2}}a^{\nu}b^{-\nu-1}y^{-\nu-\frac{1}{2}}(z^2-1)^{-\frac{1}{2}(\mu+\frac{1}{2})}\cdot$ $\cdot\, e^{-i\pi(\frac{1}{2}+\mu)}\, q_{\nu-\frac{1}{2}}^{\mu+\frac{1}{2}}(z)$ $z = \dfrac{b^2-a^2+y^2}{2ab}$
4.75	$x^{\frac{1}{2}+\nu}I_{\mu}(ax)I_{\nu}(bx)$ Re $\nu>-1$, Re $(\nu+\mu)>-1$ $y > a+b$	$(2\pi)^{-\frac{1}{2}}b^{\mu}a^{-\mu-1}y^{-\mu-\frac{1}{2}}(z^2-1)^{-\frac{1}{2}(\nu+\frac{1}{2})}\cdot$ $\cdot\, e^{-i\pi(\nu+\frac{1}{2})}\, q_{\mu-\frac{1}{2}}^{\nu+\frac{1}{2}}(z)$ $z = \dfrac{a^2-b^2+y^2}{2ab}$

	$f(x)$	$g(y) = \int\limits_0^\infty f(x)(xy)^{\frac{1}{2}}K_\nu(xy)\,dx$
4.76	$x^{-\frac{1}{2}}[K_\mu(ax)]^2$	$\dfrac{\pi^2 y^{\frac{1}{2}}}{16a}\sec(\pi\mu)\sec^2(\tfrac{1}{2}\pi\nu)\cdot$ $\cdot\ \{P^{\frac{1}{2}\nu}_{\mu-\frac{1}{2}}(z)\,[Q^{-\frac{1}{2}\nu}_{\mu-\frac{1}{2}}(z)+Q^{-\frac{1}{2}\nu}_{-\mu-\frac{1}{2}}(z)]$ $+\ P^{-\frac{1}{2}\nu}_{\mu-\frac{1}{2}}[Q^{\frac{1}{2}\nu}_{\mu-\frac{1}{2}}(z)\ +\ Q^{\frac{1}{2}\nu}_{-\mu-\frac{1}{2}}(z)]\}$ $y < 2a,\quad z=(1-\dfrac{y^2}{4a^2})^{\frac{1}{2}}$ $\dfrac{\pi^2}{4}y^{-\frac{1}{2}}\sec(\tfrac{1}{2}\pi\nu)\csc(2\pi\mu)\cdot$ $\cdot\ \{P^{\mu}_{-\frac{1}{2}-\frac{1}{2}\nu}(z)\,[Q^{-\mu}_{-\frac{1}{2}-\frac{1}{2}\nu}(z)\ +\ Q^{-\mu}_{-\frac{1}{2}+\frac{1}{2}\nu}(z)]$ $-\ P^{-\mu}_{-\frac{1}{2}-\frac{1}{2}\nu}(z)\,[Q^{\mu}_{-\frac{1}{2}-\frac{1}{2}\nu}(z)\ +\ Q^{\mu}_{-\frac{1}{2}+\frac{1}{2}\nu}(z)]\}$ $y > 2a,\quad z=[1-\dfrac{4a^2}{y^2}]^{\frac{1}{2}}$ For both, $-1 < \mathrm{Re}\ \nu < 1$ $-\tfrac{1}{2} < \mathrm{Re}\ \mu < \tfrac{1}{2}$
4.77	$x^{\sigma+\frac{1}{2}}J_\mu(ax)K_\lambda(bx)$ $x^{\sigma+\frac{1}{2}}K_\mu(ax)K_\lambda(bx)$	Bailey, W. N., 1936 J. London Math. Soc. 11, 16; Proc. London Math. Soc. 40, 37.

	$f(x)$	$g(y) = \int_0^\infty f(x)(xy)^{1/2} K_\nu(xy)\,dx$
4.78	$x^{1/2} K_{\frac{1}{2}\nu}(ax^2)$ $-1 < \mathrm{Re}\,\nu < 1$	$\tfrac{1}{8}\pi a^{-1} y^{1/2}\{\sec(\tfrac{1}{2}\pi\nu) K_{\frac{1}{2}\nu}(\tfrac{1}{4}a^{-1}y^2) +$ $+ \pi\csc(\pi\nu)[\mathbf{L}_{-\frac{1}{2}\nu}(\tfrac{1}{4}a^{-1}y^2) - \mathbf{L}_{\frac{1}{2}\nu}(\tfrac{1}{4}a^{-1}y^2)]\}$
4.79	$x^{2\mu+\nu+\frac{1}{2}}\exp(-\tfrac{1}{2}ax^2) \cdot$ $\cdot\, I_\mu(\tfrac{1}{2}ax^2)$ $\mathrm{Re}\,\mu > -\tfrac{1}{2},\ \mathrm{Re}(2\mu+\nu) > -1$	$\pi^{-\frac{1}{2}}2^{\mu-\frac{1}{2}}a^{-\frac{1}{2}\mu-\frac{1}{2}\nu-\frac{1}{4}}\Gamma(\nu+2\mu+1)\,\Gamma(\tfrac{1}{2}+\mu) \cdot$ $\cdot\, \exp(\tfrac{1}{8}a^{-1}y^2)\, W_{k,m}(\tfrac{1}{4}a^{-1}y^2)$ $2k = -3\mu-\nu-\tfrac{1}{2},\quad 2m = \mu+\nu+\tfrac{1}{2}$
4.80	$x^{-\frac{1}{2}} K_\nu(ax^{-1})$	$\pi y^{-\frac{1}{2}} K_{2\nu}[2(ay)^{\frac{1}{2}}]$
4.81	$x^{-5/2} K_\nu(ax^{-1})$	$\pi a^{-1} y^{\frac{1}{2}} K_{2\nu}[2(ay)^{\frac{1}{2}}]$
4.82	$x^{2\nu} K_{\nu+\frac{1}{2}}(ax^{-1})$	$(2\pi)^{\frac{1}{2}}(\tfrac{y}{a})^{-\nu-\frac{1}{2}} K_{2\nu+1}[(2iay)^{\frac{1}{2}}] K_{2\nu+1}[(-2iay)^{\frac{1}{2}}]$
4.83	$x^{2\nu-2} K_{\nu+\frac{1}{2}}(ax^{-1})$	$(2\pi)^{\frac{1}{2}}(\tfrac{y}{a})^{\frac{1}{2}-\nu} K_{2\nu}[(2iay)^{\frac{1}{2}}] K_{2\nu}[(-2iay)^{\frac{1}{2}}]$
4.84	$x^{-\frac{1}{2}} I_{2\nu}(ax^{\frac{1}{2}})$ $\mathrm{Re}\,\nu > -\tfrac{1}{2}$	$\tfrac{1}{2}\pi y^{-\frac{1}{2}}[I_\nu(\tfrac{1}{4}a^2 y^{-1}) + \mathbf{L}_\nu(\tfrac{1}{4}a^2 y^{-1})]$

	$f(x)$	$g(y) = \int_0^\infty f(x)(xy)^{\frac{1}{2}}K\nu(xy)\,dx$
4.85	$x^{-\frac{1}{2}}[J_{2\nu}(ax^{\frac{1}{2}})+I_{2\nu}(ax^{\frac{1}{2}})]$ $\mathrm{Re}\ \nu > -\tfrac{1}{2}$	$\pi y^{-\frac{1}{2}}I_\nu(\tfrac{1}{4}a^2y^{-1})$
4.86	$x^{-\frac{1}{2}}[I_{2\nu}(ax^{\frac{1}{2}})-J_{2\nu}(ax^{\frac{1}{2}})]$ $\mathrm{Re}\ \nu > -\tfrac{1}{2}$	$\pi y^{-\frac{1}{2}}\mathbf{L}_\nu(\tfrac{1}{4}a^2y^{-1})$
4.87	$x^{-\frac{1}{2}}K_{2\nu}(ax^{\frac{1}{2}})$ $-\tfrac{1}{2} < \mathrm{Re}\ \nu < \tfrac{1}{2}$	$\tfrac{1}{4}\pi\sec(\pi\nu)y^{-\frac{1}{2}}\{K_\nu(\tfrac{1}{4}a^2y^{-1})+\tfrac{1}{2}\pi\csc(\pi\nu)\ \cdot$ $\cdot\ [\mathbf{L}_{-\nu}(\tfrac{1}{4}a^2y^{-1}) - \mathbf{L}_\nu(\tfrac{1}{4}a^2y^{-1})]\}$
4.88	$x^{\nu+\frac{1}{2}}I_{2\nu}(ax^{\frac{1}{2}})J_{2\nu}(ax^{\frac{1}{2}})$ $\mathrm{Re}\ \nu > -\tfrac{1}{2}$	$\pi^{\frac{1}{2}}2^{-\nu-1}a^{2\nu+1}y^{-2\nu-2}J_{\nu-\frac{1}{2}}(\tfrac{1}{2}a^2y^{-1})$
4.89	$x^{\nu-\frac{1}{2}}I_{2\nu-1}(ax^{\frac{1}{2}})J_{2\nu-1}(ax^{\frac{1}{2}})$ $\mathrm{Re}\ \nu > 0$	$\pi^{\frac{1}{2}}2^{-\nu}a^{2\nu-1}y^{-2\nu}J_{\nu-\frac{1}{2}}(\tfrac{1}{2}a^2y^{-1})$
4.90	$x^{\nu-\frac{1}{2}}I_{2\nu-1}(ax^{\frac{1}{2}})Y_{2\nu-1}(ax^{\frac{1}{2}})$ $\mathrm{Re}\ \nu > 0$	$\pi^{\frac{1}{2}}2^{-\nu-1}a^{2\nu-1}y^{-2\nu}\csc(\pi\nu)[\mathbf{H}_{\frac{1}{2}-\nu}(\tfrac{1}{2}a^2y^{-1})+$ $+\cos(\pi\nu)J_{\nu-\frac{1}{2}}(\tfrac{1}{2}a^2y^{-1})+$ $+ \sin(\pi\nu)Y_{\nu-\frac{1}{2}}(\tfrac{1}{2}a^2y^{-1})]$

	$f(x)$	$g(y) = \int_0^\infty f(x)(xy)^{\frac{1}{2}} K_\nu(xy)\,dx$
4.91	$x^{\nu-\frac{1}{2}} J_{2\nu-1}(ax^{\frac{1}{2}}) K_{2\nu-1}(ax^{\frac{1}{2}})$ $\operatorname{Re}\nu > 0$	$\pi^{3/2} 2^{-\nu-2} a^{2\nu-1} y^{-2\nu}\ \cdot$ $\cdot\, \csc(\pi\nu)[\mathbf{H}_{\frac{1}{2}-\nu}(\tfrac{1}{2}a^2 y^{-1}) - Y_{\frac{1}{2}-\nu}(\tfrac{1}{2}a^2 y^{-1})]$
4.92	$x^{-\nu-\frac{1}{2}} I_{2\nu+1}(ax^{\frac{1}{2}}) J_{-2\nu-1}(ax^{\frac{1}{2}})$ $\operatorname{Re}\nu < \tfrac{1}{2}$	$-\pi^{\frac{1}{2}} 2^\nu a^{-2\nu-1} y^{2\nu}\ \cdot$ $\cdot\,[\cos(\pi\nu)\mathbf{H}_{\nu+\frac{1}{2}}(\tfrac{1}{2}a^2 y^{-1}) +$ $+ \sin(\pi\nu) J_{\nu+\frac{1}{2}}(\tfrac{1}{2}a^2 y^{-1})]$
4.93	$x^{-\nu-\frac{1}{2}} I_{-2\nu-1}(ax^{\frac{1}{2}}) J_{2\nu+1}(ax^{\frac{1}{2}})$ $\operatorname{Re}\nu < \tfrac{1}{2}$	$\pi^{\frac{1}{2}} 2^\nu a^{-2\nu-1} y^{2\nu}\ \cdot$ $\cdot\,[\cos(\pi\nu)\mathbf{H}_{\nu+\frac{1}{2}}(\tfrac{1}{2}a^2 y^{-1}) -$ $- \sin(\pi\nu) J_{\nu+\frac{1}{2}}(\tfrac{1}{2}a^2 y^{-1})]$
4.94	$x^{\frac{1}{2}-\nu}[I_{2\nu}(ax^{\frac{1}{2}}) J_{-2\nu}(ax^{\frac{1}{2}}) -$ $- J_{2\nu}(ax^{\frac{1}{2}}) I_{-2\nu}(ax^{\frac{1}{2}})]$ $\operatorname{Re}\nu < {}^3\!/_2$	$-\pi^{\frac{1}{2}} 2^\nu a^{1-2\nu} \sin(\pi\nu) y^{2\nu-2} J_{\nu+\frac{1}{2}}(\tfrac{1}{2}a^2 y^{-1})$
4.95	$x^{-\frac{1}{2}} K_\mu(ax^{\frac{1}{2}})\ \cdot$ $\cdot\,[\sin(\tfrac{1}{2}\pi\mu) J_\mu(ax^{\frac{1}{2}}) +$ $+\cos(\tfrac{1}{2}\pi\mu) Y_\mu(ax^{\frac{1}{2}})]$ $-1 < \operatorname{Re}\mu < 1$ $\nu = 0$	$-{}^1\!/_{16}\pi^2 \sec(\tfrac{1}{2}\pi\mu) y^{-\frac{1}{2}}$ $H^{(1)}_{\frac{1}{2}\mu}(\tfrac{1}{4}a^2 y^{-1}) H^{(2)}_{\frac{1}{2}\mu}(\tfrac{1}{4}a^2 y^{-1})$

	$f(x)$	$g(y) = \int\limits_{0}^{\infty} f(x)(xy)^{\frac{1}{2}}K_{\nu}(xy)\,dx$
4.96	$x^{-\frac{1}{2}}K_{\mu}(ax^{\frac{1}{2}}) \cdot$ $\cdot \{\sin[\frac{\pi}{2}(\mu-\nu)]J_{\mu}(ax^{\frac{1}{2}})+$ $+\cos[\frac{\pi}{2}(\pi-\nu)]Y_{\mu}(ax^{\frac{1}{2}})\}$ $\mathrm{Re}(\pm\mu\pm\nu) < 1$	$-\frac{1}{2}a^{-2}y^{\frac{1}{2}}\Gamma(\frac{1}{2}+\frac{1}{2}\mu-\frac{1}{2}\nu)\Gamma(\frac{1}{2}-\frac{1}{2}\mu-\frac{1}{2}\nu) \cdot$ $\cdot W_{\frac{1}{2}\nu,\frac{1}{2}\mu}(\frac{i}{2}a^2y^{-1})W_{\frac{1}{2}\nu,\frac{1}{2}\mu}(-\frac{i}{2}a^2y^{-1})$
4.97	$x^{\frac{1}{2}}\mathbf{H}_{\nu}(ax)$ $\mathrm{Re}\ \nu > -\,^{3}/_{2}$	$a^{\nu+1}y^{-\nu-\frac{1}{2}}(a^2+y^2)^{-1}$
4.98	$x^{\mu+\nu+\frac{1}{2}}\mathbf{H}_{\mu}(ax)$ $\mathrm{Re}\ \mu>-^{3}/_{2},\ \mathrm{Re}(\mu+\nu)>-^{3}/_{2}$	$\pi^{-\frac{1}{2}}2^{\mu+\nu+1}a^{\mu+1}\Gamma(\mu+\nu+^{3}/_{2})y^{-2\mu-\nu-^{5}/_{2}} \cdot$ $\cdot\,_{2}F_{1}(^{3}/_{2}+\mu+\nu,1;^{3}/_{2};-a^2y^{-2})$
4.99	$x^{\frac{1}{2}}\mathbf{H}_{\frac{1}{2}\nu}(ax^2)$ $\mathrm{Re}\ \nu > -2$	$2^{\frac{1}{2}\nu-1}(a\pi)^{-1}\Gamma(1+\frac{1}{2}\nu) \cdot$ $\cdot\, y^{\frac{1}{2}}S_{-1-\frac{1}{2}\nu,\frac{1}{2}\nu}(\frac{1}{4}a^{-1}y^2)$
4.100	$x^{3/2}\mathbf{H}_{\frac{1}{2}\nu+\frac{1}{2}}(ax^2)$ $\mathrm{Re}\ \nu > -3$	$2^{\frac{1}{2}+\frac{1}{2}\nu}(\pi a^2)^{-1}\Gamma(^{3}/_{2}+\frac{1}{2}\nu) \cdot$ $\cdot\, y^{3/2}S_{-\frac{1}{2}\nu-^{5}/_{2},\frac{1}{2}\nu-\frac{1}{2}}(\frac{1}{4}a^{-1}y^2)$
4.101	$x^{5/2}\mathbf{H}_{\frac{1}{2}\nu}(ax^2)$ $\mathrm{Re}\ \nu > -3$	$2^{\frac{1}{2}\nu-1}a^{-3}\pi^{-1}\Gamma(2+\frac{1}{2}\nu) \cdot$ $\cdot\, y^{5/2}S_{-3-\frac{1}{2}\nu,\frac{1}{2}\nu}(\frac{1}{4}a^{-1}y^2)$

	$f(x)$	$g(y) = \int_0^\infty f(x)(xy)^{\frac{1}{2}}K_\nu(xy)\,dx$
4.102	$x^{\frac{1}{2}}S_{\mu,\frac{1}{2}\nu}(ax^2)$ $\mathrm{Re}(2\mu\pm\nu) > -2$	$(4a)^{-1}\Gamma(1+\mu+\tfrac{1}{2}\nu)\Gamma(1+\mu-\tfrac{1}{2}\nu)\;\cdot$ $\cdot\; y^{\frac{1}{2}}S_{-\mu-1,\frac{1}{2}\nu}(\tfrac{1}{4}a^{-1}y^2)$
4.103	$x^{3/2}S_{\mu,\frac{1}{2}+\frac{1}{2}\nu}(ax^2)$ $\mathrm{Re}(2\mu\pm\nu) > -5$	$\tfrac{1}{8}a^{-2}(\mu+\tfrac{3}{2}-\tfrac{3}{2}\nu)\Gamma(\tfrac{1}{2}+\mu-\tfrac{1}{2}\nu)\Gamma(\tfrac{3}{2}+\mu+\tfrac{1}{2}\nu)\;\cdot$ $\cdot\; y^{3/2}S_{-\mu-2,\frac{1}{2}\nu-\frac{1}{2}}(\tfrac{1}{4}a^{-1}y^2)$
4.104	$D_{\nu-\frac{1}{2}}(ax^{-\frac{1}{2}})D_{-\nu-\frac{1}{2}}(ax^{-\frac{1}{2}})$	$\tfrac{1}{2}\pi y^{-1}\exp[-a(2y)^{\frac{1}{2}}]$
4.105	$x^{2\mu+\nu-\frac{1}{2}}\exp(-\tfrac{1}{2}ax^2)\;\cdot$ $\cdot M_{k,\mu}(ax^2)$ $\mathrm{Re}\ \mu>-\tfrac{1}{2},\mathrm{Re}(2\mu+\nu)>-1$	$2^{\mu-k-\frac{1}{2}}a^{\frac{1}{4}-\frac{1}{2}\mu-\frac{1}{2}\nu-\frac{1}{2}k}\Gamma(1+2\mu)\Gamma(1+2\mu+\nu)\;\cdot$ $\cdot\;\exp(\tfrac{1}{8}a^{-1}y^2)W_{k,m}(\tfrac{1}{4}a^{-1}y^2)$ $2k = -3\mu-\nu-k-\tfrac{1}{2}$ $2m = \mu+\nu-k+\tfrac{1}{2}$
4.106	$x^{-3/2}M_{k,0}(iax^2)\;\cdot$ $\cdot\; M_{k,0}(-iax^2)$ $\nu = 0$	$\tfrac{1}{16}\pi y^{\frac{1}{2}}\{[J_k(\tfrac{1}{8}a^{-1}y^2)]^2 + [Y_k(\tfrac{1}{8}a^{-1}y^2)]^2\}$

	$f(x)$	$g(y) = \int\limits_0^\infty f(x)(xy)^{\frac{1}{2}} K_\nu(xy)\,dx$
4.107	$x^{-3/2} M_{k,\mu}(iax^2) \cdot$ $\cdot M_{k,\mu}(-iax^2)$ $\text{Re } \mu > -\tfrac{1}{2}, \ \nu = 0$	$a[\Gamma(\mu+1)]^2 \cdot$ $\cdot y^{-3/2} W_{-\mu,k}(\tfrac{i}{4}a^{-1}y^2) W_{-\mu,k}(-\tfrac{i}{4}a^{-1}y^2)$
4.108	$x^{\frac{1}{2}} W_{\frac{1}{2}\nu,\mu}(ax^{-1}) \cdot$ $\cdot W_{-\frac{1}{2}\nu,\mu}(ax^{-1})$	$2ay^{-\frac{1}{2}} K_{2\mu}[(2iay)^{\frac{1}{2}}] K_{2\mu}[(-2iay)^{\frac{1}{2}}]$
4.109	$x^{\nu+\frac{1}{2}} {}_2F_1(a,b;\nu+1;-\lambda^2 x^2)$ $\text{Re } \lambda > 0, \ \text{Re } \nu > -1$	$2^{\nu+1} \lambda^{-a-b} \Gamma(1+\nu) \cdot$ $\cdot y^{a+b-\nu-3/2} S_{1-a-b,a-b}(\lambda^{-1}y)$
4.110	$x^{\nu+2\gamma-3/2} \cdot$ $\cdot {}_3F_2(1,a,b;\gamma,\gamma+\nu;-\lambda^2 x^2)$ $\text{Re } \lambda > 0, \ \text{Re } \gamma > 0$ $\text{Re}(\gamma+\lambda) > 0$	$2^{\nu+2\gamma-2} \lambda^{-a-b} \Gamma(\gamma)\Gamma(\gamma+\nu) \cdot$ $\cdot y^{a+b-2\gamma-\nu+\frac{1}{2}} S_{1-a-b,a-b}(\lambda^{-1}y)$
4.111	$x^{\mu-3/2} \cdot$ $\cdot {}_pF_q(a_1,\cdots a_p;$ $b_1\cdots b_q; -\lambda x^2)$ $p \le q-1, \ \text{Re}(\mu\pm\nu) > 0$	$2^{\mu-2} \Gamma(\tfrac{1}{2}\mu+\tfrac{1}{2}\nu)\Gamma(\tfrac{1}{2}\mu-\tfrac{1}{2}\nu) y^{\frac{1}{2}-\mu} \cdot$ $\cdot {}_{p+2}F_q(a_1,\cdots a_p, \tfrac{1}{2}\mu+\tfrac{1}{2}\nu, \tfrac{1}{2}\mu-\tfrac{1}{2}\nu;$ $b_1,\cdots b_q; \ 4\lambda y^{-2})$

Chapter III. Integral Transforms with Neumann Functions as Kernel

    If in the Hankel transform the Bessel function kernel is replaced by a Neumann function, the corresponding pair of inversion formulas is not symmetric.

$$(1) \quad g(y;\nu) = \int_0^\infty f(x)(xy)^{\frac{1}{2}} Y_\nu(xy)\,dx$$

$$(2) \quad f(x) = \int_0^\infty g(y;\nu)(xy)^{\frac{1}{2}} \mathbf{H}_\nu(xy)\,dy$$

Here $\mathbf{H}_\nu(z)$ is the S t r u v e function of order ν. For the special case $\nu = \pm\frac{1}{2}$

$$Y_{\frac{1}{2}}(z) = -(\tfrac{1}{2}\pi z)^{-\frac{1}{2}}\cos z, \quad Y_{-\frac{1}{2}}(z) = (\tfrac{1}{2}\pi z)^{-\frac{1}{2}}\sin z$$

$$\mathbf{H}_{\frac{1}{2}}(z) = (\tfrac{1}{2}\pi z)^{-\frac{1}{2}}(1-\cos z), \quad \mathbf{H}_{-\frac{1}{2}}(z) = (\tfrac{1}{2}\pi z)^{-\frac{1}{2}}\sin z.$$

The formulas (1) and (2) become for $\nu = -\frac{1}{2}$ the inversion formulas for the Fourier sine-transform.

Reference   Titchmarsh, E. C., 1937: Introduction to the theory of Fourier integrals, Oxford.

3.1 General Formulas

	$f(x)$	$g(y;\nu) = \int\limits_0^\infty f(x)(xy)^{\frac{1}{2}}Y_\nu(xy)\,dx \quad y > 0$
1.1	$\int\limits_0^\infty g(y)\mathbf{H}_\nu(xy)(xy)^{\frac{1}{2}}dy$	$g(y)$
1.2	$f(ax) \qquad , \qquad a > 0$	$a^{-1}g(ya^{-1};\nu)$
1.3	$x^m f(x), \quad m = 0,1,2,\cdots$	$y^{\frac{1}{2}-\nu}\left(\dfrac{d}{ydy}\right)^m [y^{m+\nu-\frac{1}{2}}g(y;\nu+m)]$
1.4	$x^m f(x), \; m = 0,1,2,\cdots$	$(-1)^m y^{\frac{1}{2}+\nu}\left(\dfrac{d}{ydy}\right)^m [y^{m-\nu-\frac{1}{2}}g(y;\nu-m)]$
1.5	$x^{-1}f(x)$	$\frac{1}{2}y\nu^{-1}[g(y;\nu-1) + g(y;\nu+1)]$
1.6	$x^{-\mu}f(x)$ $\mathrm{Re}\ \mu>0,\ \mathrm{Re}\ \nu>-{}^3\!/_2$	$2^{1-\mu}[\Gamma(\mu)]^{-1}y^{\nu+\frac{1}{2}} \cdot$ $\cdot \int\limits_y^\infty \tau^{-\nu-\mu+\frac{1}{2}}(\tau^2-y^2)^{\mu-1}g(\tau;\nu+\mu)\,d\tau$
1.7	$f'(x)$	$\frac{1}{2}\nu^{-1}[(\nu-\frac{1}{2})yg(y;\nu+1) - (\nu+\frac{1}{2})yg(y;\nu-1)]$

3.2 Transforms of Order Zero

	$f(x)$	$g(y) = \int\limits_0^\infty f(x)(xy)^{\frac{1}{2}} Y_0(xy)\,dx$
2.1	$x^{-\frac{1}{2}}(a+x)^{-1}$	$\mathbf{H}_0(ay) - Y_0(ay)$
2.2	$x^{-\frac{1}{2}}e^{-ax}$	$-2\pi^{-1}y^{\frac{1}{2}}(a^2+y^2)^{-\frac{1}{2}}\log[ay^{-1}+(1+a^2y^{-2})^{\frac{1}{2}}]$
2.3	$x^{\frac{1}{2}}(b^2+x^2)^{-\frac{1}{2}}$	$\pi^{-1}y^{-\frac{1}{2}}[e^{-by}\,\overline{Ei}(by) - e^{by}Ei(-by)]$
2.4	$x^{\frac{1}{2}}(b^2+x^2)^{-\frac{1}{2}} \cdot$ $\cdot\exp[-a(b^2+x^2)^{\frac{1}{2}}]$	$\pi^{-1}y^{\frac{1}{2}}(a^2+y^2)^{-\frac{1}{2}} \cdot$ $\cdot\{\exp[-b(a^2+y^2)^{\frac{1}{2}}]\overline{Ei}[b((a^2+y^2)^{\frac{1}{2}}-a)] -$ $-\exp[b(a^2+y^2)^{\frac{1}{2}}]Ei[-b((a^2+y^2)^{\frac{1}{2}}+a)]\}$
2.5	$x^{-\frac{1}{2}}(a^2+x^2)^{-1}\log x$	$-\frac{1}{4}\pi^2 y^{\frac{1}{2}}a^{-1}[\mathbf{L}_0(ay) - I_0(ay)] -$ $- a^{-1}y^{\frac{1}{2}}\log a\, K_0(ay)$
2.6	$x^{-\frac{1}{2}}(a^2+x^2)^{-1}\log(\frac{x}{a})$	$-\frac{1}{4}\pi^2 a^{-1}y^{\frac{1}{2}}[\mathbf{L}_0(ay) - I_0(ay)]$
2.7	$x^{-\frac{1}{2}}\sin(ax)$	$2\pi^{-1}y^{\frac{1}{2}}(a^2-y^2)^{-\frac{1}{2}}\log[\frac{a}{y} - (\frac{a^2}{y^2}-1)^{\frac{1}{2}}]\quad y < a$ $2\pi^{-1}y^{\frac{1}{2}}(y^2-a^2)^{-\frac{1}{2}}\arcsin(\frac{a}{y})\qquad y > a$

	$f(x)$	$g(y) = \int\limits_0^\infty f(x)(xy)^{\frac{1}{2}}Y_0(xy)\,dx$
2.8	$x^{-\frac{1}{2}}\cos(ax)$	$-y^{\frac{1}{2}}(a^2-y^2)^{-\frac{1}{2}} \qquad\qquad y < a$ $0 \qquad\qquad\qquad\qquad y > a$
2.9	$x^{-3/2}\sin(ax)$	$y^{\frac{1}{2}}\log[\frac{a}{y} - (\frac{a^2}{y^2}-1)^{\frac{1}{2}}] \qquad y < a$ $0 \qquad\qquad\qquad\qquad y > a$
2.10	$x^{-\frac{1}{2}}\cos(ax)\log(bx)$	$y^{\frac{1}{2}}(a^2-y^2)^{-\frac{1}{2}}[\gamma+\log(a^2-y^2)-\log(\tfrac{1}{2}by)]$ $\qquad\qquad\qquad\qquad y < a$ $\tfrac{1}{2}\pi y^{\frac{1}{2}}(y^2-a^2)^{-\frac{1}{2}} \qquad y > a$
2.11	$x^{\frac{1}{2}}\cos(ax^2)$	$\tfrac{1}{2}(\pi a)^{-1}y^{\frac{1}{2}}\{Ci(\tfrac{1}{4}y^2a^{-1})\sin(\tfrac{1}{4}y^2a^{-1}) -$ $- [\pi+si(\tfrac{1}{4}y^2a^{-1})]\cos(\tfrac{1}{4}y^2a^{-1})\}$
2.12	$x^{-\frac{1}{2}}\cos(ax^2)$	$\tfrac{1}{4}\pi^{\frac{1}{2}}a^{-\frac{1}{2}}y^{\frac{1}{2}}[J_0(\tfrac{1}{8}y^2a^{-1})\sin(\tfrac{1}{8}y^2a^{-1}-\tfrac{1}{4}\pi) +$ $+ Y_0(\tfrac{1}{8}y^2a^{-1})\cos(\tfrac{1}{8}y^2a^{-1}-\tfrac{1}{4}\pi)]$
2.13	$x^{-1}\cos(ax^{\frac{1}{2}})$	$-\tfrac{1}{4}\pi ay^{-\frac{1}{2}}\{[J_{\frac{1}{4}}(\tfrac{1}{8}y^2a^{-1})]^2+[J_{-\frac{1}{4}}(\tfrac{1}{8}y^2a^{-1})]^2\}$

	$f(x)$	$g(y) = \int_0^\infty f(x)(xy)^{\frac{1}{2}}Y_0(xy)\,dx$
2.14	$x^{\frac{1}{2}}(a^2-x^2)^{-\frac{1}{2}} \cdot$ $\cdot \cos[b(a^2-x^2)^{\frac{1}{2}}]$ $\qquad\qquad x < a$ $\qquad 0 \quad x > a$	$\pi^{-1}y^{\frac{1}{2}}(y^2+b^2)^{-\frac{1}{2}}\{\sin\alpha[\text{Ci}(z_1)+\text{Ci}(z_2)] -$ $- \cos\alpha[\text{Si}(z_1)+\text{Si}(z_2)]\}$ $\alpha = a(b^2+y^2)^{\frac{1}{2}}, \quad z_{\frac{1}{2}} = \alpha\pm ab$
2.15	$x^{-\frac{1}{2}}\exp(\tfrac{1}{2}ax^2) \cdot$ $\cdot K_0(\tfrac{1}{2}ax^2)$	$-\tfrac{1}{2}\pi a^{-\frac{1}{2}}y^{\frac{1}{2}}\exp(\tfrac{1}{8}y^2a^{-1})K_0(\tfrac{1}{8}y^2a^{-1})$
2.16	$x^{-\frac{1}{2}-2\mu}\exp(\tfrac{1}{2}ax^2) \cdot$ $\cdot K_\mu(\tfrac{1}{2}ax^2)$ $-\,^3\!/_4\,\text{Re}<\mu < \tfrac{1}{4}$	$-\pi^{\frac{1}{2}}a^\mu[\Gamma(\tfrac{1}{2}-2\mu)]^2[\Gamma(1-2_\mu)]^{-1}y^{-\frac{1}{2}}$ $\exp(\tfrac{1}{8}y^2a^{-1})W_{2\mu,0}(\tfrac{1}{4}y^2a^{-1})$
2.17	$x^{-\frac{1}{2}}[K_\mu(ax^{-1})]^2$ $-\tfrac{1}{4} < \text{Re }\mu < \tfrac{1}{4}$	$2\pi y^{-\frac{1}{2}}K_{2\mu}[2(ay)^{\frac{1}{2}}] \cdot \{\cos(\pi\mu) \cdot$ $\cdot J_{2\mu}[2(ay)^{\frac{1}{2}}]-\sin(\pi\mu)Y_{2\mu}[2(ay)^{\frac{1}{2}}]\}$
2.18	$x^{\frac{1}{2}-\mu}\mathbf{H}_\mu(ax)$ $\text{Re }\mu > 0$	$2^{1-\mu}[\Gamma(\mu)]^{-1}a^{-\mu}y^{\frac{1}{2}}(a^2-y^2)^{\mu-1} \quad y < a$ $\qquad\qquad 0 \qquad\qquad y > a$
2.19	$\qquad 0 \quad x < a$ $x^{-\frac{1}{2}}\arccos(ax^{-1})$ $\qquad\qquad x > a$	$y^{-3/2}\cos(ay)$

Transforms of General Order

3.3 Elementary Functions

	$f(x)$	$g(y) = \int\limits_0^\infty f(x)(xy)^{\frac{1}{2}}Y_\nu(xy)\,dx$
3.1	$x^{-\frac{1}{2}}$ $-1 < \mathrm{Re}\ \nu < 1$	$-\tan(\tfrac{1}{2}\pi\nu)$
3.2	$x^{\frac{1}{2}+\nu}$ $\quad\begin{array}{ll}0 & x < a \\ \\ & x > a\end{array}$ $\mathrm{Re}\ \nu < -\tfrac{1}{2}$	$-a^{\nu+1}y^{-\frac{1}{2}}Y_{\nu+1}(ay)$
3.3	x^μ $\mathrm{Re}\ \mu<0,\mathrm{Re}(\mu\pm\nu)>-\tfrac{3}{2}$	$2^{\mu+\frac{1}{2}}\cot(\tfrac{1}{2}\pi\nu-\tfrac{1}{2}\pi\mu+\tfrac{\pi}{4})\ \cdot$ $\cdot\ \Gamma(\tfrac{3}{4}+\tfrac{1}{2}\nu+\tfrac{1}{2}\mu)\,[\Gamma(\tfrac{1}{4}+\tfrac{1}{2}\nu-\tfrac{1}{2}\mu)]^{-1}y^{-\mu-1}$
3.4	x^μ $\quad\begin{array}{ll}0 & x < a \\ \\ & x > a\end{array}$ $\mathrm{Re}\ \mu < 0$	$ay^{-\mu}[Y_{\nu-1}(ay)S_{\mu+\frac{1}{2},\nu}(ay)\ -$ $-\ (\mu+\nu-\tfrac{1}{2})Y_\nu(ay)S_{\mu-\frac{1}{2},\nu-1}(ay)]$
3.5	$x^{-\frac{1}{2}}(a+x)^{-1}$ $-1 < \mathrm{Re}\ \nu < 1$	$\pi y^{\frac{1}{2}}\csc(\pi\nu)[\mathbf{E}_\nu(ay) + Y_\nu(ay)]\ +$ $+\ 2\cot(\pi\nu)[\mathbf{J}_\nu(ay) - J_\nu(ay)]$
3.6	$x^{\nu-\frac{1}{2}}(a+x)^{-1}$ $-\tfrac{1}{2} < \mathrm{Re}\ \nu < \tfrac{3}{2}$	$-2^{\nu+1}\pi^{-1}a^\nu y^{\frac{1}{2}}\Gamma(\nu+1)S_{-\nu-1,\nu}(ay)$

	$f(x)$	$g(y) = \int\limits_{0}^{\infty} f(x)(xy)^{\frac{1}{2}} Y_{\nu}(xy)\,dx$
3.7	$x^{-\nu-\frac{1}{2}}(a+x)^{-1}$ $-\frac{3}{2} < \mathrm{Re}\ \nu < \frac{1}{2}$	$a^{-\nu}y^{\frac{1}{2}}\{\frac{1}{2}\pi\tan(\pi\nu)\,[Y_{\nu}(ay)-\mathbf{H}_{\nu}(ay)]$ $-2^{1-\nu}\pi^{-1}\cos(\pi\nu)\,\Gamma(1-\nu)\,S_{\nu-1,\nu}(ay)\}$
3.8	$x^{\mu-\frac{1}{4}}(a+x)^{-1}$ $\mathrm{Re}(\mu\pm\nu) > -1$ $\mathrm{Re}\ \mu < \frac{3}{2}$	$\pi^{-1}(2a)^{\mu}y^{\frac{1}{2}}[\sin(\frac{1}{2}\pi\mu-\frac{1}{2}\pi\nu)\,\Gamma(\frac{1}{2}+\frac{1}{2}\mu+\frac{1}{2}\nu)\ \cdot$ $\cdot\ \Gamma(\frac{1}{2}+\frac{1}{2}\mu-\frac{1}{2}\nu)\,S_{-\mu,\nu}(ay)\ -$ $-2\cos(\frac{1}{2}\pi\mu-\frac{1}{2}\pi\nu)\,\Gamma(1+\frac{1}{2}\mu+\frac{1}{2}\nu)\ \cdot$ $\cdot\ \Gamma(1+\frac{1}{2}\mu-\frac{1}{2}\nu)\,S_{-\mu-1,\nu}(ay)]$
3.9	$x^{-\frac{1}{2}}(x-a)^{-1}$ $-1 < \mathrm{Re}\ \nu < 1$ Cauchy principal value.	$\pi y^{\frac{1}{2}}\{\cot(\pi\nu)\,[Y_{\nu}(ay)+\mathbf{E}_{\nu}(ay)]\ +$ $+\ \mathbf{J}_{\nu}(ay)\ +\ 2[\cot(\pi\nu)]^{2}\ \cdot$ $\cdot\ [\mathbf{J}_{\nu}(ay)\ -\ J_{\nu}(ay)]\}$
3.10	$x^{\nu-\frac{1}{2}}(x-a)^{-1}$ $-\frac{1}{2} < \mathrm{Re}\ \nu < \frac{3}{2}$ Cauchy principal value.	$a^{\nu}y^{\frac{1}{2}}[\pi J_{\nu}(ay)-2^{\nu+1}\pi^{-1}\ \cdot$ $\cdot\ \Gamma(\nu+1)\,S_{-\nu-1,\nu}(ay)]$
3.11	$x^{-\nu-\frac{1}{2}}(x-a)^{-1}$ $-\frac{3}{2} < \mathrm{Re}\ \nu < \frac{1}{2}$ Cauchy principal value	$a^{-\nu}y^{\frac{1}{2}}\{\frac{1}{2}\pi\tan(\pi\nu)\,[\mathbf{H}_{\nu}(ay)-Y_{\nu}(ay)]\ +$ $+\ \pi J_{\nu}(ay)-2^{1-\nu}\pi^{-1}\cos(\pi\nu)\,\Gamma(1-\nu)\ \cdot$ $\cdot\ S_{\nu-1,\nu}(ay)\}$

	$f(x)$	$g(y) = \int\limits_0^\infty f(x)(xy)^{\frac{1}{2}} Y_\nu(xy)\,dx$
3.12	$x^{\mu-\frac{1}{2}}(x-a)^{-1}$ Re$(\mu\pm\nu)>-1$, Re $\mu<\frac{3}{2}$ Cauchy principal value	$\pi a^\mu y^{\frac{1}{2}} J_\nu(ay) - \pi^{-1}(2a)^\mu y^{\frac{1}{2}} \cdot$ $\cdot\ [\sin(\frac{1}{2}\pi\mu-\frac{1}{2}\pi\nu)\Gamma(\frac{1}{2}+\frac{1}{2}\mu+\frac{1}{2}\nu)\ \cdot$ $\cdot\ \Gamma(\frac{1}{2}+\frac{1}{2}\mu-\frac{1}{2}\nu)S_{-\mu,\nu}(ay)\ +$ $+\ 2\ \cos(\frac{1}{2}\pi\mu-\frac{1}{2}\pi\nu)\Gamma(1+\frac{1}{2}\mu+\frac{1}{2}\nu)\ \cdot$ $\cdot\ \Gamma(1+\frac{1}{2}\mu-\frac{1}{2}\nu)S_{-\mu-1,\nu}(ay)]$
3.13	$x^{-\frac{1}{2}}(a^2+x^2)^{-1}$ $-1 < $ Re $\nu < 1$	$\frac{1}{2}\pi a^{-1}\tan(\frac{1}{2}\pi\nu)I_\nu(ay)+a^{-1}K_\nu(ay)\ -$ $-\ y(1-\nu^2)^{-1}\sin(\frac{1}{2}\pi\nu)\ \cdot$ $\cdot\ {}_1F_2(1;\frac{3}{2}-\frac{1}{2}\nu,\frac{3}{2}+\frac{1}{2}\nu;\frac{1}{4}a^2y^2)$
3.14	$x^{\nu-\frac{1}{2}}(a^2+x^2)^{-1}$ $-\frac{1}{2} < $ Re $\nu < \frac{5}{2}$	$-a^{\nu-1}y^{\frac{1}{2}}K_\nu(ay)$
3.15	$x^{\nu+\frac{3}{2}}(a^2+x^2)^{-1}$ $-\frac{3}{2} < $ Re $\nu < \frac{1}{2}$	$a^{\nu+1}y^{\frac{1}{2}}K_\nu(ay)$
3.16	$x^{-\nu-\frac{1}{2}}(a^2+x^2)^{-1}$ $-\frac{5}{2} < $ Re $\nu < \frac{1}{2}$	$a^{-\nu-1}y^{\frac{1}{2}}\{\frac{1}{2}\pi\tan(\pi\nu)[\mathbf{L}_\nu(ay)-I_\nu(ay)]$ $-\ \sec(\pi\nu)K_\nu(ay)\}$

	$f(x)$	$g(y) = \int\limits_0^\infty f(x)(xy)^{\frac{1}{2}}Y_\nu(xy)\,dx$
3.17	$x^{\mu-3/2}(a^2+x^2)^{-1}$ $\text{Re}(\mu\pm\nu)>0$ $\text{Re }\mu < 7/2$	$\pi^{-1}2^{\mu-3}y^{5/2-\mu}\cos(\tfrac{1}{2}\pi\mu-\tfrac{1}{2}\pi\nu)\Gamma(\tfrac{1}{2}\mu+\tfrac{1}{2}\nu-1)\cdot$ $\cdot\,\Gamma(\tfrac{1}{2}\mu-\tfrac{1}{2}\nu-1)\,{}_1F_2(1;2-\tfrac{1}{2}\mu-\tfrac{1}{2}\nu,2-\tfrac{1}{2}\mu+\tfrac{1}{2}\nu;\tfrac{1}{4}a^2y^2)$ $-\tfrac{1}{2}\pi a^{\mu-2}y^{\frac{1}{2}}\csc(\tfrac{1}{2}\pi\mu+\tfrac{1}{2}\pi\nu)\cot(\tfrac{1}{2}\pi\mu-\tfrac{1}{2}\pi\nu)\cdot$ $\cdot\,I_\nu(ay)-a^{\mu-2}y^{\frac{1}{2}}\csc(\tfrac{1}{2}\pi\mu-\tfrac{1}{2}\pi\nu)K_\nu(ay)$
3.18	$x^{-\frac{1}{2}}(a^2+x^2)^{-\frac{1}{2}}$ $-1 < \text{Re }\nu < 1$	$-\pi^{-1}y^{\frac{1}{2}}\sec(\tfrac{1}{2}\pi\nu)K_{\frac{1}{2}\nu}(ay)\cdot$ $\cdot\,[K_{\frac{1}{2}\nu}(\tfrac{1}{2}ay)+\pi\sin(\tfrac{1}{2}\pi\nu)I_{\frac{1}{2}\nu}(\tfrac{1}{2}ay)]$
3.19	$x^{\frac{1}{2}+\nu}(a^2+x^2)^\mu$ $-1<\text{Re }\nu<-2\text{ Re }\mu$	$\pi^{-1}2^{\nu-1}a^{2\mu+2}(1+\mu)^{-1}\Gamma(\nu)y^{\frac{1}{2}-\nu}\cdot$ $\cdot\,{}_1F_2(1;1-\nu,2+\mu;\tfrac{1}{4}a^2y^2)-2^\mu a^{\mu+\nu+1}\csc(\pi\nu)$ $\cdot\,\Gamma(\mu+1)y^{-\frac{1}{2}-\mu}[I_{\mu+\nu+1}(ay)-2\cos(\pi\mu)K_{\mu+\nu+1}(ay)]$
3.20	$x^{\frac{1}{2}-\nu}(a^2+x^2)^\mu$ $\text{Re }\nu<1,\text{Re}(\nu-2\mu)>\tfrac{1}{2}$	$2^\mu a^{\mu-\nu+1}y^{-\mu-1}[\pi^{-1}\cos(\pi\nu)\Gamma(\nu)\Gamma(1+\mu)\cdot$ $\cdot\,I_{\nu-\mu-1}(ay)-2[\Gamma(-\mu)]^{-1}\csc(\pi\nu)K_{\nu-\mu-1}(ay)]$ $-\,[(1+\mu)\Gamma(1+\nu)]^{-1}2^{-\nu-1}a^{2\mu+2}\cot(\pi\nu)y^\nu\cdot$ $\cdot\,{}_1F_2(1;\nu+1,\mu+2;\tfrac{1}{4}a^2y^2)$

	$f(x)$	$g(y) = \int\limits_{0}^{\infty} f(x)(xy)^{\frac{1}{2}} Y_{\nu}(xy)\,dx$
3.21	$x^{-\frac{1}{2}}(x^2-a^2)^{-1}$ $-1 < \mathrm{Re}\ \nu < 1$ Cauchy principle value	$\frac{1}{2}\pi a^{-1} y^{\frac{1}{2}}\{J_{\nu}(ay)+\tan(\frac{1}{2}\pi\nu)\,[\mathbf{J}_{\nu}(ay) -$ $- J_{\nu}(ay) - \mathbf{E}_{\nu}(ay) - Y_{\nu}(ay)]\}$
3.22	$x^{\nu-\frac{1}{2}}(x^2-a^2)^{-1}$ $-\frac{1}{2} < \mathrm{Re}\ \nu < \frac{5}{2}$ Cauchy principle value	$\frac{1}{2}\pi a^{\nu-1} y^{\frac{1}{2}} J_{\nu}(ay)$
3.23	$x^{\nu+\frac{1}{2}}(x^2-a^2)^{-1}$ $-1 < \mathrm{Re}\ \nu < \frac{3}{2}$ Cauchy principle value	$\frac{1}{2}\pi a^{\nu} y^{\frac{1}{2}} J_{\nu}(ay) - 2^{\nu+1}\pi^{-1}\Gamma(1+\nu)\ \cdot$ $\cdot\ y^{\frac{1}{2}} S_{-\nu-1,\nu}(ay)$
3.24	$x^{-\nu-\frac{1}{2}}(x^2-a^2)^{-1}$ $-\frac{5}{2} < \mathrm{Re}\ \nu < \frac{1}{2}$ Cauchy principle value	$\frac{1}{2}\pi a^{-\nu-1} y^{\frac{1}{2}}\sec(\pi\nu)\,[J_{-\nu}(ay) +$ $+ \sin(\pi\nu)\mathbf{H}_{\nu}(ay)]$

	$f(x)$	$g(y) = \int\limits_0^\infty f(x)(xy)^{\frac{1}{2}}Y_\nu(xy)\,dx$
3.25	$x^{\mu-3/2}(x^2-a^2)^{-1}$ $\mathrm{Re}(\mu\pm\nu)>0,\ \mathrm{Re}\ \mu < {}^7\!/_2$ Cauchy principle value	$\frac{1}{2}\pi a^{\mu-2}y^{\frac{1}{2}}J_\nu(ay)\ +$ $+\ 2^{\mu-1}\pi^{-1}a^{\mu-2}y^{\frac{1}{2}}\cos(\frac{1}{2}\pi\mu-\frac{1}{2}\pi\nu)\ \cdot$ $\cdot\ \Gamma(\frac{1}{2}\mu-\frac{1}{2}\nu)\Gamma(\frac{1}{2}\mu+\frac{1}{2}\nu)S_{1-\mu,\nu}(ay)$
3.26	$x^{-\frac{1}{2}+\nu}(a^2-x^2)^{-\frac{1}{2}+\nu}\quad x < a$ $0\qquad\qquad x > a$ $\mathrm{Re}\ \nu > -\frac{1}{2}$	$\pi^{\frac{1}{2}}2^{\nu-1}a^{2\nu}\Gamma(\frac{1}{2}+\nu)\ \cdot$ $\cdot\ J_\nu(\frac{1}{2}ay)Y_\nu(\frac{1}{2}ay)$
3.27	$x^{\frac{1}{2}-\nu}(a^2-x^2)^{-\frac{1}{2}}\qquad x < a$ $0\qquad\qquad x > a$ $\mathrm{Re}\ \nu < 1$	$(\frac{1}{2}\pi)^{\frac{1}{2}}a^{\frac{1}{2}-\nu}\{\cot(\pi\nu)[\mathbf{H}_{\nu-\frac{1}{2}}(ay)\ -$ $-\ Y_{\nu-\frac{1}{2}}(ay)]\ -\ J_{\nu-\frac{1}{2}}(ay)\}$
3.28	$x^{\frac{1}{2}-\nu}(a^2-x^2)^{\mu}\qquad x < a$ $0\qquad\qquad x > a$ $\mathrm{Re}\ \mu > -1,\ \mathrm{Re}\ \nu < 1$	$a^{\mu-\nu+1}y^{-\mu-\frac{1}{2}}[\pi^{-1}2^{1-\nu}\cos(\pi\nu)\Gamma(1-\nu)\ \cdot$ $\cdot\ S_{\mu+\nu,\mu-\nu+1}(ay)\ -$ $-\ 2^\mu\Gamma(1+\mu)\csc(\pi\nu)J_{\mu-\nu+1}(ay)]$
3.29	$x^{\frac{1}{2}+\nu}(a^2-x^2)^{\mu}$ $\qquad\qquad x < a$ $0\qquad\qquad x > a$ $\mathrm{Re}\ \mu > -1,\ \mathrm{Re}\ \nu > -1$	$a^{\mu+\nu+1}y^{-\mu-\frac{1}{2}}[2^\mu\Gamma(1+\mu)Y_{\mu+\nu+1}(ay)\ +$ $+\ \pi^{-1}a^{\nu+1}\Gamma(1+\nu)S_{\mu-\nu,\mu+\nu+1}(ay)]$

	$f(x)$	$g(y) = \int_0^\infty f(x)(xy)^{\frac{1}{2}} Y_\nu(xy)\,dx$
3.30	$\begin{array}{ll} 0 & x < a \\ x^{-\frac{1}{2}}(x^2-a^2)^{-\frac{1}{2}} & x > a \end{array}$	$\tfrac{1}{4}\pi y^{\frac{1}{2}}\{[J_{\frac{1}{2}\nu}(\tfrac{1}{2}ay)]^2 - [Y_{\frac{1}{2}\nu}(\tfrac{1}{2}ay)]^2\}$
3.31	$\begin{array}{ll} 0 & x < a \\ x^{\nu+\frac{1}{2}}(x^2-a^2)^{-\frac{1}{2}} & x > a \\ \mathrm{Re}\ \nu < \tfrac{1}{2} \end{array}$	$(\tfrac{1}{2}\pi)^{\frac{1}{2}} a^{\nu+\frac{1}{2}} J_{\nu+\frac{1}{2}}(ay)$
3.32	$\begin{array}{ll} 0 & x < a \\ x^{\nu-\frac{1}{2}}(x^2-a^2)^{\nu-\frac{1}{2}} & x > a \\ -\tfrac{1}{2} < \mathrm{Re}\ \nu < \tfrac{1}{2} \end{array}$	$\pi^{\frac{1}{2}} 2^{\nu-2} a^{2\nu} y^{\frac{1}{2}-\nu} \Gamma(\tfrac{1}{2}+\nu)\ \cdot$ $\cdot\ [J_\nu(\tfrac{1}{2}ay) J_{-\nu}(\tfrac{1}{2}ay) - Y_\nu(\tfrac{1}{2}ay) Y_{-\nu}(\tfrac{1}{2}ay)]$
3.33	$\begin{array}{ll} 0 & x < a \\ x^{\frac{1}{2}-\nu}(x^2-a^2)^{\nu-\frac{1}{2}} & x > a \\ -\tfrac{1}{2} < \mathrm{Re}\ \nu < \tfrac{1}{2} \end{array}$	$\pi^{-\frac{1}{2}} 2^\nu \Gamma(\tfrac{1}{2}+\nu) \sin(ay)\, y^{-\nu-\frac{1}{2}}$
3.34	$\begin{array}{ll} 0 & x < a \\ x^{-\nu-\frac{1}{2}}(x^2-a^2)^{-\nu-\frac{1}{2}} & x > a \\ -\tfrac{1}{2} < \mathrm{Re}\ \nu < \tfrac{1}{2} \end{array}$	$\pi^{\frac{1}{2}} 2^{-\nu-2} a^{-2\nu} \Gamma(\tfrac{1}{2}-\nu)\ \cdot$ $\cdot\ y^{\frac{1}{2}}\{[J_\nu(\tfrac{1}{2}ay)]^2 - [Y_\nu(\tfrac{1}{2}ay)]^2\}$
3.35	$\begin{array}{ll} 0 & x < a \\ x^{\frac{1}{2}+\nu}(x^2-a^2)^{\mu} & x > a \\ -2 < 2\mathrm{Re}\ \mu < -\tfrac{1}{2}-\mathrm{Re}\ \nu \end{array}$	$-2^\mu a^{\mu+\nu+1} \Gamma(1+\mu) y^{-\mu-\frac{1}{2}}\ \cdot$ $\cdot\ [\sin(\pi\mu) J_{\mu+\nu+1}(ay) + \cos(\pi\mu) Y_{\mu+\nu+1}(ay)]$

	$f(x)$	$g(y) = \int\limits_0^\infty f(x)(xy)^{\frac{1}{2}} Y_\nu(xy)\,dx$
3.36	$\begin{array}{ll} 0 & x < a \\ x^{\frac{1}{2}-\nu}(x^2-a^2)^\mu & x > a \end{array}$ $-1 < \mathrm{Re}\ \mu < \tfrac{1}{2}\mathrm{Re}\ \nu - \tfrac{1}{4}$	$2^\mu a^{\mu-\nu+1}\Gamma(1+\mu)y^{-\mu-\frac{3}{2}}Y_{\nu-\mu-1}(ay)$
3.37	$x^{-\frac{1}{2}}(a^2+x^2)^{-\frac{1}{2}} \cdot$ $\cdot[(a^2+x^2)^{\frac{1}{2}}-x]^\mu$ $\mathrm{Re}\ \mu > -\tfrac{3}{2},\ -1 < \mathrm{Re}\ \nu < 1$	$a^\mu y^{\frac{1}{2}}[\cot(\pi\nu)I_{\frac{1}{2}\mu+\frac{1}{2}\nu}(\tfrac{1}{2}ay)K_{\frac{1}{2}\mu-\frac{1}{2}\nu}(\tfrac{1}{2}ay) -$ $-\csc(\pi\nu)I_{\frac{1}{2}\mu-\frac{1}{2}\nu}(\tfrac{1}{2}ay)K_{\frac{1}{2}\mu+\frac{1}{2}\nu}(\tfrac{1}{2}ay)]$
3.38	$x^{-\frac{1}{2}}(a^2+x^2)^{-\frac{1}{2}} \cdot$ $\cdot\{[(a^2+x^2)^{\frac{1}{2}}+x]^\mu +$ $+[(a^2+x^2)^{\frac{1}{2}}-x]^\mu\}$ $\nu=0,\ -\tfrac{3}{2} < \mathrm{Re}\ \mu < \tfrac{3}{2}$	$-2\pi^{-1}a^\mu\cos(\tfrac{1}{2}\pi\mu)y^{\frac{1}{2}}[K_{\frac{1}{2}\mu}(\tfrac{1}{2}ay)]^2$
3.39	$x^{-\frac{1}{2}}(a^2+x^2)^{-\frac{1}{2}} \cdot$ $\cdot[(a^2+x^2)^{\frac{1}{2}}-a]^{2k}$ $-\tfrac{1}{2}-\mathrm{Re}\ k < \mathrm{Re}\ \nu < \tfrac{1}{2}+\mathrm{Re}\ k$	$-a^{-1}y^{-\frac{1}{2}}W_{-k,\frac{1}{2}\nu}(ay)\cdot\{\tan(\tfrac{1}{2}\pi\nu-\pi k) \cdot$ $\cdot\Gamma(\tfrac{1}{2}+\tfrac{1}{2}\nu+k)[\Gamma(1+\nu)]^{-1}M_{k,\frac{1}{2}\nu}(ay) +$ $+ \sec(\tfrac{1}{2}\pi\nu-\pi k)W_{k,\frac{1}{2}\nu}(ay)\}$
3.40	$\begin{array}{ll} 0 & x < a \end{array}$ $x^{-\frac{1}{2}}(x^2-a^2)^{-\frac{1}{2}} \cdot$ $\cdot\{[x+(x^2-a^2)^{\frac{1}{2}}]^\mu +$ $+[x-(x^2-a^2)^{\frac{1}{2}}]^\mu\ \ x > a$ $-\tfrac{3}{2} < \mathrm{Re}\ \mu < \tfrac{3}{2}$	$\tfrac{1}{2}\pi a^\mu y^{\frac{1}{2}}[J_{\frac{1}{2}\nu+\frac{1}{2}\mu}(\tfrac{1}{2}ay)J_{\frac{1}{2}\nu-\frac{1}{2}\mu}(\tfrac{1}{2}ay) -$ $- Y_{\frac{1}{2}\nu+\frac{1}{2}\mu}(\tfrac{1}{2}ay)Y_{\frac{1}{2}\nu-\frac{1}{2}\mu}(\tfrac{1}{2}ay)$

	$f(x)$	$g(y) = \int\limits_0^\infty f(x)(xy)^{\frac{1}{2}}Y_\nu(xy)\,dx$
3.41	$x^{-\frac{1}{2}}e^{-ax}$ $-1 < \mathrm{Re}\ \nu < 1$	$y^{\frac{1}{2}}(y^2+a^2)^{-\frac{1}{2}}\csc(\pi\nu)\ \cdot$ $\cdot\{y^\nu[(a^2+y^2)^{\frac{1}{2}}+a]^{-\nu}\cos(\pi\nu)\ -$ $-y^{-\nu}[(a^2+y^2)^{\frac{1}{2}}+a]^\nu\}$
3.42	$x^{\mu-3/2}e^{-ax}$ $-\mathrm{Re}\ \mu < \mathrm{Re}\ \nu < \mathrm{Re}\ \mu$	$-2\pi^{-1}\Gamma(\nu+\mu)y^{\frac{1}{2}}(a^2+y^2)^{-\frac{1}{2}\mu}\ \cdot$ $\cdot Q_{\mu-1}^{-\nu}[a(a^2+y^2)^{-\frac{1}{2}}]$
3.43	$x^{-\frac{1}{2}}e^{-ax^2}$ $-1 < \mathrm{Re}\ \nu < 1$	$-\tfrac{1}{2}(\tfrac{\pi y}{a})^{\frac{1}{2}}\exp(-\tfrac{1}{8}y^2a^{-1})\ \cdot$ $\cdot[\tan(\tfrac{1}{2}\pi\nu)I_{\frac{1}{2}\nu}(\tfrac{1}{8}y^2a^{-1})\ +$ $+\ \pi^{-1}\sec(\tfrac{1}{2}\pi\nu)K_{\frac{1}{2}\nu}(\tfrac{1}{8}y^2a^{-1})]$
3.44	$x^{\mu-\frac{1}{2}}e^{-ax^2}$ $\mathrm{Re}(\mu\pm\nu) > -1$	$-a^{-\frac{1}{2}\mu}\sec(\tfrac{1}{2}\pi\nu-\tfrac{1}{2}\pi\mu)y^{-\frac{1}{2}}\exp(-\tfrac{1}{8}y^2a^{-1})\ \cdot$ $\cdot\{\Gamma(\tfrac{1}{2}+\tfrac{1}{2}\mu+\tfrac{1}{2}\nu)[\Gamma(1+\nu)]^{-1}\sin(\tfrac{1}{2}\pi\nu-\tfrac{1}{2}\pi\mu)\ \cdot$ $\cdot\ M_{\frac{1}{2}\mu,\frac{1}{2}\nu}(\tfrac{1}{4}y^2a^{-2})\ +\ W_{\frac{1}{2}\mu,\frac{1}{2}\nu}(\tfrac{1}{4}y^2a^{-1})\}$
3.45	$x^{-3/2}e^{-ax^{-1}}$	$2y^{\frac{1}{2}}Y_\nu[(2ay)^{\frac{1}{2}}]K_\nu[(2ay)^{\frac{1}{2}}]$

	$f(x)$	$g(y) = \int\limits_0^\infty f(x)(xy)^{\frac{1}{2}} Y_\nu(xy)dx$
3.46	$x^{-3/2}\exp(-ax-bx^{-1})$	$2y^{\frac{1}{2}}Y_\nu\{(2b)^{\frac{1}{2}}[(a^2+y^2)^{\frac{1}{2}}-a]^{\frac{1}{2}}\}\cdot$ $\cdot K_\nu\{(2b)^{\frac{1}{2}}[(a^2+y^2)^{\frac{1}{2}}+a]^{\frac{1}{2}}\}$
3.47	$(a^2+x^2)^{-\frac{1}{2}}[(a^2+x^2)^{\frac{1}{2}}+a]^k\cdot$ $\cdot x^{-k-\frac{1}{2}}\exp[-b(a^2+x^2)^{\frac{1}{2}}]$ $\mathrm{Re}(-k\pm\nu) > -1$	$-a^{-1}y^{-\frac{1}{2}}\sec(\tfrac{1}{2}\pi\nu+\tfrac{1}{2}\pi k)\cdot$ $\cdot W_{\frac{1}{2}k,\frac{1}{2}\nu}\{a[(b^2+y^2)^{\frac{1}{2}}+b]\}\cdot$ $\cdot[W_{-\frac{1}{2}k,\frac{1}{2}\nu}\{a[(b^2+y^2)^{\frac{1}{2}}-b]\}+\Gamma(\tfrac{1}{2}+\tfrac{1}{2}\nu-\tfrac{1}{2}k)\cdot$ $\cdot\sin(\tfrac{1}{2}\pi\nu+\tfrac{1}{2}\pi k)[\Gamma(1+\nu)]^{-1}\cdot$ $\cdot M_{-\frac{1}{2}k,\frac{1}{2}\nu}\{a[(b^2+y^2)^{\frac{1}{2}}-b]\}$
3.48	$x^{-\frac{1}{2}}(a^2+x^2)^{-\frac{1}{2}}\cdot$ $\cdot\exp[-b(a^2+x^2)^{\frac{1}{2}}]$ $-1 < \mathrm{Re}\ \nu < 1$	$-y^{\frac{1}{2}}\sec(\tfrac{1}{2}\pi\nu)K_{\frac{1}{2}\nu}\{\tfrac{1}{2}a[(b^2+y^2)^{\frac{1}{2}}+b]\}\cdot$ $\cdot[\pi^{-1}K_{\frac{1}{2}\nu}\{\tfrac{1}{2}a[(b^2+y^2)^{\frac{1}{2}}-b]\}+\sin(\tfrac{1}{2}\pi\nu)\cdot$ $\cdot I_{\frac{1}{2}\nu}\{\tfrac{1}{2}a[(b^2+y^2)^{\frac{1}{2}}-b]\}]$
3.49	$x^{-\frac{1}{2}}\sin(ax)$ $-2<\mathrm{Re}\ \nu<2$	$\tfrac{1}{2}y^{\frac{1}{2}}\csc(\tfrac{1}{2}\pi\nu)\cdot(a^2-y^2)^{-\frac{1}{2}}\cdot$ $\cdot\{y^{-\nu}\cos(\pi\nu)[a-(a^2-y^2)^{\frac{1}{2}}]^\nu -$ $- y^\nu[a-(a^2-y^2)^{\frac{1}{2}}]^{-\nu}\}\qquad y < a$ $y^{\frac{1}{2}}\cot(\tfrac{1}{2}\pi\nu)(y^2-a^2)^{-\frac{1}{2}}\cdot$ $\cdot\sin[\nu\ \arcsin(ay^{-1})\qquad y > a$

	$f(x)$	$g(y) = \int\limits_0^\infty f(x)(xy)^{\frac{1}{2}} Y_\nu(xy)\,dx$
3.50	$x^{-\frac{1}{2}}\cos(ax)$ $-1 < \mathrm{Re}\ \nu < 1$	$-\sin(\tfrac{1}{2}\pi\nu)y^{\frac{1}{2}}(a^2-y^2)^{-\frac{1}{2}}\ \cdot$ $\cdot\{\cot(\pi\nu)y^{-\nu}[a-(a^2-y^2)^{\frac{1}{2}}]^\nu +$ $+\ y^\nu[a-(a^2-y^2)^{\frac{1}{2}}]^{-\nu}\csc(\pi\nu)$ $y < a$ $-\ \tan(\tfrac{1}{2}\pi\nu)y^{\frac{1}{2}}(y^2-a^2)^{-\frac{1}{2}}\ \cdot$ $\cdot\ \cos[\nu\ \arcsin(ay^{-1})]$ $y > a$
3.51	$x^{-\frac{3}{2}}\sin(ax)$ $-1 < \mathrm{Re}\ \nu < 1$	$\tfrac{1}{2}\nu^{-1}\sec(\tfrac{1}{2}\pi\nu)y^{\frac{1}{2}}\{y^{-\nu}\cos(\pi\nu)\ \cdot$ $\cdot\ [a-(a^2-y^2)^{\frac{1}{2}}]^\nu - y^\nu[a-(a^2-y^2)^{\frac{1}{2}}]^{-\nu}\}$ $y < a$ $-\nu^{-1}\tan(\tfrac{1}{2}\pi\nu)y^{\frac{1}{2}}\sin[\nu\ \arcsin(ay^{-1})]$ $y > a$
3.52	$x^{\frac{1}{2}+\nu}\sin(ax)$ $-\tfrac{3}{2} < \mathrm{Re}\ \nu < -\tfrac{1}{2}$	$\pi^{\frac{1}{2}}2^{\nu+1}[\Gamma(-\tfrac{1}{2}-\nu)]^{-1}ay^{\nu+\frac{1}{2}}(a^2-y^2)^{-\nu-\frac{3}{2}}$ $y < a$ 0 $y > a$
3.53	$x^{\frac{1}{2}-\nu}\sin(ax)$ $-\tfrac{1}{2} < \mathrm{Re}\ \nu < \tfrac{3}{2}$	$-2\pi^{-\frac{1}{2}}\Gamma(\tfrac{3}{2}-\nu)ay^{\frac{1}{2}}(2y)^{-\nu}(a^2-y^2)^{\nu-\frac{3}{2}}$ $y < a$ $2\pi^{-\frac{1}{2}}\sin(\pi\nu)\Gamma(\tfrac{3}{2}-\nu)ay^{\frac{1}{2}}(2y)^{-\nu}(y^2-a^2)^{\nu-\frac{3}{2}}$ $y > a$
3.54	$x^{\nu-\frac{1}{2}}\cos(ax)$ $-\tfrac{1}{2} < \mathrm{Re}\ \nu < \tfrac{1}{2}$	$-2^\nu\pi^{\frac{1}{2}}[\Gamma(\tfrac{1}{2}-\nu)]^{-1}y^{\nu+\frac{1}{2}}(a^2-y^2)^{-\nu-\frac{1}{2}}$ $y < a$ 0 $y > a$

	$f(x)$	$g(y) = \int\limits_0^\infty f(x)(xy)^{\frac{1}{2}} Y_\nu(xy)\,dx$
3.55	$x^{-\nu-\frac{1}{2}}\cos(ax)$ $-\frac{1}{2} < \text{Re } \nu < \frac{1}{2}$	$-\pi^{-\frac{1}{2}}\Gamma(\frac{1}{2}-\nu)y^{\frac{1}{2}}(2y)^{-\nu}(a^2-y^2)^{\nu-\frac{1}{2}}$ $\qquad y < a$ $-\pi^{\frac{1}{2}}\Gamma(\frac{1}{2}-\nu)\sin(\pi\nu)y^{\frac{1}{2}}(2y)^{-\nu}(y^2-a^2)^{\nu-\frac{1}{2}}$ $\quad y>a$
3.56	$x^{-\frac{1}{2}}\sin(ax^2)$ $-3 < \text{Re } \nu < 3$	$-\frac{1}{4}(\frac{\pi}{a}y)^{\frac{1}{2}}\sec(\frac{1}{2}\pi\nu)\,[\cos(\frac{1}{8}y^2a^{-1}-\frac{3}{4}\pi\nu-\frac{1}{4}\pi)\;\cdot$ $J_{\frac{1}{2}\nu}(\frac{1}{8}y^2a^{-1})-\sin(\frac{1}{8}y^2a^{-1}+\frac{1}{4}\pi\nu-\frac{1}{4}\pi)$ $\qquad\qquad Y_{\frac{1}{2}\nu}(\frac{1}{8}y^2a^{-1})\,]$
3.57	$x^{-\frac{1}{2}}\cos(ax^2)$ $-1 < \text{Re } \nu < 1$	$\frac{1}{4}(\frac{\pi}{a}y)^{\frac{1}{2}}\sec(\frac{1}{2}\pi\nu)\;\cdot$ $\cdot\,[\sin(\frac{1}{8}y^2a^{-1}-\frac{3}{4}\pi\nu-\frac{1}{4}\pi)J_{\frac{1}{2}\nu}(\frac{1}{8}y^2a^{-1})\;+$ $+\,\cos(\frac{1}{8}y^2a^{-1}+\frac{1}{4}\pi\nu-\frac{1}{4}\pi)Y_{\frac{1}{2}\nu}(\frac{1}{8}y^2a^{-1})\,]$
3.58	$x^{-3/2}\sin(ax)\sin(bx^{-1})$ $-2 < \text{Re } \nu < 2$	$\frac{1}{2}\pi y^{\frac{1}{2}}Y_\nu(z_1)\,[\cos(\frac{1}{2}\pi\nu)Y_\nu(z_2)\;+$ $+\,\sin(\frac{1}{2}\pi\nu)J_\nu(z_2)]\,-\,y^{\frac{1}{2}}K_\nu(z_2)\;\cdot$ $\cdot\,[I_\nu(z_1)\sin(\frac{1}{2}\pi\nu)+2\pi^{-1}\cos(\frac{1}{2}\pi\nu)K_\nu(z_1)]$ $z_{\frac{1}{2}} = b^{\frac{1}{2}}[(a+y)^{\frac{1}{2}}\pm(a-y)^{\frac{1}{2}}]$ $\qquad\qquad y < a$

	$f(x)$	$g(y) = \int\limits_0^\infty f(x)(xy)^{\frac{1}{2}} Y_\nu(xy)\,dx$
3.59	$x^{-3/2}\sin(ax)\cos(bx^{-1})$ $-2 < \mathrm{Re}\ \nu < 2$	$\frac{1}{2}\pi y^{\frac{1}{2}} Y_\nu(z_1)[\cos(\frac{1}{2}\pi\nu)J_\nu(z_2) -$ $- \sin(\frac{1}{2}\pi\nu)Y_\nu(z_2)] + y^{\frac{1}{2}}K_\nu(z_2)\ \cdot$ $\cdot\ [I_\nu(z_1)\cos(\frac{1}{2}\pi\nu)+2\pi^{-1}\sin(\frac{1}{2}\pi\nu)K_\nu(z_1)]$ $z_{\frac{1}{2}} = b^{\frac{1}{2}}[(a+y)^{\frac{1}{2}} \pm (a-y)^{\frac{1}{2}}]\quad y < a$
3.60	$x^{-3/2}\cos(ax)\sin(bx^{-1})$ $-1 < \mathrm{Re}\ \nu < 1$	$-\frac{1}{2}\pi y^{\frac{1}{2}} Y_\nu(z_1)[\cos(\frac{1}{2}\pi\nu)J_\nu(z_2) +$ $+ \sin(\frac{1}{2}\pi\nu)Y_\nu(z_2)] - y^{\frac{1}{2}}K_\nu(z_2)\ \cdot$ $\cdot\ [\cos(\frac{1}{2}\pi\nu)I_\nu(z_1)+2\pi^{-1}\sin(\frac{1}{2}\pi\nu)K_\nu(z_1)]$ $z_{\frac{1}{2}} = b^{\frac{1}{2}}[(a+y)^{\frac{1}{2}} \pm (a-y)^{\frac{1}{2}}]\quad y < a$
3.61	$x^{-3/2}\cos(ax)\cos(bx^{-1})$ $-1 < \mathrm{Re}\ \nu < 1$	$-\frac{1}{2}\pi y^{\frac{1}{2}} Y_\nu(z_1)[\sin(\frac{1}{2}\pi\nu)J_\nu(z_2) +$ $+ \cos(\frac{1}{2}\pi\nu)Y_\nu(z_2)] - y^{\frac{1}{2}}K_\nu(z_2)\ \cdot$ $\cdot\ [\sin(\frac{1}{2}\pi\nu)I_\nu(z_1)+2\pi^{-1}\cos(\frac{1}{2}\pi\nu)K_\nu(z_1)]$ $z_{\frac{1}{2}} = b^{\frac{1}{2}}[(a+y)^{\frac{1}{2}} \pm (a-y)^{\frac{1}{2}}]\quad y < a$

	$f(x)$	$g(y) = \int\limits_0^\infty f(x)(xy)^{\frac{1}{2}} Y_\nu(xy)\, dx$
3.62	$x^{-\frac{1}{2}}(a^2+x^2)^{-\frac{1}{2}} \cdot$ $\cdot \cos[b(a^2+x^2)^{\frac{1}{2}}]$ $-1 < \mathrm{Re}\ \nu < 1$	$-\tfrac{1}{4}\pi y^{\frac{1}{2}}\sec(\tfrac{1}{2}\pi\nu)\{Y_{\frac{1}{2}\nu}(z_1)Y_{\frac{1}{2}\nu}(z_2)\ +$ $+J_{\frac{1}{2}\nu}(z_2)[\cos(\pi\nu)J_{\frac{1}{2}\nu}(z_1)-\sin(\pi\nu)Y_{\frac{1}{2}\nu}(z_1)]$ $z_{\frac{1}{2}} = \tfrac{1}{2}a[b\pm(b^2-y^2)^{\frac{1}{2}}] \qquad\qquad y < b$
3.63	$x^{-\frac{1}{2}}(a^2+x^2)^{-\frac{1}{2}} \cdot$ $\cdot \sin[b(a^2+x^2)^{\frac{1}{2}}]$ $-1 < \mathrm{Re}\ \nu < 1$	$\tfrac{1}{4}\pi y^{\frac{1}{2}}\sec(\tfrac{1}{2}\pi\nu)\{J_{\frac{1}{2}\nu}(z_1)Y_{\frac{1}{2}\nu}(z_2)\ -$ $-J_{\frac{1}{2}\nu}(z_2)[\cos(\pi\nu)Y_{\frac{1}{2}\nu}(z_1)+\sin(\pi\nu)J_{\frac{1}{2}\nu}(z_1)]\}$ $z_{\frac{1}{2}} = \tfrac{1}{2}a[b\pm(b^2-y^2)^{\frac{1}{2}}] \qquad\qquad y < b$

3.4 Higher Transcendental Functions

	$f(x)$	$g(y) = \int\limits_0^\infty f(x)(xy)^{\frac{1}{2}} Y_\nu(xy)\,dx$
4.1	$\begin{array}{ll} 0 & x < a \\ P_{\nu-\frac{1}{2}}(a^{-1}x) & x > a \\ \operatorname{Re}\nu < \frac{1}{2} \end{array}$	$(\frac{1}{2}ay^{-1})^{\frac{1}{2}}[\cos(\frac{1}{2}ay)J_\nu(\frac{1}{2}ay) - \\ \quad - \sin(\frac{1}{2}ay)Y_\nu(\frac{1}{2}ay)]$
4.2	$\begin{array}{ll} 0 & x < a \\ x^{-\mu}(x^2-a^2)^{-\frac{1}{2}\mu} \cdot \\ \cdot P_{\nu-\frac{1}{2}}^{\mu}(a^{-1}x) & x > a \\ -\frac{1}{4} < \operatorname{Re}\mu < 1 \\ \operatorname{Re}(2\mu-\nu) > -\frac{1}{2} \end{array}$	$2^{-\frac{3}{2}}\pi^{\frac{1}{2}}a^{1-\mu}y^{\mu} \cdot \\ \cdot [J_\nu(\frac{1}{2}ay)J_{\mu-\frac{1}{2}}(\frac{1}{2}ay) - Y_\nu(\frac{1}{2}ay)Y_{\mu-\frac{1}{2}}(\frac{1}{2}ay)]$
4.3	$\begin{array}{ll} 0 & x < a \\ (x^2-a^2)^{\frac{1}{2}\nu-\frac{1}{4}} \\ P_\mu^{\frac{1}{2}-\nu}(2x^2a^{-2}-1) \\ & x > a \\ \operatorname{Re}\nu > -\frac{1}{2} \\ \operatorname{Re}\nu+|2\operatorname{Re}\mu+1| < \frac{3}{2} \end{array}$	$\pi^{\frac{1}{2}}2^{\nu-2}ay^{\frac{1}{2}-\nu}[J_{\mu+\frac{1}{2}}(ay)J_{-\mu-\frac{1}{2}}(ay) - \\ \quad - Y_{\mu+\frac{1}{2}}(\frac{1}{2}ay)Y_{-\mu-\frac{1}{2}}(\frac{1}{2}ay)]$
4.4	$\begin{array}{l} x^{\frac{1}{2}+\nu+\mu}J_\mu(ax) \\ -1 < \operatorname{Re}(\nu+\mu) < 0 \\ \operatorname{Re}\mu > -1 \end{array}$	$-\pi^{-1}2^{\nu+\mu+1}\cos(\pi\nu)\,\Gamma(\nu+\mu+1)a^{\mu}y^{\nu+\frac{1}{2}} \cdot \\ \quad \cdot (a^2-y^2)^{-\mu-\nu-1}, \quad y < a \\ \pi^{-1}2^{\nu+\mu+1}\cos(\pi\mu)\,\Gamma(\nu+\mu+1)a^{\mu}y^{\nu+\frac{1}{2}} \cdot \\ \quad \cdot (y^2-a^2)^{-\mu-\nu-1}, \quad y > a \end{array}$

	$f(x)$	$g(y) = \int\limits_0^\infty f(x)(xy)^{\frac12} Y_\nu(xy)\,dx$
4.5	$x^{\frac12+\mu-\nu} J_\mu(ax)$ $\mathrm{Re}\,\mu > -1$ $-1 < \mathrm{Re}(\mu-\nu) < 0$	$-\pi^{-1} 2^{\mu-\nu+1} \Gamma(\mu-\nu+1) a^\mu y^{\frac12-\nu} \cdot$ $\cdot (a^2-y^2)^{\nu-\mu-1}$, $\quad y < a$ $\pi^{-1} 2^{\mu-\nu+1} \cos[\pi(\mu-\nu)]\, \Gamma(\mu-\nu+1) a^\mu y^{\frac12-\nu} \cdot$ $\cdot (y^2-a^2)^{\nu-\mu-1}$, $\quad y > a$
4.6	$x^{-\lambda-\frac12} J_\mu(ax)$ $\mathrm{Re}(\mu\pm\nu-\lambda+1) > 0$	$\int\limits_0^\infty x^{-\lambda-\frac12} [\mathrm{ctn}(\pi\nu) J_\nu(xy) - \csc(\pi\nu) J_{-\nu}(xy)](xy)^{\frac12} \cdot$ $\cdot J_\mu(ax)\,dx$ $\quad\quad\quad y < a$ For the integral see (Chapter I, 10.26). $\dfrac{2^{-\lambda} a^\nu}{\pi\Gamma(1+\mu)} \sin[\tfrac{\pi}{2}(\mu-\nu-\lambda)] \cdot$ $\cdot \Gamma(\tfrac12+\tfrac12\mu+\tfrac12\nu-\tfrac12\lambda)\,\Gamma(\tfrac12+\tfrac12\mu-\tfrac12\nu-\tfrac12\lambda)\, y^{\nu-\mu-\frac12} \cdot$ $\cdot\, {}_2F_1(\tfrac12+\tfrac12\mu+\tfrac12\nu-\tfrac12\lambda, \tfrac12+\tfrac12\mu-\tfrac12\nu-\tfrac12\lambda; \mu+1; \dfrac{a^2}{y^2})$ $\quad\quad\quad y > a$

	$f(x)$	$g(y) = \int\limits_0^\infty f(x)(x)^{\frac{1}{2}}Y_\nu(xy)\,dx$
4.7	$x^{-\lambda-\frac{1}{2}}Y_\mu(ax)$ $\mathrm{Re}(\pm\mu\pm\nu-\lambda+1) > 0$	$\int\limits_0^\infty x^{-\lambda-\frac{1}{2}}(xy)^{\frac{1}{2}}\{J_\mu(ax)J_\nu(xy) +$ $+ 4\pi^{-2}\sin[\frac{\pi}{2}(\lambda+\mu+\nu)]K_\mu(ax)K_\nu(xy)\}dx$ For the integrals see Chapter I, 10.26 and Chapter II, 4.70.
4.8	$x^{\frac{1}{2}-\nu}J_\mu(ax)J_\mu(bx)$ $\mathrm{Re}\ \nu > -\frac{1}{2},\ \mathrm{Re}\ \mu > -1$ $\mathrm{Re}(\mu-\nu) > -1$	$-(\tfrac{1}{2}\pi^3)^{-\frac{1}{2}}(ab)^{\nu-1}y^{\frac{1}{2}-\nu} \cdot$ $\cdot(z_1^2-1)^{\frac{1}{2}\nu-\frac{1}{4}}e^{-i\pi(\frac{1}{2}-\nu)}\ q_{\mu-\frac{1}{2}}^{\frac{1}{2}-\nu}(z_1)$ $\qquad\qquad\qquad y < |a-b|$ $-(\tfrac{1}{2}\pi^3)^{-\frac{1}{2}}(ab)^{\nu-1}y^{\frac{1}{2}-\nu}(1-z_1^2)^{\frac{1}{2}\nu-\frac{1}{4}} \cdot$ $\cdot Q_{\mu-\frac{1}{2}}^{\frac{1}{2}-\nu}(z_1),\ |a-b| < y < a+b$ $(\tfrac{1}{2}\pi^3)^{-\frac{1}{2}}\cos[\pi(\nu-\mu)](ab)^{\nu-1}y^{\frac{1}{2}-\nu} \cdot$ $\cdot(z_2^2-1)^{\frac{1}{2}\nu-\frac{1}{4}}e^{-i\pi(\frac{1}{2}-\nu)}q_{\mu-\frac{1}{2}}^{\frac{1}{2}-\nu}(z_2)$ $\qquad\qquad\qquad\qquad y > a + b$ $z_{\frac{1}{2}} = \pm(\dfrac{a^2+b^2-y^2}{2ab})$
4.9	$x^{-\frac{1}{2}}J_\mu(ax)J_{-\mu}(ax)$ $-1 < \mathrm{Re}\ \nu < 1$	$-\tan(\tfrac{1}{2}\pi\nu)y^{-\frac{1}{2}}P_{\frac{1}{2}\nu-\frac{1}{2}}^\mu[(1-4a^2y^{-2})^{\frac{1}{2}}] \cdot$ $\cdot P_{\frac{1}{2}\nu-\frac{1}{2}}^{-\mu}[(1-4a^2y^{-2})^{\frac{1}{2}}]\qquad y > 2a$

	$f(x)$	$g(y) = \int_0^\infty f(x)(xy)^{\frac{1}{2}} Y_\nu(xy)\,dx$
4.10	$x^{\frac{1}{2}+\nu} J_\mu(ax) J_\mu(bx)$ $\mathrm{Re}\ \nu < \frac{1}{2},\ \mathrm{Re}\ \mu > -1$ $\mathrm{Re}(\nu+\mu) > -1$	$-(\frac{1}{2}\pi^3)^{-\frac{1}{2}}\cos(\pi\nu)(ab)^{-\nu-1}y^{\nu+\frac{1}{2}}(z_1^2-1)^{-\frac{1}{4}-\frac{1}{2}\nu}\cdot$ $\cdot e^{-i\pi(\nu+\frac{1}{2})} q_{\mu-\frac{1}{2}}^{\nu+\frac{1}{2}}(z_1)\qquad,\qquad y < |a-b|$ $-(2\pi)^{-\frac{1}{2}}(ab)^{-\nu-1}y^{\nu+\frac{1}{2}}(1-z_1^2)^{-\frac{1}{4}-\frac{1}{2}\nu}\cdot$ $\cdot[P_{\mu-\frac{1}{2}}^{\nu+\frac{1}{2}}(z_1)\sin(\pi\nu)+\frac{2}{\pi}Q_{\mu-\frac{1}{2}}^{\nu+\frac{1}{2}}(z_1)\cos(\pi\nu)]$ $\qquad |a-b| < y < a+b$ $(\frac{1}{2}\pi^3)^{-\frac{1}{2}}\cos(\pi\mu)(ab)^{-\nu-1}y^{\frac{1}{2}+\nu}(z_2^2-1)^{-\frac{1}{4}-\frac{1}{2}\nu}\cdot$ $\cdot e^{-i\pi(\nu+\frac{1}{2})} q_{\mu-\frac{1}{2}}^{\nu+\frac{1}{2}}(z_2)\qquad,\qquad y > a+b$ $z_{\frac{1}{2}} = \pm\left(\dfrac{a^2+b^2-y^2}{2ab}\right)$
4.11	$x^{-\frac{1}{2}}[J_\mu(ax)]^2$ $\mathrm{Re}(2\mu\pm\nu) > -1$	$y^{-\frac{1}{2}}\tan(\pi\mu-\tfrac{1}{2}\pi\nu)\,\Gamma(\tfrac{1}{2}+\tfrac{1}{2}\nu+\mu)\cdot$ $\cdot[\Gamma(\tfrac{1}{2}+\tfrac{1}{2}\nu-\mu)]^{-1}\{P_{\frac{1}{2}\nu-\frac{1}{2}}^{-\mu}[(1-4a^2y^{-2})]^{\frac{1}{2}}\}^2$ $\qquad y > 2a$
4.12	$x^{\frac{1}{2}+\nu}[J_\nu(ax)]^2$ $-\frac{1}{2} < \mathrm{Re}\ \nu < \frac{1}{2}$	$0\qquad\qquad y < 2a$ $2^{3\nu+1}\pi^{-\frac{1}{2}}a^{2\nu}[\Gamma(\tfrac{1}{2}-\nu)]^{-1}y^{-\nu-\frac{1}{2}}(y^2-4a^2)^{-\nu-\frac{1}{2}}$ $\qquad y > 2a$

	$f(x)$	$g(y) = \int\limits_0^\infty f(x)\,(xy)^{\frac12}Y_\nu(xy)\,dx$
4.13	$x^{\frac12}J_{\frac12\nu}(ax^2)$ $\mathrm{Re}\ \nu > -1$	$\tfrac14 y^{\frac12}a^{-1}[Y_{\frac12\nu}(\tfrac14 y^2 a^{-1}) -$ $-\tan(\tfrac12\pi\nu)J_{\frac12\nu}(\tfrac14 y^2 a^{-1})+\sec(\tfrac12\pi\nu)\mathbf{H}_{-\frac12\nu}(\tfrac14 y^2 a^{-1})\,]$
4.14	$x^{-5/2}J_{\frac12\nu-\frac12}(ax^2)$ $\mathrm{Re}\ \nu > -\,^3/_2$	$a^{-2}y^{\frac12}J_{\frac12\nu+\frac12}(\tfrac14 y^2 a^{-1})$
4.15	$x^{\frac12}J_{\frac14\nu}(ax^2)J_{-\frac14\nu}(ax^2)$ $-2 < \mathrm{Re}\ \nu < 2$	$^1/_{16}a^{-1}\sec(\tfrac14\pi\nu)y^{\frac12}\{2\cos^2(\tfrac14\pi\nu)[J_{\frac14\nu}(^1/_{16}y^2 a^{-1})]^2$ $+2\sin(\tfrac12\pi\nu)J_{\frac14\nu}(^1/_{16}y^2 a^{-1})Y_{\frac14\nu}(^1/_{16}y^2 a^{-1}) -$ $-\,[Y_{\frac14\nu}(^1/_{16}y^2 a^{-1})]^2\}$
4.16	$x^{-\frac12}J_\nu(a^2 x^{-1})$ $-\tfrac12 < \mathrm{Re}\ \nu < \,^3/_2$	$y^{-\frac12}[Y_{2\nu}(2a y^{\frac12}) + 2\pi^{-1}K_{2\nu}(2a y^{\frac12})\,]$
4.17	$x^{-5/2}J_\nu(a^2 x^{-1})$ $-\tfrac12 < \mathrm{Re}\ \nu < \tfrac12$	$a^{-2}y^{\frac12}[Y_{2\nu}(2a y^{\frac12}) - 2\pi^{-1}K_{2\nu}(2a y^{\frac12})\,]$
4.18	$x^{-\frac12}Y_\nu(a^2 x^{-1})$ $-\tfrac12 < \mathrm{Re}\ \nu < \tfrac12$	$-y^{-\frac12}J_{2\nu}(2a y^{\frac12})$

	$f(x)$	$g(y) = \int\limits_0^\infty f(x)(xy)^{\frac{1}{2}} Y_\nu(xy)\,dx$
4.19	$x^{-5/2} Y_\nu(a^2 x^{-1})$ $-\tfrac{1}{2} < \mathrm{Re}\ \nu < \tfrac{1}{2}$	$-a^{-2} y^{\frac{1}{2}} J_{2\nu}(2ay^{\frac{1}{2}})$
4.20	$x^{-3/2} Y_{\nu+1}(a^2 x^{-1})$ $-\tfrac{3}{2} < \mathrm{Re}\ \nu < \tfrac{1}{2}$	$-a^{-1} J_{2\nu+1}(2ay^{\frac{1}{2}})$
4.21	$J_{2\nu-1}(ax^{\frac{1}{2}})$ $\mathrm{Re}\ \nu > -\tfrac{1}{2}$	$-\tfrac{1}{2} ay^{-3/2} \mathbf{H}_{\nu-1}(\tfrac{1}{4}a^2 y^{-1})$
4.22	$x^{-\frac{1}{2}} J_{2\nu}(ax^{\frac{1}{2}})$ $\mathrm{Re}\ \nu > -\tfrac{1}{2}$	$-y^{-\frac{1}{2}} \mathbf{H}_\nu(\tfrac{1}{4}a^2 y^{-1})$
4.23	$x^{-\frac{1}{2}} Y_{2\nu}(ax^{\frac{1}{2}})$ $-\tfrac{1}{2} < \mathrm{Re}\ \nu < \tfrac{1}{2}$	$\tfrac{1}{2} y^{-\frac{1}{2}} [\sec(\pi\nu) J_{-\nu}(\tfrac{1}{4}a^2 y^{-1}) +$ $+ \csc(\pi\nu)\mathbf{H}_{-\nu}(\tfrac{1}{4}a^2 y^{-1}) - 2\cot(2\pi\nu)\mathbf{H}_\nu(\tfrac{1}{4}a^2 y^{-1})]$
4.24	$x^{\nu+2n-\frac{1}{2}}(x^2+k^2)^{-1} \cdot$ $\cdot (a^2+x^2)^{-\frac{1}{2}\mu} \cdot$ $\cdot J_\mu[b(a^2+x^2)^{\frac{1}{2}}]$ $n = 0,1,2,\cdots$ $-\tfrac{1}{2}-n\ \mathrm{Re}\ \mu < 3-2n+\mathrm{Re}\ \nu$	$(-1)^{n+1} k^{\nu+2n-1} y^{\frac{1}{2}} K_\nu(ky) \cdot$ $\cdot (k^2-a^2)^{-\frac{1}{2}\mu} I_\mu[b(k^2-a^2)^{\frac{1}{2}}]$ $y > b$

	$f(x)$	$g(y) = \int\limits_0^\infty f(x)(xy)^{\frac{1}{2}}Y_\nu(xy)\,dx$
4.25	$\begin{array}{ll}0 & x < b\\[4pt] x^{\frac{1}{2}-\nu}(x^2-b^2)^{\frac{1}{2}\mu}\;\cdot\\ \quad\cdot J_\mu[a(x^2-b^2)^{\frac{1}{2}}]\\ & x > b\\[4pt] R\,\nu > \mathrm{Re}\,\mu > -1\end{array}$	$-\dfrac{2}{\pi}a^\mu b^{\mu-\nu+1}y^{\frac{1}{2}-\nu}(a^2-y^2)^{\frac{1}{2}\nu-\frac{1}{2}\mu-\frac{1}{2}}\;\cdot$ $\cdot\,K_{\nu-\mu-1}[b(a^2-y^2)^{\frac{1}{2}}]\quad,\quad y < a$ $a^\mu b^{\mu-\nu+1}y^{\frac{1}{2}-\nu}(y^2-a^2)^{\frac{1}{2}\nu-\frac{1}{2}\mu-\frac{1}{2}}\;\cdot$ $\cdot\,Y_{\nu-\mu-1}[b(y^2-a^2)^{\frac{1}{2}}]\quad,\quad y > a$
4.26	$\begin{array}{ll}0 & x < b\\[4pt] x^{\frac{1}{2}+\nu}(x^2-b^2)^{\frac{1}{2}\mu}\;\cdot\\ \quad\cdot J_\mu[a(x^2-b^2)^{\frac{1}{2}}]\\ & x > b\\[4pt] \mathrm{Re}\,\mu>-1,\ \mathrm{Re}(\nu+\mu)<0\end{array}$	$-2\pi^{-1}\cos(\pi\nu)a^\mu b^{\mu+\nu+1}y^{\frac{1}{2}+\nu}(a^2-y^2)^{-\frac{1}{2}\nu-\frac{1}{2}\mu-\frac{1}{2}}\;\cdot$ $\cdot\,K_{\mu+\nu+1}[b(a^2-y^2)^{\frac{1}{2}}]\qquad y < a$ $-a^\mu b^{\nu+\mu+1}y^{\frac{1}{2}+\nu}(y^2-a^2)^{-\frac{1}{2}\mu-\frac{1}{2}\nu-\frac{1}{2}}\;\cdot$ $\cdot\,\{\sin(\pi\nu)J_{\nu+\mu+1}[b(y^2-a^2)^{\frac{1}{2}}]\;+$ $+\;\cos(\pi\mu)Y_{\nu+\mu+1}[b(y^2-a^2)^{\frac{1}{2}}]\}$ $\qquad\qquad\qquad\qquad y > a$
4.27	$x^{-\lambda-\frac{1}{2}}K_\mu(ax)$ $\mathrm{Re}(\pm\nu\pm\mu-\lambda+1) > 0$	$\int\limits_0^\infty x^{-\lambda-\frac{1}{2}}(xy)^{\frac{1}{2}}K_\mu(ax)\;\cdot$ $\cdot\,[\mathrm{ctn}(\pi\nu)J_\nu(xy)-\csc(\pi\nu)J_{-\nu}(xy)]\,dx$ For the integrals see
4.28	$x^{\frac{1}{2}}K_{\frac{1}{2}\nu}(ax^2)$ $-1 < \mathrm{Re}\,\nu < 1$	$\frac{1}{4}\pi a^{-1}y^{\frac{1}{2}}[\csc(\pi\nu)\mathbf{L}_{-\frac{1}{2}\nu}(\frac{1}{4}y^2a^{-1})\;-$ $-\cot(\pi\nu)\mathbf{L}_{\frac{1}{2}\nu}(\frac{1}{4}y^2a^{-1})-\tan(\frac{1}{2}\pi\nu)I_{\frac{1}{2}\nu}(\frac{1}{4}y^2a^{-1})$ $-\pi^{-1}\sec(\frac{1}{2}\pi\nu)K_{\frac{1}{2}\nu}(\frac{1}{4}y^2a^{-1})]$

	$f(x)$	$g(y) = \int\limits_0^\infty f(x)(xy)^{\frac{1}{2}} Y_\nu(xy)\,dx$
4.29	$x^{-\frac{1}{2}} K_\nu(ax^{-1})$ $-\frac{1}{2} < \mathrm{Re}\ \nu < \frac{1}{2}$	$-2y^{-\frac{1}{2}}[\sin(\tfrac{3}{2}\pi\nu)\,\mathrm{ker}_{2\nu}(2a^{\frac{1}{2}}y^{\frac{1}{2}}) +$ $+ \cos(\tfrac{3}{2}\pi\nu)\,\mathrm{kei}_{2\nu}(2a^{\frac{1}{2}}y^{\frac{1}{2}})]$
4.30	$x^{-\frac{5}{2}} K_\nu(ax^{-1})$ $-\frac{5}{2} < \mathrm{Re}\ \nu < \frac{5}{2}$	$2a^{-1}y^{\frac{1}{2}}[\sin(\tfrac{3}{2}\pi\nu)\,\mathrm{kei}_{2\nu}(2a^{\frac{1}{2}}y^{\frac{1}{2}}) -$ $- \cos(\tfrac{3}{2}\pi\nu)\,\mathrm{ker}_{2\nu}(2a^{\frac{1}{2}}y^{\frac{1}{2}})]$
4.31	$x^{-2\nu} K_{\nu-\frac{1}{2}}(ax^{-1})$ $\mathrm{Re}\ \nu > \frac{1}{6}$	$(2\pi a)^{\frac{1}{2}} a^{-\nu} y^{\nu-\frac{1}{2}} Y_{2\nu-1}[(2ay)^{\frac{1}{2}}]\ \cdot$ $\cdot\ K_{2\nu-1}[(2ay)^{\frac{1}{2}}]$
4.32	$x^{-2\nu-2} K_{\nu-\frac{1}{2}}(ax^{-1})$ $\mathrm{Re}\ \nu > -\frac{1}{2}$	$(2\pi)^{\frac{1}{2}} a^{-\nu-\frac{1}{2}} y^{\nu+\frac{1}{2}} Y_{2\nu}[(2ay)^{\frac{1}{2}}]K_{2\nu}[(2ay)^{\frac{1}{2}}]$
4.33	$x^{2\nu-2} K_{\nu+\frac{1}{2}}(ax^{-1})$ $\mathrm{Re}\ \nu < \frac{1}{2}$	$(\tfrac{1}{2}\pi)^{\frac{1}{2}} \csc(\pi\nu)\, a^{\nu-\frac{1}{2}} y^{\frac{1}{2}-\nu} K_{2\nu}[(2ay)^{\frac{1}{2}}]\ \cdot$ $\cdot\ \{J_{2\nu}[(2ay)^{\frac{1}{2}}] - J_{-2\nu}[(2ay)^{\frac{1}{2}}]\}$
4.34	$x^{-\frac{1}{2}} K_{2\nu}(ax^{\frac{1}{2}})$ $-\frac{1}{2} < \mathrm{Re}\ \nu < \frac{1}{2}$	$-\tfrac{1}{4}\pi y^{-\frac{1}{2}}[\sec(\pi\nu)\,J_{-\nu}(\tfrac{1}{4}a^2 y^{-1}) -$ $- \csc(\pi\nu)\mathbf{H}_{-\nu}(\tfrac{1}{4}a^2 y^{-1}) +$ $+ 2\csc(2\pi\nu)\mathbf{H}_\nu(\tfrac{1}{4}a^2 y^{-1})]$

	$f(x)$	$g(y) = \int_0^\infty f(x)(xy)^{\frac{1}{2}} Y_\nu(xy)\,dx$
4.35	$x^{\nu-\frac{1}{2}} J_{2\nu-1}(ax^{\frac{1}{2}}) K_{2\nu-1}(ax^{\frac{1}{2}})$ $\mathrm{Re}\ \nu > 0$	$\pi^{\frac{1}{2}} 2^{-\nu-1} a^{2\nu-2} \csc(\pi\nu)\ \cdot$ $\cdot\ [\mathbf{L}_{\frac{1}{2}-\nu}(\tfrac{1}{2}a^2 y^{-1}) - I_{\nu-\frac{1}{2}}(\tfrac{1}{2}a^2 y^{-1})]$
4.36	$x^{-\frac{1}{2}} \mathbf{H}_{\nu-1}(ax)$ $-\tfrac{1}{2} < \mathrm{Re}\ \nu < \tfrac{1}{2}$	$-a^{\nu-1} y^{\frac{1}{2}-\nu} \qquad\qquad\qquad y < a$ $0 \qquad\qquad y > a$
4.37	$x^{\nu-\mu+\frac{1}{2}} \mathbf{H}_\mu(ax)$ $\mathrm{Re}\ \mu > \mathrm{Re}\ \nu$ $-\tfrac{3}{2} < \mathrm{Re}\ \nu < \tfrac{1}{2}$	$a^{-\mu}[\Gamma(\mu-\nu)]^{-1} 2^{\nu-\mu+1} y^{\nu+\frac{1}{2}}\ \cdot$ $\cdot\ (a^2-y^2)^{\mu-\nu-1} \qquad\qquad y < a$ $0 \qquad\qquad y > a$
4.38	$x^{2\mu+\nu-\frac{1}{2}} \exp(-\tfrac{1}{4}a^2 x^2)$ $x^{\frac{1}{2}-\mu} S_{\mu+\nu,\,\mu-\nu}(ax)$ $\mathrm{Re}\ \mu > 0,\ -\tfrac{3}{2} < \mathrm{Re}\ \nu < \tfrac{1}{2}$	$2^\nu \Gamma(\tfrac{1}{2}+\nu) \Gamma(\tfrac{1}{2}+\mu)\ [\Gamma(\mu)]^{-1} a^{\nu-\mu}\ \cdot$ $\cdot\ y^{\frac{1}{2}-\nu}(a^2-y^2)^{\mu-1} \qquad\qquad y < a$ $0 \qquad\qquad y > a$
4.39	$x^{-\frac{5}{2}} S_{-\nu-3,\,\nu}(a^2 x^{-1})$ $-\tfrac{3}{2} < \mathrm{Re}\ \nu < \tfrac{1}{2}$	$\pi 2^{-\nu-2} a^{-2} [\Gamma(\nu+2)]^{-1} K_{2\nu}(2ay^{\frac{1}{2}})$
4.40	$x^{\nu-\frac{1}{2}} \exp(\tfrac{1}{4}a^2 x^2)\ \cdot$ $\cdot\ D_{\frac{1}{2}\nu-\frac{1}{2}}(ax)$ $-\tfrac{1}{2} < \mathrm{Re}\ \nu < \tfrac{2}{3}$	$-\pi^{-1} 2^{\frac{3}{4}+\frac{3}{4}\nu} a^{-\nu} \Gamma(1+\nu) y^{-\frac{1}{2}} \exp(\tfrac{1}{4}a^{-2}y^2)\ \cdot$ $\cdot\ W_{\frac{1}{2}\nu-\frac{1}{2},\,\frac{1}{2}\nu}(\tfrac{1}{2}y^2 a^{-2})$

	$f(x)$	$g(y) = \int_0^\infty f(x)(xy)^{\frac{1}{2}} Y_\nu(xy)\, dx$
4.41	$D_{\nu-\frac{1}{2}}(ax^{-\frac{1}{2}}) D_{-\nu-\frac{1}{2}}(ax^{-\frac{1}{2}})$	$y^{-1}\exp(-ay^{\frac{1}{2}})\sin(ay^{\frac{1}{2}}-\tfrac{1}{2}\pi\nu+\tfrac{1}{4}\pi)$
4.42	$x^{\nu-m}\exp(-\tfrac{1}{4}x^2)\cdot$ $\cdot M_{k,\frac{1}{4}-\frac{1}{2}m}(\tfrac{1}{2}x^2)$ m integer $\mathrm{Re}\ \nu > m - \tfrac{3}{2}$ $\mathrm{Re}(2k-\nu) > m \geq -1$	$(-1)^m\Gamma(\tfrac{3}{2}-m)[\Gamma(\tfrac{3}{4}+k-\tfrac{1}{2}m)]^{-1}\cdot$ $\cdot\,2^{\frac{1}{2}\nu-\frac{1}{2}m}(\tfrac{1}{2}y^2)^{\lambda}\exp(-\tfrac{1}{4}y^2)W_{\alpha,\beta}(\tfrac{1}{2}y^2)$ $2\alpha = k + \nu+\tfrac{1}{2}m + \tfrac{5}{4}$ $2\beta = k-\nu+\tfrac{1}{2}m - \tfrac{3}{4}$ $2\lambda = k + \tfrac{1}{2}m - \tfrac{5}{4}$
4.43	$x^{-m}\exp(-\tfrac{1}{4}x^2)\cdot$ $\cdot M_{k,\frac{1}{4}+\frac{1}{2}\nu-\frac{1}{2}m}(\tfrac{1}{2}x^2)$ m integer $\mathrm{Re}\ \nu > m - \tfrac{3}{2}$ $2\mathrm{Re}\ k > -m \geq -1$	$(-1)^m\Gamma(\nu-m+\tfrac{3}{2})[\Gamma(k+\tfrac{1}{2}\nu-\tfrac{1}{2}m+\tfrac{3}{4})]^{-1}\cdot$ $\cdot\,2^{-\frac{1}{2}m}(\tfrac{1}{2}y^2)^{\lambda}\exp(-\tfrac{1}{4}y^2)W_{\alpha,\beta}(\tfrac{1}{2}y^2)$ $2\alpha = k +\tfrac{1}{2}\nu - \tfrac{3}{2}m+\tfrac{5}{4}$ $2\beta = k + \tfrac{1}{2}\nu + \tfrac{1}{2}m - \tfrac{3}{4}$ $2\lambda = k - \tfrac{1}{2}\nu + \tfrac{1}{2}m - \tfrac{5}{4}$
4.44	$x^{2\mu+\nu-\frac{1}{2}}\exp(-\tfrac{1}{4}x^2)\cdot$ $\cdot M_{k,\mu}(\tfrac{1}{2}x^2)$ $\mathrm{Re}(2\mu+\nu) > -1$ $-1<2\mathrm{Re}\ \mu<\tfrac{1}{2}+\mathrm{Re}(2k-\nu)$	$\pi^{-1}2^{\mu+\beta}\Gamma(2\mu+1)[\Gamma(\tfrac{1}{2}-\mu-k)]^{-1}y^{k-\mu-1}\cdot$ $\cdot\exp(-\tfrac{1}{4}y^2)\{\Gamma(2\mu+\nu+1)[\Gamma(\tfrac{3}{2}+\mu+\nu-k)]^{-1}\cdot$ $\cdot\cos(2\pi\mu)M_{\alpha,\beta}(\tfrac{1}{2}y^2)+\sin(\pi\mu-\pi k)\cdot$ $\cdot W_{\alpha,\beta}(\tfrac{1}{2}y^2)\}$ $2\alpha = 3\mu+\nu+k+\tfrac{1}{2}$ $2\beta = \mu+\nu-k+\tfrac{1}{2}$

	$f(x)$	$g(y) = \int\limits_0^\infty f(x)(xy)^{\frac{1}{2}} Y_\nu(xy)\,dx$
4.45	$x^{2\mu-\nu-\frac{1}{2}} \exp(-\frac{1}{4}x^2) \cdot$ $\cdot M_{k,\mu}(\frac{1}{2}x^2)$ $\mathrm{Re}(2\mu-\nu) > -1$ $-1 < 2\mathrm{Re}\ \mu < \frac{1}{2}+\mathrm{Re}(2k+\nu)$	$\pi^{-1}2^{\mu+\beta}\Gamma(2\mu+1)\Gamma(\frac{1}{2}-k-\mu)y^{k-\mu-1}\exp(-\frac{1}{4}y^2)\cdot$ $\cdot\{\Gamma(2\mu-\nu-1)[\Gamma(2\beta+1)]^{-1}\cos(\pi\nu-2\pi\mu)\cdot$ $\cdot M_{\alpha,\beta}(\frac{1}{2}y^2)-\sin(\pi\nu+\pi k-\pi\mu)W_{\alpha,\beta}(\frac{1}{2}y^2)\}$ $2\alpha = 3\mu-\nu+k+\frac{1}{2},\ \ 2\beta = \mu-\nu-k+\frac{1}{2}$
4.46	$x^{\frac{1}{2}}W_{\frac{1}{2}\nu,\mu}(ax^{-1}) \cdot$ $\cdot W_{-\frac{1}{2}\nu,\mu}(ax^{-1})$ $-\frac{1}{2} < \mathrm{Re}\ \mu < \frac{1}{4}$	$2ay^{-\frac{1}{2}}K_{2\mu}[(2ay)^{\frac{1}{2}}] \cdot$ $\cdot\{\cos(\pi\mu-\frac{1}{2}\pi\nu)J_{2\mu}[(2ay)^{\frac{1}{2}}] -$ $- \sin[\pi\mu-\frac{1}{2}\pi\nu)Y_{2\mu}[(2ay)^{\frac{1}{2}}]\}$

Chapter IV. Integral Transforms with Struve Functions as Kernel

4.1 General Formulas

	$f(x)$	$g(y;\nu) = \int\limits_0^\infty f(x)\,(xy)^{\frac{1}{2}}\mathbf{H}_\nu(xy)\,dx$
1.1	$\int\limits_0^\infty g(y)\,(xy)^{\frac{1}{2}}Y_\nu(xy)\,dy$	$g(y)$
1.2	$f(ax)$, $a>0$	$a^{-1}g(ya^{-1};\nu)$
1.3	$x^m f(x)$, $m=0,1,2,\cdots$	$y^{\frac{1}{2}-\nu}\left(\dfrac{d}{ydy}\right)^m [y^{m+\nu-\frac{1}{2}}g(y;m+\nu)]$
1.4	$x^{-\mu}f(x)$ Re $\mu>0$, Re $\nu+\sfrac{3}{2}>$Re μ	$2^{1-\mu}[\Gamma(\mu)]^{-1}y^{\frac{1}{2}-\nu}\int\limits_0^y \tau^{\frac{1}{2}-\mu+\nu}(y^2-\tau^2)^{\mu-1}\,\cdot$ $\cdot\,g(\tau;\nu-\mu)\,d\tau$

4.2 Transforms of Order Zero

	$f(x)$	$g(y) = \int_0^\infty f(x)(xy)^{\frac{1}{2}}\mathbf{H}_0(xy)\,dx$
2.1	$x^{\frac{1}{2}}(a^2-x^2)^{\mu-1} \quad y < a$ $\qquad\qquad 0 \quad y > a$ $\mathrm{Re}\ \mu > 0$	$2^{\mu-1}\Gamma(\mu)a^\mu y^{\frac{1}{2}-\mu}\mathbf{H}_\mu(ay)$
2.2	$x^{-\frac{1}{2}}e^{-ax}$	$2\pi^{-1}y^{\frac{1}{2}}(a^2+y^2)^{-\frac{1}{2}}\log[ya^{-1}+(1+y^2a^{-2})^{\frac{1}{2}}]$
2.3	$x^{\nu+\frac{1}{2}}e^{-ax^2}$ $\qquad \mathrm{Re}\ \nu > -\tfrac{3}{2}$	$-i2^{-\nu-1}a^{-\nu-1}y^{\nu+\frac{1}{2}}\mathrm{Erf}(\tfrac{1}{2}iya^{-\frac{1}{2}})$
2.4	$x^{\frac{1}{2}}(a^2+x^2)^{-\frac{1}{2}} \cdot$ $\cdot\log[\tfrac{a}{x} +(1+\tfrac{a^2}{x^2})^{\frac{1}{2}}]$	$-\tfrac{1}{2}\pi y^{-\frac{1}{2}}e^{-ay}$
2.5	$x^{\frac{1}{2}}\log[\tfrac{a}{x} -(\tfrac{a^2}{x^2}-1)^{\frac{1}{2}}]$ $\qquad\qquad x < a$ $\qquad\qquad 0 \quad x > a$	$y^{-\frac{3}{2}}\sin(ay)$
2.6	$x^{-\frac{1}{2}}\sin(ax)$	$\qquad\qquad 0 \quad y < a$ $y^{\frac{1}{2}}(y^2-a^2)^{-\frac{1}{2}} \quad y > a$

	$f(x)$	$g(y) = \int_0^\infty f(x)(xy)^{\frac{1}{2}} \mathbf{H}_0(xy)\,dx$
2.7	$x^{-\frac{1}{2}}\cos(ax)$	$-2\pi^{-1}y^{\frac{1}{2}}(a^2-y^2)^{-\frac{1}{2}}\arcsin\left(\frac{y}{a}\right)\quad y < a$ $2\pi^{-1}y^{\frac{1}{2}}(y^2-a^2)^{-\frac{1}{2}}\log\left[\frac{y}{a}+\left(\frac{y^2}{a^2}-1\right)^{\frac{1}{2}}\right]$ $\hspace{6cm} y > a$
2.8	$x^{-3/2}\cos(ax)$	$0 \hspace{4cm} y > a$ $y^{-\frac{1}{2}}\arccos(ay^{-1}) \hspace{2cm} y > a$
2.9	$x^{-\nu-3/2}\cos(ax)$ $\mathrm{Re}\ \nu > -\,3/2$	$0 \hspace{4cm} y < a$ $(2\pi)^{\frac{1}{2}}(y^2-a^2)^{\frac{1}{2}\nu+\frac{1}{4}}P_{\nu-\frac{1}{2}}^{-\nu-\frac{1}{2}}(ay^{-1}) \quad y < a$
2.10	$x^{\frac{1}{2}}\sin(ax^2)$	$\frac{1}{2}a^{-1}y^{\frac{1}{2}}\{\sin z[C(z)+S(z)] +$ $\quad + \cos z[C(z) - S(z)]\}$ $z = \frac{1}{4}y^2a^{-1}$
2.11	$x^{-1}\cos(ax^{\frac{1}{2}})$	$\frac{1}{4}\pi a y^{-\frac{1}{2}}\{[J_{\frac{1}{4}}(\tfrac{1}{8}a^2y^{-2})]^2-[Y_{-\frac{1}{4}}(\tfrac{1}{8}a^2y^{-2})]^2\}$
2.12	$\begin{array}{l} 0 \quad x < a \\ x^{-\frac{1}{2}}(x^2-a^2)^{-\frac{1}{2}} \cdot \\ \cdot\cos[b(x^2-a^2)^{\frac{1}{2}}] \\ \hspace{2cm} x > a \end{array}$	$0 \hspace{3cm} y < b$

	$f(x)$	$g(y) = \int\limits_0^\infty f(x)(xy)^{\frac{1}{2}}\mathbf{H}_0(xy)\,dx$
2.13	$x^{\frac{1}{2}}(a^2-x^2)^{-\frac{1}{2}}\cdot$ $\cdot\log[\frac{a}{x}-(\frac{a^2}{x^2}-1)^{\frac{1}{2}}]$ $x < a$ $x^{\frac{1}{2}}(x^2-a^2)^{-\frac{1}{2}}\arcsin(ax^{-1})$ $x > a$	$\frac{1}{2}\pi y^{-\frac{1}{2}}\sin(ay)$
2.14	$x^{-\frac{1}{2}}J_0(a^2x^{-1})$	$y^{-\frac{1}{2}}[2\pi^{-1}K_0(2ay^{\frac{1}{2}}) - Y_0(2ay^{\frac{1}{2}})]$
2.15	$x^{-\frac{1}{2}}K_0(ax^{\frac{1}{2}})J_0(ax^{\frac{1}{2}})$	$(2\pi)^{-1}y^{-\frac{1}{2}}[K_0(\frac{1}{4}a^2y^{-1})]^2$
2.16	$x^{\frac{1}{2}}[K_\mu(ax)]^2$ $-\frac{3}{2} < \mathrm{Re}\ \mu < \frac{3}{2}$	$-\pi 2^{-\mu-1}a^{-2\mu}\sec(\pi\mu)y^{-\frac{1}{2}}z^{-1}\cdot$ $\cdot[(z+y)^{2\mu}+(z-y)^{2\mu}]$ $z = (4a^2+y^2)$
2.17	$x^{-\frac{1}{2}}K_1[a(ix)^{\frac{1}{2}}]\cdot$ $\cdot K_1[a(-ix)^{\frac{1}{2}}]$	$2^{-5/4}a^{-\frac{1}{2}}y^{-3/4}[\sin z\,\mathrm{Ci}(z)-\cos z\,\mathrm{si}(z)]$ $z = a(2y)^{\frac{1}{2}}$
2.18	$x^{\frac{1}{2}}K_0\{a[(b^2+x^2)^{\frac{1}{2}}+b]\}$ $K_0\ a[(b^2+x^2)^{\frac{1}{2}}-b]\}$	$-\pi y^{-\frac{1}{2}}(4a^2+y^2)^{-\frac{1}{2}}\exp[-b(4a^2+y^2)^{\frac{1}{2}}]$

	$f(x)$	$g(y) = \int\limits_0^\infty f(x)(xy)^{\frac{1}{2}}\mathbf{H}_0(xy)\,dx$
2.19	$\mathbf{H}_0(ax) - Y_0(ax)$	$y^{-\frac{1}{2}}(a+y)^{-1}$
2.20	$x^{\frac{1}{2}}[Y_0(ax^2)-\mathbf{H}_0(ax^2)]$	$a^{-1}y^{\frac{1}{2}}J_0(\frac{1}{4}y^2a^{-1})$
2.21	$x^{\frac{1}{2}}\exp(ax^2)\,\cdot$ $\cdot K_0(ax^2)$	$-\pi^{-1}2^{-\frac{1}{2}}a^{-\frac{1}{2}}y^{-\frac{1}{2}}\exp(\frac{1}{16}y^2a^{-1})\,\cdot$ $\cdot K_0(\frac{1}{16}y^2a^{-1})$
2.22	$x^{\frac{1}{2}}[\mathbf{L}_0(ax)-I_0(ax)]$	$-4\pi^{-2}ay^{-\frac{1}{2}}(a^2+y^2)^{-1}\log(ya^{-1})$

4.3 Elementary Functions

	$f(x)$	$g(y) = \int\limits_0^\infty f(x)\,(xy)^{\frac{1}{2}}\mathbf{H}_\nu(xy)\,dx$
3.1	$x^{-\frac{1}{2}}$ $-2 < \mathrm{Re}\ \nu < 0$	$-\cot(\tfrac{1}{2}\pi\nu)\,y^{\frac{1}{2}}$
3.2	$x^{\nu+\frac{1}{2}} \quad x < a$ $0 \qquad x > a$ $\mathrm{Re}\ \nu > -\tfrac{3}{2}$	$a^{\nu+1}y^{-\frac{1}{2}}\,\mathbf{H}_{\nu+1}(ay)$
3.3	$x^{\frac{1}{2}-\nu} \qquad x < a$ $0 \qquad\quad x > a$	$\pi^{-\frac{1}{2}}2^{1-\nu}[\Gamma(\tfrac{1}{2}+\nu)]^{-1}ay^{\nu-\frac{1}{2}} -$ $-a^{1-\nu}y^{-\frac{1}{2}}\mathbf{H}_{\nu-1}(ay)$
3.4	$x^{\mu-\frac{1}{2}}$ $\mathrm{Re}\ \mu<\tfrac{1}{2},\ -2<\mathrm{Re}\,(\mu+\nu)<0$	$2^\mu y^{-\mu-\frac{1}{2}}\tan[\tfrac{1}{2}\pi(\mu+\nu+1)]\ \cdot$ $\cdot\,[\Gamma(\tfrac{1}{2}+\tfrac{1}{2}\nu-\tfrac{1}{2}\mu)]^{-1}\Gamma(\tfrac{1}{2}+\tfrac{1}{2}\nu+\tfrac{1}{2}\mu)$
3.5	$x^{\mu-\frac{1}{2}} \qquad x < a$ $0 \qquad\quad x > a$ $\mathrm{Re}\,(\nu+\mu) > -2$	$\pi^{-\frac{1}{2}}2^{-\nu}[\Gamma(\tfrac{3}{2}+\nu)(2+\nu+\mu)]^{-1}a^{\mu+\nu+2}y^{\frac{3}{2}+\nu}\ \cdot$ $\cdot\,{}_2F_3(1,1+\tfrac{1}{2}\mu+\tfrac{1}{2}\nu;\tfrac{3}{2},\tfrac{3}{2}+\nu,2+\tfrac{1}{2}\mu+\tfrac{1}{2}\nu;-\tfrac{1}{4}a^2y^2)$
3.6	$x^{-\frac{1}{2}}(a^2+x^2)^{-1}\ ;\quad \nu = 1$	$\tfrac{1}{2}\pi a^{-1}y^{\frac{1}{2}}[I_1(ay) - \mathbf{L}_1(ay)]$

	$f(x)$	$g(y) = \int_0^\infty f(x)(xy)^{\frac{1}{2}}\mathbf{H}_\nu(xy)\,dx$
3.7	$x^{-\frac{1}{2}}(a^2+x^2)^{-1}$ $-2 < \mathrm{Re}\ \nu < 2$	$-\tfrac{1}{2}\pi a^{-1}\csc(\tfrac{1}{2}\pi\nu)y^{\frac{1}{2}}\mathbf{L}_\nu(ay)\ +$ $+\ (1-\nu^2)^{-1}\cot(\tfrac{1}{2}\pi\nu)y^{3/2}\ \cdot$ $\cdot\ {}_1F_2(1;\ \tfrac{3}{2}-\tfrac{1}{2}\nu,\ \tfrac{3}{2}+\tfrac{1}{2}\nu;\ \tfrac{1}{4}a^2y^2)$
3.8	$x^{\nu+\frac{1}{2}}(a^2+x^2)^{\mu-1}$ $\mathrm{Re}(\nu+\mu)<\tfrac{1}{2},\mathrm{Re}\ \nu>-\tfrac{3}{2}$ $\mathrm{Re}(\nu+2\mu) < \tfrac{3}{2}$	$2^{\mu-1}\pi a^{\mu+\nu}\{\Gamma(1-\mu)\cos[\pi(\mu+\nu)]\}^{-1}\ \cdot$ $\cdot\ y^{\frac{1}{2}-\mu}[I_{-\mu-\nu}(ay)\ -\ \mathbf{L}_{\mu+\nu}(ay)]$
3.9	$x^{\frac{1}{2}-\nu}(a^2+x^2)^{\mu-1}$ $\mathrm{Re}\ \mu<\tfrac{1}{2},\mathrm{Re}(2\mu-\nu) < \tfrac{3}{2}$	$\pi 2^{\mu-1}a^{\mu-\nu}[\Gamma(1-\mu)\cos(\pi\mu)]^{-1}y^{\frac{1}{2}-\mu}I_{\nu-\mu}(ay)\ +$ $+\ 2^{-\nu-2}a^{1+2\mu}\Gamma(-\tfrac{1}{2}-\mu)[\Gamma(1-\mu)\Gamma(\tfrac{3}{2}+\nu)]^{-1}\ \cdot$ $\cdot\ y^{3/2+\nu}{}_1F_2(1;\ \tfrac{3}{2}+\mu,\ \tfrac{3}{2}+\nu;\ \tfrac{1}{4}a^2y^2)$
3.10	$(a^2+x^2)^{-\frac{1}{2}}x^{-\frac{1}{2}}\ \cdot$ $\cdot\ [x+(a^2+x^2)^{\frac{1}{2}}]^{\nu+1}$ $-\tfrac{5}{2} < \mathrm{Re}\ \nu < 0$	$\pi^{\frac{1}{2}}a^{\nu+\frac{1}{2}}\csc(\pi\nu)\ \cdot$ $\cdot\ [\sinh(\tfrac{1}{2}ay)I_{\nu+\frac{1}{2}}(\tfrac{1}{2}ay)\ -$ $-\ \cosh(\tfrac{1}{2}ay)I_{-\nu-\frac{1}{2}}(\tfrac{1}{2}ay)]$
3.11	$x^{\frac{1}{2}+\nu}(a^2-x^2)^{-\nu-\frac{1}{2}},x<a$ $\qquad\qquad 0\qquad,x>a$ $-\tfrac{3}{2} < \mathrm{Re}\ \nu < \tfrac{1}{2}$	$-\pi^{-\frac{1}{2}}2^{-\nu}\Gamma(\tfrac{1}{2}-\nu)y^{\nu-\frac{1}{2}}\cos(ay)$

	$f(x)$	$g(y) = \int\limits_0^\infty f(x)(xy)^{\frac{1}{2}}\mathbf{H}_\nu(xy)\,dx$
3.12	$x^{\nu+\frac{1}{2}}(a^2-x^2)^{\mu-1} \quad x < a$ $0 \qquad x > a$ $\mathrm{Re}\ \mu > 0, \mathrm{Re}\ \nu > -\,{}^3/_2$	$2^{\mu-1}a^{\mu+\nu}y^{\frac{1}{2}-\mu}\Gamma(\mu)\mathbf{H}_{\mu+\nu}(ay)$
3.13	$x^{\frac{1}{2}+\nu}(a^2-x^2)^{-\nu-{}^3/_2} \quad x<a$ $0 \qquad x>a$ $-\,{}^3/_2 < \mathrm{Re}\ \nu < -\frac{1}{2}$	$\pi^{-\frac{1}{2}}2^{-1-\nu}\Gamma(-\frac{1}{2}-\nu)a^{-1}y^{\frac{1}{2}+\nu}\sin(ay)$
3.14	$x^{\frac{1}{2}-\nu}(a^2-x^2)^{\mu-1}, \quad x < a$ $0 \quad, \quad x > a$ $\mathrm{Re}\ \mu > 0$	$2^{-\nu}a^{\mu-\nu}\Gamma(\mu)[\Gamma(\frac{1}{2}+\nu)\Gamma(\frac{1}{2}+\mu)]^{-1}\cdot$ $\cdot\ y^{\frac{1}{2}-\mu}s_{\mu+\nu,\mu-\nu}(ay)$
3.15	$x^{\lambda-\frac{1}{2}}(a^2-x^2)^{\mu-1}, \quad x < a$ $0 \quad, \quad x > a$ $\mathrm{Re}\ \mu>0,\ \mathrm{Re}(\lambda+\nu)>-2$	$\pi^{-\frac{1}{2}}2^{-\nu-1}a^{2\mu+\nu+\lambda}\Gamma(\mu)\Gamma(1+\frac{1}{2}\lambda+\frac{1}{2}\mu)\cdot$ $\cdot[\Gamma({}^3/_2+\nu)\Gamma(1+\mu+\frac{1}{2}\lambda+\frac{1}{2}\nu)]^{-1}y^{\nu+{}^3/_2}\cdot$ $\cdot\ {}_2F_3(1,\frac{1}{2}\lambda+\frac{1}{2}\nu+1;{}^3/_2,{}^3/_2+\nu,\frac{1}{2}\lambda+\frac{1}{2}\nu+\mu+1;-\frac{1}{4}a^2y^2)$
3.16	$0 \qquad x < a$ $x^{-\nu-\frac{1}{2}}(x^2-a^2)^{-\nu-\frac{1}{2}} \quad x>a$ $-\frac{1}{2} < \mathrm{Re}\ \nu < \frac{1}{2}$	$2^{-\nu-1}\pi^{\frac{1}{2}}a^{-2\nu}\Gamma(\frac{1}{2}-\nu)[J_\nu(\frac{1}{2}ay)]^2y^{\frac{1}{2}+\nu}$

	$f(x)$	$g(y) = \int\limits_0^\infty f(x)(xy)^{\frac{1}{2}}\mathbf{H}_\nu(xy)\,dx$
3.17	$\begin{array}{ll} 0 & x < a \\ x^{\nu+\frac{1}{2}}(x^2-a^2)^m & x > a \\ m = 0, 1, 2, \cdots \\ \operatorname{Re} \nu < -2m-\frac{1}{2} \end{array}$	$(-1)^{m+1}2^m a^{m+\nu+1} y^{-m-\frac{1}{2}} m!\;\cdot$ $\cdot\;\mathbf{H}_{\nu+m+1}(ay)$
3.18	$\begin{array}{l} \begin{array}{ll} 0 & x < a \\ x^{\frac{1}{2}+\nu}(x^2-a^2)^{\mu-1}, & x>a \end{array} \\ \operatorname{Re}\mu>0, \operatorname{Re}(\mu+\nu) < \frac{1}{2} \\ \operatorname{Re}(2\mu+\nu) < \frac{3}{2} \end{array}$	$2^{\mu-1}a^{\mu+\nu}\Gamma(\mu)\sec[\pi(\mu+\nu)]\;\cdot$ $\cdot\,[\sin(\pi\mu)J_{-\mu-\nu}(ay)+\cos(\pi\nu)\mathbf{H}_{\mu+\nu}(ay)]$
3.19	$\begin{array}{l} \begin{array}{ll} 0 & x < a \\ x^{-\frac{1}{2}-\nu}(x^2-a^2)^{\mu-1} & x>a \end{array} \\ 0 < \operatorname{Re}\mu < 1 \\ \operatorname{Re}\nu>-\frac{5}{2}+2\operatorname{Re}\mu \end{array}$	$\pi 2^{2\mu-2-\nu}y^{\nu-2\mu+\frac{3}{2}}\;\cdot$ $\cdot\,[\Gamma(\tfrac{3}{2}-\mu)\Gamma(\tfrac{3}{2}+\nu-\mu)\sin(\pi\nu)]^{-1}\;\cdot$ $\cdot\,{}_1F_2(1-\mu;\tfrac{3}{2}-\mu,\tfrac{3}{2}+\nu-\mu;-\tfrac{1}{4}a^2y^2)$
3.20	$\begin{array}{ll} x^{\frac{1}{2}-\nu}(x^2-a^2)^\mu & x > a \\ 0 & x < a \end{array}$ $\operatorname{Re}\mu>\frac{1}{2}, \operatorname{Re}(2\mu-\nu)<-\frac{1}{2}$	$\pi^{-1}2^{-\nu}\Gamma(1+\mu)a^{1+\mu-\nu}y^{-\mu-\frac{1}{2}}\;\cdot$ $\cdot\,[\frac{\Gamma(-\frac{1}{2}-\mu)}{\Gamma(\frac{1}{2}+\nu)}\,\mathbf{s}_{\mu+\nu+1,\mu-\nu+1}(ay)\;+$ $+\;\pi 2^{\mu+\nu}\tan(\pi\mu)J_{\nu-\mu+1}(ay)]$

	$f(x)$	$g(y) = \int_0^\infty f(x)(xy)^{\frac{1}{2}} \mathbf{H}_\nu(xy)\,dx$
3.21	$x^{\mu-\frac{1}{2}} e^{-ax}$ $\mathrm{Re}(\mu+\nu) > -2$	$2^{-\nu}\pi^{-\frac{1}{2}} a^{-\mu-\nu-2} [\Gamma(\tfrac{3}{2}+\nu)]^{-1}\Gamma(2+\nu+\mu) y^{\nu+\frac{3}{2}} \cdot$ $\cdot {}_3F_2(1,\tfrac{1}{2}\mu+\tfrac{1}{2}\nu+1,\tfrac{1}{2}\mu+\tfrac{1}{2}\nu+\tfrac{3}{2};\tfrac{3}{2},\tfrac{3}{2}+\nu;-y^2 a^{-2})$
3.22	$x^{\mu+\frac{1}{2}}\exp(-ax^2)$ $\mathrm{Re}(\mu+\nu) > -3$	$\pi^{-\frac{1}{2}}2^{-\nu-1} a^{-\frac{1}{2}\mu-\frac{1}{2}\nu-\frac{3}{2}}\Gamma(\tfrac{1}{2}\mu+\tfrac{1}{2}\nu+\tfrac{3}{2}) \cdot$ $\cdot\, [\Gamma(\tfrac{3}{2}+\nu)]^{-1} \cdot$ $\cdot\, {}_2F_2(1,\tfrac{1}{2}\mu+\tfrac{1}{2}\nu+\tfrac{3}{2};\tfrac{3}{2},\tfrac{3}{2}+\nu;-\tfrac{1}{4}y^2 a^{-1})$
3.23	$x^{\nu+\frac{1}{2}} e^{-ax^2}$ $\mathrm{Re}\,\nu > -\tfrac{3}{2}$	$-i(2a)^{-\nu-1} y^{\nu+\frac{1}{2}}\mathrm{Erf}(\tfrac{1}{2}iya^{-\frac{1}{2}})$
3.24	$x^{-\nu-\frac{1}{2}}\sin(ax)$ $\mathrm{Re}\,\nu > -\tfrac{1}{2}$	$\quad\quad\quad 0 \qquad y < a$ $\pi^{\frac{1}{2}}2^{-\nu}[\Gamma(\tfrac{1}{2}+\nu)]^{-1} y^{\frac{1}{2}-\nu}(y^2-a^2)^{\nu-\frac{1}{2}}\quad y > a$
3.25	$x^{-\nu-\frac{3}{2}}\cos(ax)$ $\mathrm{Re}\,\nu > -\tfrac{3}{2}$	$\quad\quad\quad 0 \qquad y < a$ $(2\pi)^{\frac{1}{2}}(y^2-a^2)^{\frac{1}{2}\nu+\frac{1}{4}} P^{-\nu-\frac{1}{2}}_{\nu-\frac{1}{2}}(ay^{-1})\quad y > a$

	$f(x)$	$g(y) = \int\limits_0^\infty f(x)(xy)^{\frac{1}{2}} \mathbf{H}_\nu(xy)\,dx$
3.26	$x^{\frac{1}{2}}(a^2-x^2)^{-\frac{1}{2}}\cos[(\nu+1)\arccos\frac{x}{a}]$ $x < a$ $\quad 0 \qquad\qquad x > a$ $\mathrm{Re}\ \nu > -2$	$\pi^{\frac{1}{2}}a^{-\frac{1}{2}}\sin(\tfrac{1}{2}ay)\,J_{\nu+\frac{1}{2}}(\tfrac{1}{2}ay)$
3.27	$x^{-\nu-\frac{1}{2}}(b^2+x^2)^{-1}\sin(ax)$ $\mathrm{Re}\ \nu > -\tfrac{5}{2}$	$\tfrac{1}{2}b^{-\nu-1}y^{\frac{1}{2}}e^{-ab}\mathbf{L}_\nu(by) \quad y > a$

4.4 Higher Transcendental Functions

	$f(x)$	$g(y) = \int_0^\infty f(x)(xy)^{\frac{1}{2}}\mathbf{H}_\nu(xy)\,dx$
4.1	$x^{\frac{1}{2}}(a^2+x^2)^{-\frac{1}{2}-\frac{1}{2}\mu} \cdot$ $\cdot Q_\mu^{-\nu}[a(a^2+x^2)^{\frac{1}{2}}]$ $\mathrm{Re}(\nu+\mu) > -1$	$-\frac{1}{2}\pi\,[\Gamma(\nu+\mu+1)]^{-1}y^{\mu-\frac{1}{2}}e^{-ay}$
4.2	$J_{\nu+\frac{1}{2}}(ax)$ $-\frac{3}{2} < \mathrm{Re}\ \nu < 1$	$\begin{array}{ll} 0 & y < a \\ (\frac{1}{2}\pi)^{-\frac{1}{2}}(ya^{-1})^{\nu+\frac{1}{2}}(y^2-a^2)^{-\frac{1}{2}} & y > a \end{array}$
4.3	$x^{-\frac{1}{2}}Y_{\nu+1}(ax)$ $-\frac{3}{2} < \mathrm{Re}\ \nu < \frac{3}{2}$	$\begin{array}{ll} 0 & y < a \\ -a^{-\nu-1}y^{\frac{1}{2}+\nu} & y > a \end{array}$
4.4	$x^{\mu-\nu+\frac{1}{2}}Y_\mu(ax)$ $-\frac{3}{2} < \mathrm{Re}\ \mu < \frac{1}{2}$ $\mathrm{Re}(\mu-\nu) < 0$	$\begin{array}{ll} 0 & y < a \\ 2^{1+\mu-\nu}a^\mu[\Gamma(\nu-\mu)]^{-1}y^{\frac{1}{2}-\nu}(y^2-a^2)^{\nu-\mu-1} & \end{array}$
4.5	$x^{\frac{1}{2}-\mu}[\sin(\pi\mu)J_{\mu+\nu}(ax) +$ $+\cos(\pi\mu)Y_{\mu+\nu}(ax)]$ $1 < \mathrm{Re}\ \mu < \frac{3}{2}$ $\mathrm{Re}\ \nu > -\frac{3}{2},\ \mathrm{Re}(\nu-\mu) < \frac{1}{2}$	$\begin{array}{ll} 0 & y < a \\ 2^{1-\mu}a^{-\mu-\nu}[\Gamma(\mu)]^{-1}y^{\frac{1}{2}+\nu}(y^2-a^2)^{\mu-1} & \\ & y > a \end{array}$

	$f(x)$	$g(y) = \int\limits_{0}^{\infty} f(x)(xy)^{\frac{1}{2}}\mathbf{H}_{\nu}(xy)\,dx$
4.6	$x^{-\frac{1}{2}}[\cos(ax)J_{\nu}(ax) -$ $-\sin(ax)Y_{\nu}(ax)]$ $-1 < \mathrm{Re}\ \nu < 1$	$\begin{array}{ll} 0 & y < 2a \\ a^{-\frac{1}{2}}P_{\nu-\frac{1}{2}}(\tfrac{1}{2}ya^{-1}) & y > 2a \end{array}$
4.7	$J_{\nu}(ax)Y_{\nu}(ax)$ $-\tfrac{5}{6} < \mathrm{Re}\ \nu < \tfrac{1}{2}$	$\begin{array}{l} \pi^{-\frac{1}{2}}[\Gamma(\tfrac{1}{2}+\nu)]^{-1}a^{-2\nu}2^{1-3\nu}y^{\nu-\frac{1}{2}}(4a^2-y^2)^{\nu-\frac{1}{2}} \\ \qquad\qquad\qquad\qquad\qquad\qquad y < 2a \\ 0 \qquad\qquad\qquad\qquad\qquad\quad y > 2a \end{array}$
4.8	$x^{\mu+\frac{1}{2}}[J_{\nu}(ax)J_{\mu}(ax) -$ $-Y_{\nu}(ax)Y_{\mu}(ax)]$ $\mathrm{Re}\ \nu > -\tfrac{3}{2},\ -\tfrac{3}{2} < \mathrm{Re}\ \mu < \tfrac{1}{2}$ $-\tfrac{3}{2} < \mathrm{Re}(\nu+\mu) < 0$	$\begin{array}{l} 0 \qquad\qquad\qquad\qquad y < 2a \\ 2^{\mu+1}\pi^{-\frac{1}{2}}a^{\mu-\frac{1}{2}}y^{-\mu-\frac{1}{2}}\ \cdot \\ \quad\cdot\ (y^2-4a^2)^{-\frac{1}{2}\mu-\frac{1}{4}}P_{\nu-\frac{1}{2}}^{\mu+\frac{1}{2}}(\tfrac{1}{2}ya^{-1}) \quad y > 2a \end{array}$
4.9	$x^{\frac{1}{2}-\nu}[J_{\nu}(ax)J_{-\nu}(ax) -$ $-Y_{\nu}(ax)Y_{-\nu}(ax)]$ $-\tfrac{1}{2} < \mathrm{Re}\ \nu < \tfrac{1}{2}$	$\begin{array}{l} 0 \qquad\qquad\qquad\qquad y < 2a \\ \pi^{-\frac{1}{2}}2^{2-3\nu}a^{-2\nu}[\Gamma(\tfrac{1}{2}+\nu)]^{-1}\ \cdot \\ \quad\cdot\ y^{\nu-\frac{1}{2}}(y^2-4a^2)^{\nu-\frac{1}{2}} \qquad y > 2a \end{array}$
4.10	$x^{\nu+\frac{1}{2}}J_{\nu}(ax)Y_{\nu}(ax)$ $-1 < \mathrm{Re}\ \nu < \tfrac{1}{2}$	$\begin{array}{l} \pi^{-\frac{3}{2}}2^{-\nu-2}a^{-2\nu-3}\Gamma(2\nu+\tfrac{3}{2})[\Gamma(\nu+2)]^{-1}y^{\nu+\frac{3}{2}}\ \cdot \\ \quad\cdot\ {}_2F_1(1,2\nu+\tfrac{3}{2};\nu+2;\tfrac{1}{4}y^2a^{-1}) \quad y < 2a \end{array}$

	$f(x)$	$g(y) = \int\limits_0^\infty f(x)(xy)^{\frac{1}{2}}\mathbf{H}_\nu(xy)\,dx$
4.11	$x^{\nu+\frac{1}{2}}\{[J_\nu(ax)]^2-[Y_\nu(ax)]^2\}$ $-1 < \operatorname{Re}\nu < \tfrac{1}{2}$	$\begin{aligned}&0 \qquad y < 2a\\[4pt] &\pi^{-\frac{1}{2}}2^{2+3\nu}a^{2\nu}[\Gamma(\tfrac{1}{2}-\nu)]^{-1}y^{-\nu-\frac{1}{2}}(y^2-4a^2)^{-\nu-\frac{1}{2}}\\[4pt] &\qquad\qquad y > 2a\end{aligned}$
4.12	$x^{\frac{1}{2}}\{[J_{\frac{1}{2}\nu}(ax)]^2-[Y_{\frac{1}{2}\nu}(ax)]^2\}$ $-\tfrac{3}{2} < \operatorname{Re}\nu < 1$	$\begin{aligned}&0 \qquad y < 2a\\[4pt] &4\pi^{-1}y^{-\frac{1}{2}}(y^2-4a^2)^{-\frac{1}{2}} \qquad y > 2a\end{aligned}$
4.13	$x^{\frac{1}{2}}[J_{\frac{1}{2}\nu+\frac{1}{2}\mu}(ax)\cdot$ $\cdot\,J_{\frac{1}{2}\nu-\frac{1}{2}\mu}(ax)-$ $-Y_{\frac{1}{2}\nu+\frac{1}{2}\mu}(ax)Y_{\frac{1}{2}\nu-\frac{1}{2}\mu}(ax)]$ $-\tfrac{3}{2} < \operatorname{Re}\nu < 1$	$\begin{aligned}&0 \qquad y < 2a\\[4pt] &2\pi^{-1}(2a)^{-\mu}y^{-\frac{1}{2}}(y^2-4a^2)^{-\frac{1}{2}}\\[4pt] &\cdot\{[y+(y^2-4a^2)^{\frac{1}{2}}]^\mu+[y-(y^2-4a^2)^{\frac{1}{2}}]^\mu\} \quad y>2a\end{aligned}$
4.14	$x^{\frac{1}{2}}J_{\frac{1}{2}\nu+\frac{1}{2}}(ax^2)$ $-2 < \operatorname{Re}\nu < 1$	$\tfrac{1}{16}\,a^{-2}y^{-\frac{5}{2}}J_{\frac{1}{2}\nu-\frac{1}{2}}(\tfrac{1}{4}y^2a^{-1})$
4.15	$x^{-\frac{1}{2}}J_{2\nu}(ax^{\frac{1}{2}})$ $-1 < \operatorname{Re}\nu < \tfrac{5}{4}$	$-y^{-\frac{1}{2}}Y_{\nu+1}(\tfrac{1}{4}a^2y^{-1})$
4.16	$x^{\frac{1}{2}}(\{J_{\frac{1}{2}\nu}[b(z-a)]\}^2-$ $-x^{\frac{1}{2}}\{Y_{\frac{1}{2}\nu}[b(z+a)]\}^2$ $-\tfrac{3}{2} < \operatorname{Re}\nu < 1$ $z = (a^2+x^2)^{\frac{1}{2}}$	$\begin{aligned}&-4\pi^{-1}y^{-\frac{1}{2}}(4b^2-y^2)^{-\frac{1}{2}}\sin[a(4b^2-y^2)^{\frac{1}{2}}] \quad y<2b\\[4pt] &4\pi^{-1}y^{-\frac{1}{2}}(y^2-4b^2)^{-\frac{1}{2}}\exp[-a(y^2-4b^2)^{\frac{1}{2}}] \quad y>2b\end{aligned}$

	$f(x)$	$g(y) = \int\limits_0^\infty f(x)(xy)^{\frac{1}{2}}\mathbf{H}_\nu(xy)\,dx$
4.17	$x^{\frac{1}{2}}K_\nu(ax)$ $\mathrm{Re}\ \nu > -\ ^3\!/_2$	$-a^{1-\nu}y^{\nu-\frac{1}{2}}(a^2+y^2)^{-1}$
4.18	$x^{\mu+\nu+\frac{1}{2}}K_\mu(ax)$ $\mathrm{Re}\ \nu > -\ ^3\!/_2,\ \mathrm{Re}(\mu+\nu) > -\ ^3\!/_2$	$\pi^{-\frac{1}{2}}2^{\mu+\nu+1}a^{-\mu-2\nu-3}\Gamma(\nu+\mu+^3\!/_2)\ \cdot$ $\cdot\ y^{^3\!/_2+\nu}{}_2F_1(1,\,^3\!/_2+\mu+\nu;\,^3\!/_2;\,-y^2a^{-2})$
4.19	$x^{\mu-\nu+\frac{1}{2}}K_\mu(ax)$ $\mathrm{Re}\ \mu > -\ ^3\!/_2$	$2^{\mu-\nu}a^{-\mu-3}\Gamma(^3\!/_2+\mu)\,[\Gamma(^3\!/_2+\nu)]^{-1}y^{^3\!/_2+\nu}\ \cdot$ $\cdot\ {}_2F_1(1,\,^3\!/_2+\mu;\,^3\!/_2+\nu;\,-y^2a^{-2})$
4.20	$x^{\sigma-\frac{1}{2}}K_\mu(ax)$ $\mathrm{Re}(\sigma+\nu) > |\mathrm{Re}\ \mu|-2$	$\pi^{-\frac{1}{2}}2^\sigma a^{-\nu-\sigma-2}\Gamma(1+\tfrac{1}{2}\sigma+\tfrac{1}{2}\nu+\tfrac{1}{2}\mu)\ \cdot$ $\cdot\ \Gamma(1+\tfrac{1}{2}\sigma+\tfrac{1}{2}\nu-\tfrac{1}{2}\mu)\,[\Gamma(^3\!/_2+\nu)]^{-1}y^{^3\!/_2+\nu}\ \cdot$ $\cdot\ {}_3F_2(1,1+\tfrac{1}{2}\nu+\tfrac{1}{2}\sigma+\tfrac{1}{2}\mu,1+\tfrac{1}{2}\nu+\tfrac{1}{2}\sigma-\tfrac{1}{2}\mu;$ $^3\!/_2,\,^3\!/_2+\nu;\,-y^2a^{-2})$
4.21	$x^{\nu+\frac{1}{2}}[K_\nu(ax)]^2$ $\mathrm{Re}\ \nu > -\ ^3\!/_4$	$\pi^{\frac{1}{2}}2^{-\nu-3}a^{-2\nu-3}\Gamma(^3\!/_2+2\nu)\,[\Gamma(2+\nu)]^{-1}\ \cdot$ $\cdot\ y^{^3\!/_2+\nu}{}_2F_1(1,2\nu+^3\!/_2;\,\nu+2;\,-\tfrac{1}{4}y^2a^{-2})$
4.22	$x^{\frac{1}{2}}[K_\mu(ax)]^2$ $\nu = 0$ $-\ ^3\!/_2 < \mathrm{Re}\ \mu < \ ^3\!/_2$	$-2^{-\mu-1}\pi a^{-2\mu}\sec(\pi\mu)y^{-\frac{1}{2}}(y^2+4a^2)^{-\frac{1}{2}}\ \cdot$ $\cdot\ \{[(y^2+4a^2)^{\frac{1}{2}}+y]^{2\mu}+[(y^2+4a^2)^{\frac{1}{2}}-y]^{2\mu}\}$

	$f(x)$	$g(y) = \int\limits_0^\infty f(x)(xy)^{\frac{1}{2}}\mathbf{H}_\nu(xy)\,dx$
4.23	$x^{-\nu-\frac{1}{2}}K_0(ax)K_1(ax)$	$2^{-\nu-3}a^{-2}\pi^{\frac{3}{2}}[\Gamma(\frac{3}{2}+\nu)]^{-1}y^{\frac{3}{2}+\nu}$. $\cdot\,_2F_1(\frac{1}{2},\frac{1}{2};\frac{3}{2}+\nu;-\frac{1}{4}y^2a^{-2})$
4.24	$x^{-\nu-\frac{1}{2}}K_\nu(ax)K_{\nu+1}(ax)$ $\mathrm{Re}\ \nu < \frac{1}{2}$	$\pi^{\frac{1}{2}}2^{-\nu-2}a^{-2}\Gamma(\frac{1}{2}-\nu)y^{\frac{3}{2}+\nu}$. $\cdot\,_2F_1(\frac{1}{2},\frac{1}{2}-\nu;\frac{3}{2};-\frac{1}{4}y^2a^{-2})$
4.25	$x^{\sigma-\frac{5}{2}}K_\lambda(ax)K_\mu(ax)$ $\mathrm{Re}(\sigma+\nu)>\lvert\mathrm{Re}\lambda\rvert+\lvert\mathrm{Re}\ \mu\rvert$	$\pi^{-\frac{1}{2}}2^{\sigma-3}a^{-\sigma-\mu}[\Gamma(\frac{3}{2}+\nu)\Gamma(\sigma+\nu)]^{-1}$. $\cdot\,\Gamma(\frac{1}{2}\sigma+\frac{1}{2}\nu+\frac{1}{2}\lambda+\frac{1}{2}\mu)\Gamma(\frac{1}{2}\sigma+\frac{1}{2}\nu+\frac{1}{2}\lambda-\frac{1}{2}\mu)$. $\cdot\,\Gamma(\frac{1}{2}\sigma+\frac{1}{2}\nu-\frac{1}{2}\lambda+\frac{1}{2}\mu)\Gamma(\frac{1}{2}\sigma+\frac{1}{2}\nu-\frac{1}{2}\lambda-\frac{1}{2}\mu)$. $\cdot\,_5F_4(1,\frac{1}{2}\sigma+\frac{1}{2}\nu+\frac{1}{2}\lambda+\frac{1}{2}\mu,\frac{1}{2}\sigma+\frac{1}{2}\nu+\frac{1}{2}\lambda-\frac{1}{2}\mu,$ $\frac{1}{2}\sigma+\frac{1}{2}\nu-\frac{1}{2}\lambda+\frac{1}{2}\mu,\frac{1}{2}\sigma+\frac{1}{2}\nu-\frac{1}{2}\lambda-\frac{1}{2}\mu;\frac{3}{2},\frac{3}{2}+\nu,$ $\frac{1}{2}\sigma+\frac{1}{2}\nu,\frac{1}{2}\sigma+\frac{1}{2}\nu+\frac{1}{2};-\frac{1}{4}y^2a^{-2})$
4.26	$x^{\nu+\frac{3}{2}}K_\nu(ax)K_{\nu+1}(ax)$ $\mathrm{Re}\ \nu > -\frac{5}{4}$	$\pi^{\frac{1}{2}}2^{-\nu-3}a^{-2\nu-4}\Gamma(\frac{5}{2}+2\nu)[\Gamma(2+\nu)]^{-1}y^{\frac{3}{2}+\nu}$. $\cdot\,_2F_1(1,\frac{5}{2}+2\nu;2+\nu;-\frac{1}{4}y^2a^{-2})$
4.27	$x^{\sigma-\frac{5}{2}}\exp(-\frac{1}{2}a^2x^2)$. $\cdot\,K_\mu(\frac{1}{2}a^2x^2)$ $\mathrm{Re}(\sigma+\nu)>2\lvert\mathrm{Re}\ \mu\rvert$	$2^{-\nu-1}a^{-\nu-\sigma}\Gamma(\frac{1}{2}\nu+\frac{1}{2}\sigma+\mu)\Gamma(\frac{1}{2}\nu+\frac{1}{2}\sigma-\mu)$. $\cdot\,[\Gamma(\frac{3}{2}+\nu)\Gamma(\frac{1}{2}\nu+\frac{1}{2}\sigma)]^{-1}y^{\frac{3}{2}+\nu}$. $\cdot\,_3F_3(1,\frac{1}{2}\nu+\frac{1}{2}\sigma+\mu,\frac{1}{2}\nu+\frac{1}{2}\sigma-\mu;\frac{3}{2},\frac{3}{2}+\nu,$ $\frac{1}{2}\nu+\frac{1}{2}\sigma;-\frac{1}{4}y^2a^{-2})$

	$f(x)$	$g(y) = \int\limits_0^\infty f(x)(xy)^{\frac{1}{2}}\mathbf{H}_\nu(xy)\,dx$
4.28	$x^{\frac{1}{2}}\exp(\tfrac{1}{8}a^2x^2)\cdot$ $\cdot K_{\frac{1}{2}\nu}(\tfrac{1}{8}a^2x^2)$ $-\tfrac{3}{2} < \mathrm{Re}\ \nu < 0$	$2\pi^{-\frac{1}{2}}a^{-1-\frac{1}{2}\nu}\Gamma(-\tfrac{1}{2}\nu)\cos(\tfrac{1}{2}\pi\nu)y^{\frac{1}{2}\nu-\frac{1}{2}}\cdot$ $\cdot\exp(\tfrac{1}{2}y^2a^{-2})W_{\frac{1}{4}\nu,\,\frac{1}{2}+\frac{1}{4}\nu}(y^2a^{-2})$
4.29	$K_{2\nu-1}(ax^{\frac{1}{2}})$ $\mathrm{Re}\ \nu > -1$	$\pi^{-1}2^\nu a\Gamma(1+\nu)y^{-\frac{3}{2}}S_{-\nu-2,\,\nu-1}(\tfrac{1}{4}a^2y^{-1})$
4.30	$x^{-\frac{1}{2}}K_{2\nu}(2ax^{\frac{1}{2}})$ $\mathrm{Re}\ \nu > -1$	$\pi^{-1}2^\nu\Gamma(1+\nu)y^{-\frac{1}{2}}S_{-\nu-1,\,\nu}(a^2y^{-1})$
4.31	$x^{\frac{1}{2}}K_{2\nu}(2ax^{\frac{1}{2}})$ $\mathrm{Re}\ \nu > -2$	$\pi^{-1}2^\nu a^2\Gamma(2+\nu)y^{-\frac{5}{2}}S_{-\nu-3,\,\nu}(a^2y^{-1})$
4.32	$x^{-\frac{1}{2}}[2\pi^{-1}K_{2\nu}(2ax^{\frac{1}{2}}) +$ $+Y_{2\nu}(2ax^{\frac{1}{2}})]$ $-\tfrac{1}{2} < \mathrm{Re}\ \nu < \tfrac{1}{2}$	$y^{-\frac{1}{2}}J_\nu(a^2y^{-1})$
4.33	$x^{\frac{1}{2}-\nu}[J_{2\nu}(ax^{\frac{1}{2}}) -$ $-J_{-2\nu}(ax^{\frac{1}{2}})]K_{2\nu}(ax^{\frac{1}{2}})$ $-\tfrac{3}{2} < \mathrm{Re}\ \nu < \tfrac{3}{2}$	$\pi^{-\frac{1}{2}}2^\nu a^{1-2\nu}y^{2\nu-2}\sin(\pi\nu)\cdot$ $\cdot K_{\nu+\frac{1}{2}}(\tfrac{1}{2}a^2y^{-1})$

	$f(x)$	$g(y) = \int\limits_{0}^{\infty} f(x) \, x^{\frac{1}{2}} \mathbf{H}_\nu(xy)\,dx$
4.34	$x^{\frac{1}{2}} Y_\nu(ax^{\frac{1}{2}}) K_\nu(ax^{\frac{1}{2}})$ $\mathrm{Re}\ \nu > -\,{}^{3}\!/_{2}$	$\tfrac{1}{2} y^{-3/2} \exp(-\tfrac{1}{2} a^2 y^{-1})$
4.35	$x^{\nu-\frac{1}{2}} Y_{2\nu-1}(ax^{\frac{1}{2}})\ \cdot$ $\cdot K_{2\nu-1}(ax^{\frac{1}{2}})$ $\mathrm{Re}\ \nu > -\tfrac{1}{4}$	$\pi^{-\frac{1}{2}} 2^{-\nu} a^{2\nu-1} y^{-2\nu} K_{\nu-\frac{1}{2}}(\tfrac{1}{2} a^2 y^{-1})$
4.36	$x^{\nu+\frac{1}{2}} Y_{2\nu}(ax^{\frac{1}{2}}) K_{2\nu}(ax^{\frac{1}{2}})$ $\mathrm{Re}\ \nu > -\,{}^{3}\!/_{4}$	$\pi^{-\frac{1}{2}} 2^{-\nu-1} a^{2\nu+1} y^{-2\nu-2} K_{\nu-\frac{1}{2}}(\tfrac{1}{2} a^2 y^{-1})$
4.37	$x^{-\frac{1}{2}} \{\cos[\tfrac{1}{2}\pi(\mu-\nu)]\ \cdot$ $\cdot J_\mu(ax^{\frac{1}{2}}) -$ $-\sin[\tfrac{1}{2}\pi(\mu-\nu)] Y_\mu(ax^{\frac{1}{2}})\}$ $\cdot K_\mu(ax^{\frac{1}{2}})$ $\mathrm{Re}\ \nu > |\mathrm{Re}\ \mu| - 2$	$a^{-2} y^{\frac{1}{2}} W_{\frac{1}{2}\nu,\,\frac{1}{2}\mu}(\tfrac{1}{2} a^2 y^{-1})\ \cdot$ $\cdot W_{-\frac{1}{2}\nu,\,\frac{1}{2}\mu}(\tfrac{1}{2} a^2 y^{-1})$
4.38	$x^{\nu-\frac{1}{2}} K_{2\nu-1}[a(ix)^{\frac{1}{2}}]\ \cdot$ $\cdot K_{2\nu-1}[a(-ix)^{\frac{1}{2}}]$ $\mathrm{Re}\ \nu > -\tfrac{1}{4}$	$\pi^{-\frac{1}{2}} 2^{3\nu-1} \Gamma(1+\nu)\,\Gamma(\tfrac{1}{2}+2\nu)\,y^{-\nu-\frac{1}{2}}\ \cdot$ $\cdot\, S_{-\frac{1}{2}-3\nu,\,\nu-\frac{1}{2}}[a(2y)^{\frac{1}{2}}]$

	$f(x)$	$g(y) = \int\limits_0^\infty f(x)(xy)^{\frac{1}{2}}\mathbf{H}_\nu(xy)\,dx$
4.39	$x^{\frac{1}{2}}Y_\nu\{b[(a^2+y^2)^{\frac{1}{2}}-a]\}\cdot$ $\cdot K_\nu\{b[(a^2+y^2)^{\frac{1}{2}}+a]\}$ $-1 < \mathrm{Re}\ \nu < 3$	$\tfrac{1}{2}y^{-3/2}\exp(-ay-\tfrac{1}{2}by^{-1})$
4.40	$x^{-\frac{1}{2}}\mathbf{H}_\nu(a^2x^{-1})$ $\mathrm{Re}\ \nu > -\,{}^3\!/_2$	$-y^{-\frac{1}{2}}J_{2\nu}(2ay^{\frac{1}{2}})$
4.41	$x^{-3/2}\mathbf{H}_{\nu-1}(a^2x^{-1})$ $\mathrm{Re}\ \nu > -\tfrac{1}{2}$	$-a^{-1}J_{2\nu-1}(2ay^{\frac{1}{2}})$
4.42	$x^{-\frac{1}{2}}[J_{-\nu}(a^2x^{-1})\ +$ $+\sin(\pi\nu)\mathbf{H}_\nu(a^2x^{-1})]$ $-{}^3\!/_2 < \mathrm{Re}\ \nu < 0$	$y^{-\frac{1}{2}}[2\pi^{-1}K_{2\nu}(2ay^{\frac{1}{2}}) - Y_{2\nu}(2ay^{\frac{1}{2}})]$
4.43	$x^{-\frac{1}{2}}J_{-\nu}(a^2x^{-1})$	$y^{-\frac{1}{2}}[\sin(\pi\nu)J_{2\nu}(2ay^{\frac{1}{2}}) - Y_{2\nu}(2ay^{\frac{1}{2}})\ +$ $+\ 2\pi^{-1}K_{2\nu}(2ay^{\frac{1}{2}})]$

	$f(x)$	$g(y) = \int\limits_0^\infty f(x)(xy)^{\frac{1}{2}}\mathbf{H}_\nu(xy)\,dx$
4.44	$x^{-\nu-\frac{1}{2}}\exp(-\tfrac{1}{4}x^2)\cdot$ $\cdot[D_\mu(x)-D_\mu(-x)]$ $\mathrm{Re}\ \mu>-1,\ \mathrm{Re}(\mu+\nu)>-\tfrac{3}{2}$	$2^{2+\frac{1}{4}\mu+\frac{1}{2}\nu}\Gamma(\tfrac{1}{2}+\tfrac{1}{2}\mu)[\Gamma(1+\nu+\tfrac{1}{2}\mu)]^{-1}\cdot$ $\cdot\sin(\tfrac{1}{2}\pi\mu)y^{\frac{1}{2}\mu-\frac{1}{2}}\exp(-\tfrac{1}{4}y^2)\cdot$ $\cdot M_{\frac{1}{4}\mu-\frac{1}{2}\nu,\,\frac{1}{4}\mu+\frac{1}{2}\nu}(\tfrac{1}{2}y^2)$
4.45	$x^{2\lambda}\exp(-\tfrac{1}{4}x^2)\cdot$ $\cdot W_{k,\mu}(\tfrac{1}{2}x^2)$ $\mathrm{Re}(2\lambda+\nu)>2|\mathrm{Re}\ \mu|-\tfrac{7}{2}$	$2^{\frac{1}{4}-\lambda-\frac{1}{2}\nu}\pi^{-\frac{1}{2}}\Gamma(\tfrac{7}{4}+\tfrac{1}{2}\nu+\lambda+\mu)\cdot$ $\cdot\Gamma(\tfrac{7}{4}+\tfrac{1}{2}\nu+\lambda-\mu)\cdot$ $\cdot[\Gamma(\tfrac{3}{2}+\nu)\Gamma(\tfrac{9}{4}+\lambda-k-\tfrac{1}{2}\nu)]^{-1}y^{\frac{3}{2}+\nu}$ $\cdot {}_3F_3(1,\tfrac{7}{4}+\tfrac{1}{2}\nu+\lambda+\mu,\tfrac{7}{4}+\tfrac{1}{2}\nu+\lambda-\mu;$ $\tfrac{3}{2},\tfrac{3}{2}+\nu,\tfrac{9}{4}+\lambda-k+\tfrac{1}{2}\nu;-\tfrac{1}{2}y^2)$
4.46	$x^{-\frac{1}{2}}\exp(\tfrac{1}{2}x^2)\cdot$ $\cdot W_{-\frac{1}{2}\nu-\frac{1}{2},\,\frac{1}{2}\nu}(x^2)$ $\mathrm{Re}\ \nu>-1$	$2^{-\nu-1}\pi y^{\frac{1}{2}+\nu}\exp(\tfrac{1}{4}y^2)\mathrm{Erfc}(\tfrac{1}{2}y)$
4.47	$x^{-\frac{1}{2}}\exp(\tfrac{1}{4}x^2)\cdot$ $\cdot W_{k,\frac{1}{2}\nu}(\tfrac{1}{2}x^2)$ $-\tfrac{3}{2}<\mathrm{Re}\ \nu<-2\,\mathrm{Re}\ k$ $\mathrm{Re}\ k<\tfrac{1}{4}$	$2^{\frac{1}{2}k-\frac{1}{4}\nu}\Gamma(-k-\tfrac{1}{2}\nu)\cdot$ $\cdot[\Gamma(\tfrac{1}{2}-k+\tfrac{1}{2}\nu)\Gamma(\tfrac{1}{2}-k-\tfrac{1}{2}\nu)]^{-1}y^{\frac{1}{2}\nu-k-\frac{1}{2}}\cdot$ $\cdot\exp(\tfrac{1}{4}y^2)W_{\frac{1}{2}k+\frac{1}{4}\nu,\,\frac{1}{2}k+\frac{1}{4}\nu+\frac{1}{2}}(\tfrac{1}{2}y^2)$

Chapter V. Kontorovich-Lebedev Transforms

The pair of inversian formulas

$$(1) \quad g(y) = \int_0^\infty f(x) K_{ix}(y) \, dx$$

$$(2) \quad f(x) = 2\pi^{-2} x \sinh(\pi x) \int_0^\infty g(y) K_{ix}(y) y^{-1} dy$$

was given by Kontorovich and Lebedev (1938, 1939) in connection with the solution of certain boundary value problems of the wave equation. The mathematical theory was developed by Lebedev (1946, 1949). Here $K_{ix}(y)$ is the modified Hankel function given for instance by

$$K_{ix}(y) = \int_0^\infty \exp(-y \cosh t) \cos(xt) \, dt$$

The inversion formulas (1), (2) can also be applied to the K-transform as displayed in Chapter II

$$g(\nu, y) = \int_0^\infty f(x) (xy)^{\frac{1}{2}} K_\nu(xy) \, dx$$

or, with ν replaced by $i\nu$ and $y = 1$

$$(3) \quad \int_0^\infty f(x) x^{\frac{1}{2}} K_{i\nu}(x) \, dx = g(i\nu, 1)$$

If now, ν is regarded to be the transformation parameter, the inversion of (3) gives by (1) and (2)

$$(4) \quad f(x) = 2\pi^{-2} y^{-3/2} \int_0^\infty \nu \sinh(\pi\nu) g(i\nu,1) K_{i\nu}(x) d\nu$$

It is assumed here that x and y are positive and real
(some of the integrals listed here are valid for complex y).
The convergence of the integral (1) can be determined by the
asymptotic behavior of $K_{ix}(y)$ for fixed y and large x

$$K_{ix}(y) \sim \left(\frac{2\pi}{x}\right)^{1/2} \exp\left(-\frac{\pi}{2}x\right) \sin\left[x \log\left(\frac{2x}{y}\right) - x + \frac{\pi}{4}\right]$$

Of importance are representations of various types of waves in
the form of (1) such as:

plane wave

$$\exp(-\gamma\rho\cos\phi) = \frac{2}{\pi} \int_0^\infty \cosh(\phi x) K_{ix}(\gamma\rho) dx$$

$$|\phi| \leq \frac{\pi}{2}$$

cylindrical wave

$$K_0 \left[\gamma(\rho^2+\rho'^2-2\rho\rho'\cos\phi)^{1/2}\right] = \frac{2}{\pi} \int_0^\infty K_{ix}(\gamma\rho) K_{ix}(\gamma\rho') \cdot$$

$$\cdot \cosh[x(\pi-|\phi|)] dx$$

$$0 \leq \phi \leq 2\pi$$

spherical wave

$$(r^2+r'^2-2rr'\cos\theta)^{-1/2} \exp[-\gamma(r^2+r'^2-2rr'\cos\theta)^{1/2}] =$$

$$= \frac{2}{\pi}(rr')^{-1/2} \int_0^\infty x \tanh(\pi x) P_{ix-\frac{1}{2}}(-\cos\theta) K_{ix}(\gamma r) K_{ix}(\gamma r') dx$$

$$0 \leq \theta \leq 2\pi$$

Generalized spherical wave

$$(r^2+r'^2-2rr'\cos\ \theta)^{-\frac{1}{2}\nu}K_\nu[\gamma(r^2+r'^2-2rr'\cos\ \theta)^{\frac{1}{2}}] =$$

$$= \frac{1}{\pi}\ (\frac{2}{\pi})^{\frac{1}{2}}(rr')^{-\nu}(\sin\ \theta)^{\frac{1}{2}-\nu}\ \cdot$$

$$\cdot \int_0^\infty x\ \sinh(\pi x)\,\Gamma(\nu+ix)\,\Gamma(\nu-ix)\,P_{ix-\frac{1}{2}}^{\frac{1}{2}-\nu}(-\cos\ \theta)\ \cdot$$

$$\cdot\ K_{ix}(\gamma r)K_{ix}(\gamma r')\,dx$$

$$\text{Re}\ \nu > -1,\quad 0 \le \theta \le 2\pi$$

References

Erdélyi et al. Higher Transcendental Functions. Vol. 2.
McGraw-Hill, 1953.

Kontorovich M. J. and N. N. Lebedev, 1938
J. Exper. Theor. Phys. USSR, 8, 1192.

Kontorovich, M. J. and N. N. Lebedev, 1939
Acad. Sci. USSR, J. Phys. 1, 229.

Kontorovich, M. J. and N. N. Lebedev, 1939
J. Exper. Theor. Phys. USSR, 9, 729.

Lebedev, N. N., 1946
Acad. Sci. USSR, Doklady 52, 655

Lebedev, N. N., 1949
Acad. Sci. USSR, Doklady, 65, 621.

Oberhettinger, F. and T. P. Higgins, 1961
Tables of Lebedev, Mehler and Generalized Mehler Transforms.
Boeing Scientific Research Laboratories, Research Report
D1-82-0136.

	$f(x)$	$g(y) = \int\limits_{0}^{\infty} f(x) K_{ix}(y)\, dx$
1	x^2	$\tfrac{1}{2} y\, \pi \exp(-y)$
2	x^{2n}, $n = 0,1,2,\cdots$	$(-1)^n \tfrac{1}{2}\pi \left[\dfrac{d^{2n}}{dz^{2n}} \exp(-y \cosh z)\right]_{z=0}$
3	$(a^2+x^2)^{-1}$	$\tfrac{1}{2}\pi a^{-1} \int\limits_{0}^{\infty} \exp(-y \cosh t - at)\, dt$
4	$(a^2+x^2)^{-\frac{1}{2}}$	$\int\limits_{0}^{\infty} \exp(-y \cosh t) K_0(ta)\, dt$
5	$\exp(-ax)$	$a \int\limits_{0}^{\infty} (a^2+t^2)^{-1} \exp(-y \cosh t)\, dt$
6	$\cos(ax)$	$\tfrac{1}{2}\pi \exp(-y \cosh a)$
7	$x \sin(ax)$	$\tfrac{1}{2}\pi y\, \sinh a \exp(-y \cosh a)$
8	$\sin(ax)\sinh(bx)$	$\tfrac{1}{2}\pi \exp(-y \cos b \cosh a) \sin(y \sin b \sinh a)$
9	$\cos(ax)\cosh(bx)$	$\tfrac{1}{2}\pi \exp(-y \cos b \cosh a) \cos(y \sin b\ \sinh a)$
10	$\sinh(ax)\sinh(bx)$	$\tfrac{1}{2}\pi \exp(-y \cos a \cos b) \sinh(y \sin a \sin b)$ $a + b \leq \tfrac{1}{2}\pi$

	$f(x)$	$g(y) = \int\limits_0^\infty f(x) K_{ix}(y)\,dx$
11	$\cosh(ax)\cosh(bx)$	$\tfrac{1}{2}\pi\exp(-y\cos a\cos b)\cosh(y\sin a\sin b)$ $a + b \leq \tfrac{1}{2}\pi$
12	$\operatorname{sech}(\tfrac{1}{2}\pi x)$	$\tfrac{1}{2}\pi\{1-y[K_0(y)\mathbf{L}_{-1}(y)+\mathbf{L}_0(y)K_1(y)]\}$
13	$x^2\operatorname{sech}(\pi x)$	$\tfrac{1}{2}e^{-y}\{(\tfrac{1}{2}\pi y)^{\frac{1}{2}}-\pi e^{2y}\operatorname{Erfc}[(2y)^{\frac{1}{2}}]\}$
14	$\operatorname{sech}(\pi x)\cosh(ax)$	$\tfrac{1}{2}\pi\exp(y\cos a)\operatorname{Erfc}[(2y)^{\frac{1}{2}}\,|\cos(\tfrac{1}{2}a)|]$ $a \leq \tfrac{3}{2}\pi$
15	$\operatorname{sech}(\tfrac{1}{2}\pi x)\cosh(ax)$	$y\int\limits_0^\infty (y^2+t^2)^{-\frac{1}{2}}\exp(-t\cos a)K_1[(y^2+t^2)^{\frac{1}{2}}]\,dt$ $a \leq \pi$
16	$\operatorname{csch}(\tfrac{1}{2}\pi x)\sinh(ax)$	$\sin a\int\limits_0^\infty \exp(-t\cos a)K_0[(y^2+t^2)^{\frac{1}{2}}]\,dt$ $a \leq \pi$
17	$\operatorname{csch}(\pi x)\sinh(ax)$	$\tfrac{1}{2}\sin a\int\limits_0^\infty \exp(-t\cos a)K_0(y+t)\,dt$ $a \leq \tfrac{3}{2}\pi$

	$f(x)$	$g(y) \int_0^\infty f(x) K_{ix}(y) dx$
18	$\tanh(\pi x)\sinh(ax)$	$\frac{1}{2}\pi \exp(-y\cos a)\,\mathrm{Erf}[(2y)^{\frac{1}{2}}\sin(\frac{1}{2}a)]$ $a \leq \frac{1}{2}\pi$
19	$\mathrm{sech}(\pi x)\sinh(ax)\cdot$ $\cdot\sinh(bx)$	$\frac{1}{4}\pi\{\exp[y\cos(a+b)]\,\mathrm{Erfc}[(2y)^{\frac{1}{2}}|\cos(\frac{1}{2}a+\frac{1}{2}b)|]-$ $-\exp[y\cos(a-b)]\,\mathrm{Erfc}[(2y)^{\frac{1}{2}}|\cos(\frac{1}{2}a-\frac{1}{2}b)|]\}$ $a+b \leq \frac{3}{2}\pi$
20	$\mathrm{sech}(\pi x)\cosh(ax)\cdot$ $\cdot\cosh(bx)$	$\frac{1}{4}\pi\{\exp[y\cos(a+b)]\,\mathrm{Erfc}[(2y)^{\frac{1}{2}}|\cos(\frac{1}{2}a+\frac{1}{2}b)|]+$ $+\exp[y\cos(a-b)\,\mathrm{Erfc}[(2y)^{\frac{1}{2}}|\cos(\frac{1}{2}a-\frac{1}{2}b)|]\}$ $a+b \leq \frac{3}{2}\pi$
21	$x\,\tanh(\frac{1}{2}\pi x)$	$y\,K_0(y)$
22	$\mathrm{csch}(bx)\sinh(\pi x)\cdot$ $\cdot\cosh(ax)$	$\frac{1}{2}\pi^2 b^{-1}\sum_{n=0}^{\infty}(-1)^n\varepsilon_n I_{n\frac{\pi}{b}}(y)\cos(n\pi\frac{a}{b})$ $b-a \geq \frac{1}{2}\pi$
23	$\tanh(\pi x)\sinh(bx)\cdot$ $\cdot\mathrm{csch}(ax)$	$\frac{1}{4}\pi\{\exp[-y\cos(a+b)]\,\mathrm{Erf}[(2y)^{\frac{1}{2}}\sin(\frac{1}{2}a+\frac{1}{2}b)]-$ $-\exp[-y\cos(a-b)]\,\mathrm{Erf}[(2y)^{\frac{1}{2}}\sin(\frac{1}{2}a-\frac{1}{2}b)]\}$ $a+b \leq \frac{1}{2}\pi$

	$f(x)$	$g(y) = \int\limits_0^\infty f(x)K_{ix}(y)\,dx$
24	$x \sinh(\pi x) \cdot$ $\cdot \Gamma(\nu+\tfrac{1}{2}ix)\Gamma(\nu-\tfrac{1}{2}ix)$ $0 \le \mathrm{Re}\ \nu \le \tfrac{1}{4}$	$2^{1-2\nu}\pi^2 y^{2\nu}$
25	$x\sinh(\pi x)\Gamma(\nu+\tfrac{1}{2}ix)\cdot$ $\cdot\Gamma(\nu-\tfrac{1}{2}ix)\Gamma(\tfrac{1}{2}-\nu+\tfrac{1}{2}ix)\cdot$ $\cdot\Gamma(\tfrac{1}{2}-\nu-\tfrac{1}{2}ix)$	$2^{3/2}\pi^{5/2}y^{\tfrac{1}{2}}K_{\tfrac{1}{2}-2\nu}(y)$
26	$x\sinh(\tfrac{1}{2}\pi x)P^{-\mu}_{\tfrac{1}{2}ix-\tfrac{1}{2}}(z)$ $-1 < z < 1$	$\pi 2^{\tfrac{3}{2}\mu-1}(z+1)^{-\tfrac{1}{2}\mu}y^{1-\mu}I_\mu[y(\tfrac{1}{2}-\tfrac{1}{2}z)^{\tfrac{1}{2}}]$
27	$x\sinh(\pi x)\Gamma(\tfrac{1}{2}+\mu+\tfrac{1}{2}ix)\cdot$ $\cdot\Gamma(\tfrac{1}{2}+\mu-\tfrac{1}{2}ix)P^{-\mu}_{\tfrac{1}{2}ix-\tfrac{1}{2}}(z)$ $-1 < z < 1$ $\mathrm{Re}\ \mu > -\tfrac{1}{2}$	$\pi^2 2^{-\tfrac{3}{2}\mu}(z+1)^{\tfrac{1}{2}\mu}y^{\mu+1}I_\mu[\tfrac{1}{2}-\tfrac{1}{2}z)^{\tfrac{1}{2}}]$
28	$x\sinh(\tfrac{1}{2}\pi x)\Gamma(\tfrac{1}{2}+\mu+\tfrac{1}{2}ix)\cdot$ $\cdot\Gamma(\tfrac{1}{2}+\mu-\tfrac{1}{2}ix)P^{-\mu}_{\tfrac{1}{2}ix-\tfrac{1}{2}}(z)$ $-1 < z < 1, \mathrm{Re}\ \mu > -\tfrac{1}{2}$	$\pi 2^{-\tfrac{3}{2}\mu}(1-z)^{\tfrac{1}{2}\mu}y^{\mu+1}K_\mu[y(\tfrac{1}{2}+\tfrac{1}{2}z)^{\tfrac{1}{2}}]$
29	$x\tanh(\tfrac{1}{2}\pi x)P_{\tfrac{1}{2}ix-\tfrac{1}{2}}(z)$ $-1 < z < 1$	$yK_0[y(\tfrac{1}{2}+\tfrac{1}{2}z)^{\tfrac{1}{2}}]$

	$f(x)$	$g(y) = \int_0^\infty f(x)\,K_{ix}(y)\,dx$
30	$x\,\tanh(\tfrac{1}{2}\pi x)\,p_{\frac{1}{2}ix-\frac{1}{2}}(z)$ $z > 1$	$y\,K_0\,[y\,(\tfrac{1}{2}+\tfrac{1}{2}z)^{\frac{1}{2}}]$
31	$x\,\tanh(\pi x)\,p_{\frac{1}{2}ix-\frac{1}{2}}(z)$ $z > 1$	$(\tfrac{1}{2}\pi)^{\frac{1}{2}}e^{-zy}$
32	$x\,\mathrm{sech}(\pi x)\,\tanh(\pi x)\,\cdot$ $\cdot\;p_{ix-\frac{1}{2}}(z)$ $z > 1$	$-\,(\tfrac{2\pi}{y})^{-\frac{1}{2}}\exp(zy)\,\mathrm{Ei}\,(-zy-y)$
33	$x\,\sinh(\pi x)\,\mathrm{sech}(2\pi x)\,\cdot$ $\cdot\;p_{2ix-\frac{1}{2}}(z)$ $z > 1$	$2^{-7/4}y^{\frac{1}{4}}\exp(\tfrac{1}{2}z^2y-\tfrac{1}{2}y)\,D_{-\frac{3}{2}}\,[z(2y)^{\frac{1}{2}}]$
34	$x\,\tanh(2\pi x)\,\mathrm{sech}(\pi x)\,\cdot$ $\cdot\;p_{2ix-\frac{1}{2}}(z)$ $z > 1$	$\tfrac{1}{2}(\tfrac{1}{2}\pi)^{\frac{1}{2}}y^{\frac{1}{4}}\exp(z^2y-y)\,D_{-\frac{3}{2}}(2zy^{\frac{1}{2}})$
35	$x\,\sinh(\tfrac{1}{2}\pi x)\,p_{\frac{1}{2}ix-\frac{1}{2}}^{\mu}(z)$ $z > 1$ $\mathrm{Re}\;\mu \leq 0$	$\pi 2^{-3/2\,\mu-1}(z+1)^{\frac{1}{2}\mu}y^{1+\mu}J_{-\mu}\,[y\,(\tfrac{1}{2}z-\tfrac{1}{2})^{\frac{1}{2}}]$

	$f(x)$	$g(y) = \int_0^\infty f(x) K_{ix}(y) dx$
36	$x \sinh(\tfrac{1}{2}\pi x) \Gamma(\tfrac{1}{2}+\mu+\tfrac{1}{2}ix) \cdot$ $\cdot \Gamma(\tfrac{1}{2}+\mu-\tfrac{1}{2}ix) p_{\frac{1}{2}ix-\frac{1}{2}}^{-\mu}(z)$ $z > 1, \ \mathrm{Re}\ \mu > -\tfrac{1}{2}$	$\pi 2^{-3/2\mu}(z-1)^{\frac{1}{2}\mu}y^{\mu+1}K_\mu[y(\tfrac{1}{2}z+\tfrac{1}{2})^{\frac{1}{2}}]$
37	$x \sinh(\pi x) \Gamma(\tfrac{1}{2}-\mu+ix) \cdot$ $\cdot \Gamma(\tfrac{1}{2}-\mu-ix) p_{ix-\frac{1}{2}}^{\mu}(z)$ $z > 1, \ \mathrm{Re}\ \mu \geq \tfrac{1}{2}$	$2^{-\frac{1}{2}}\pi^{3/2}(z^2-1)^{-\frac{1}{2}\mu}y^{\frac{1}{2}-\mu}e^{-zy}$
38	$x \sinh(\pi x) \Gamma(\tfrac{1}{2}-\mu+2ix) \cdot$ $\cdot \Gamma(\tfrac{1}{2}-\mu-2ix) p_{2ix-\frac{1}{2}}^{\mu}(z)$ $z > 1, \ \mathrm{Re}\ \mu \leq \tfrac{1}{2}$	$\pi 2^{-3/2-\mu}y^{\frac{1}{4}-\frac{1}{2}\mu}\Gamma(\tfrac{3}{2}-\mu)(z^2-1)^{-\frac{1}{2}\mu} \cdot$ $\cdot \exp(z^2 y-y) D_{\mu-3/2}(2zy^{\frac{1}{2}})$
39	$x \sinh(\pi x) \Gamma(\tfrac{1}{2}-\mu+\tfrac{1}{2}ix) \cdot$ $\cdot \Gamma(\tfrac{1}{2}-\mu-\tfrac{1}{2}ix) p_{\frac{1}{2}ix-\frac{1}{2}}^{\mu}(z)$ $z > 1, 0 \leq \mathrm{Re}\ \mu \leq \tfrac{1}{2}$	$\pi^2 2^{3/2\mu}(z+1)^{-\frac{1}{2}\mu}y^{1-\mu}J_{-\mu}[y(\tfrac{1}{2}z-\tfrac{1}{2})^{\frac{1}{2}}]$
40	$x \sinh(\pi x) \Gamma(\tfrac{1}{4}-\tfrac{1}{2}\mu+\tfrac{1}{2}ix) \cdot$ $\cdot \Gamma(\tfrac{1}{4}-\tfrac{1}{2}\mu-\tfrac{1}{2}ix) p_{ix-\frac{1}{2}}^{\mu}(z)$ $z > 1, \ \mathrm{Re}\ \mu \leq \tfrac{1}{2}$	$z^{\frac{1}{2}+\mu}\pi^2 y^{\frac{1}{2}}J_{-\mu}[y(z^2-1)^{\frac{1}{2}}]$

	$f(x)$	$g(y) = \int\limits_0^\infty f(x) K_{ix}(y)\, dx$
41	$x \sinh(\pi x) \Gamma(\tfrac{3}{4}-\tfrac{1}{2}\mu+\tfrac{1}{2}ix) \cdot$ $\cdot \Gamma(\tfrac{3}{4}-\tfrac{1}{2}\mu-\tfrac{1}{2}ix) p^\mu_{ix-\frac{1}{2}}(z)$ Re $\mu \le \tfrac{3}{2}$	$2^{\mu-\frac{1}{2}}\pi^2 z\, y^{3/2} J_{-\mu}[y(z^2-1)^{\frac{1}{2}}]$
42	$x \tanh(\pi x) \Gamma(\tfrac{1}{2}-\mu+ix) \cdot$ $\cdot \Gamma(\tfrac{1}{2}-\mu-ix) p^\mu_{ix-\frac{1}{2}}(z)$ $z > 1$	$(\tfrac{1}{2}\pi y)^{\frac{1}{2}} \Gamma(1-\mu)(z^2-1)^{-\frac{1}{2}\mu} y^{-\mu} \cdot$ $\cdot e^{yz} \Gamma(\mu, yz+y)$
43	$x \sinh(\tfrac{1}{2}\pi x) \cdot$ $\cdot [q_{\frac{1}{2}ix-\frac{1}{2}}(z) + q_{-\frac{1}{2}ix-\frac{1}{2}}(z)]$ $z > 1$	$-\tfrac{1}{2}\pi^2 y\, Y_0[y(\tfrac{1}{2}z-\tfrac{1}{2})^{\frac{1}{2}}]$
44	$x \sinh(\tfrac{1}{2}\pi x) \cdot$ $\cdot P^\mu_{\frac{1}{2}ix-\frac{1}{2}}(z) P^{-\mu}_{\frac{1}{2}ix-\frac{1}{2}}(z)$ $-1 < z < 1$	$\tfrac{1}{2}\pi y I_\mu[\tfrac{1}{2}y(1-z^2)^{\frac{1}{2}}] I_{-\mu}[\tfrac{1}{2}y(1-z^2)^{\frac{1}{2}}]$
45	$x\Gamma(\tfrac{1}{2}+\mu+\tfrac{1}{2}ix) \Gamma(\tfrac{1}{2}+\mu-\tfrac{1}{2}ix) \cdot$ $\cdot [P^{-\mu}_{\frac{1}{2}ix-\frac{1}{2}}(z)]^2$ $-1 < z < 1$	$\pi^2 y\{I_\mu[\tfrac{1}{2}y(1-z^2)^{\frac{1}{2}}]\}^2$

	$f(x)$	$g(y) = \int\limits_0^\infty f(x) K_{ix}(y)\, dx$
46	$x \sinh(\pi x)\, \Gamma(\tfrac{1}{2}+\mu+\tfrac{1}{2}ix) \cdot$ $\cdot\, \Gamma(\tfrac{1}{2}+\mu-\tfrac{1}{2}ix) \cdot$ $\cdot\, [p_{\frac{1}{2}ix-\frac{1}{2}}^{-\mu}(z)]^2$ $z > 1,\ \mathrm{Re}\ \mu \geq -\tfrac{1}{2}$	$\pi^2 y\{J_\mu[\tfrac{1}{2}y(z^2-1)^{\frac{1}{2}}]\}^2$
47	$x \sinh(\tfrac{1}{2}\pi x) \cdot$ $\cdot\, P_{\frac{1}{2}ix}^{\mu}(z) P_{\frac{1}{2}ix-\frac{1}{2}}^{-\mu}(z)\quad z > 1$	$\tfrac{1}{2}\pi y J_\mu[\tfrac{1}{2}y(z^2-1)^{\frac{1}{2}}] J_{-\mu}[\tfrac{1}{2}y(z^2-1)^{\frac{1}{2}}]$
48	$x \sinh(\pi x)\, \Gamma(\tfrac{1}{2}-\mu+ix) \cdot$ $\cdot\, \Gamma(\tfrac{1}{2}-\mu-ix)\, [p_{ix-\frac{1}{2}}^{\mu}(z)]^2$ $z > 1,\ \mathrm{Re}\ \mu \leq \tfrac{1}{2}$	$(\tfrac{1}{2}\pi)^{\frac{1}{2}} e^{-z^2 y} I_{-\mu}[y(z^2-1)]$
49	$x \tanh(\pi x) \cdot$ $\cdot\, [p_{ix-\frac{1}{2}}(z)]^2$	$(2\pi)^{-\frac{1}{2}} e^{-z^2 y} I_0[y(z^2-1)]$
50	$J_{ix}(a) + J_{-ix}(a)$	$\pi J_0[(a^2-y^2)^{\frac{1}{2}}]\qquad y < a$ $\qquad\qquad 0 \qquad\qquad y > a$

	$f(x)$	$g(y) = \int_0^\infty f(x) K_{ix}(y)\,dx$
51	$i\,\sin(ax)\cdot$ $\cdot[J_{ix}(b)-J_{-ix}(b)]$ $z_{1\atop 2}=(b^2-y^2\pm 2by\,\sinh a)^{\frac{1}{2}}$	$\tfrac{1}{2}\pi[J_0(z_1)-J_0(z_2)]\sinh a<\dfrac{b^2-y^2}{2by}$ $\tfrac{1}{2}\pi J_0(z_1)\qquad \sinh a>\dfrac{b^2-y^2}{2by}$ $y<b$ $0\qquad \sinh a<\dfrac{y^2-b^2}{2by}$ $\tfrac{1}{2}\pi J_0(z_1)\qquad \sinh a>\dfrac{y^2-b^2}{2by}$ $y>b$
52	$\cos(ax)\cdot$ $\cdot[J_{ix}(b)+J_{-ix}(b)]$ $z_{1\atop 2}=(b^2-y^2\pm 2by\,\sinh a)^{\frac{1}{2}}$	$\tfrac{1}{2}\pi[J_0(z_1)+J_0(z_2)]\ \sinh a<\dfrac{b^2-y^2}{2by}$ $\tfrac{1}{2}\pi J_0(z_1)\qquad\qquad \sinh a>\dfrac{b^2-y^2}{2by}$ $y<b$ $0\qquad\qquad \sinh a<\dfrac{y^2-b^2}{2by}$ $\tfrac{1}{2}\pi J_0(z_1)\qquad\qquad \sinh a>\dfrac{y^2-b^2}{2by}$ $y>b$
53	$[J_{\frac{1}{2}ix}(a)]^2+[Y_{\frac{1}{2}ix}(a)]^2$	$2\pi^{-1}K_0\{\tfrac{1}{2}y\,[y+(y^2-4a^2)^{\frac{1}{2}}]\}\cdot$ $\cdot K_0\{\tfrac{1}{2}y\,[y-(y^2-4a^2)^{\frac{1}{2}}]\}$

	$f(x)$	$g(y) = \int\limits_{0}^{\infty} f(x) K_{ix}(y)\,dx$
54	$x \sinh(\tfrac{1}{2}\pi x) \cdot$ $\cdot \{[J_{\frac{1}{2}ix}(a)]^{2} + [Y_{\frac{1}{2}ix}(a)]^{2}\}$	$2y(4a^{2}+y^{2})^{-\frac{1}{2}}$
55	$-ix \operatorname{sech}(\pi x) \cdot$ $\cdot [J_{ix}(a) Y_{-ix}(a) - J_{-ix}(a) Y_{ix}(a)]$	$\exp(-y+\tfrac{1}{2}a^{2}y^{-1})\operatorname{Erfc}[a(2y)^{\frac{1}{2}}]$
56	$x \sinh(\tfrac{1}{2}\pi x) \cdot$ $\cdot [J_{i\frac{1}{2}x}(a) J_{i\frac{1}{2}x}(b) +$ $+Y_{i\frac{1}{2}x}(a) Y_{i\frac{1}{2}x}(b)]$	$2y(y^{2}+4ab)^{-\frac{1}{2}} \cdot$ $\cdot \cos\{\tfrac{1}{2}[(\tfrac{a}{b})^{\frac{1}{2}} - (\tfrac{b}{a})^{\frac{1}{2}}](y^{2}+4ab)^{\frac{1}{2}}\}$
57	$x \sinh(\tfrac{1}{2}\pi x) \cdot$ $\cdot [J_{i\frac{1}{2}x}(a) Y_{i\frac{1}{2}x}(b) -$ $-J_{i\frac{1}{2}x}(b) Y_{i\frac{1}{2}x}(a)]$	$-2y(y^{2}+4ab)^{-\frac{1}{2}} \cdot$ $\cdot \sin\{\tfrac{1}{2}[(\tfrac{a}{b})^{\frac{1}{2}} - (\tfrac{b}{a})^{\frac{1}{2}}](y^{2}+4ab)^{\frac{1}{2}}\}$
58	$x \sinh(\tfrac{1}{2}\pi x) \cdot$ $\cdot [J_{i\frac{1}{2}x}(a) Y_{-\frac{1}{2}ix}(b) +$ $+ J_{-i\frac{1}{2}x}(b) Y_{\frac{1}{2}ix}(a)]$	$-2y(4ab-y^{2})^{-\frac{1}{2}} \cdot$ $\cdot \cos\{\tfrac{1}{2}[(\tfrac{a}{b})^{\frac{1}{2}} + (\tfrac{b}{a})^{\frac{1}{2}}](4ab-y^{2})^{\frac{1}{2}}\}$ $\qquad\qquad y < 2(ab)^{\frac{1}{2}}$ $0 \qquad\qquad y > 2(ab)^{\frac{1}{2}}$

	$f(x)$	$g(y) = \int\limits_0^\infty f(x)K_{ix}(y)\,dx$
59	$x[J_{\mu+\frac{1}{2}ix}(a)Y_{\mu-\frac{1}{2}ix}(a) -$ $-Y_{\mu+\frac{1}{2}ix}(a)J_{\mu-\frac{1}{2}ix}(a)]$	$i2^{2\mu+1}a^{2\mu}y(4a^2+y^2)^{-\frac{1}{2}}\cdot$ $\cdot[y+(y^2+4a^2)^{\frac{1}{2}}]^{-2\mu}$
60	$\cosh(ax)K_{ix}(b)$ $a \le \pi$	$\frac{1}{2}\pi K_0[(b^2+y^2+2by\,\cos a)^{\frac{1}{2}}]$
61	$x\sinh(\pi x)K_{i2x}(a)$	$\frac{1}{8}a\pi(2y\pi^{-1})^{-\frac{1}{2}}\exp(-y-\frac{1}{8}a^2y^{-1})$
62	$x\tanh(\pi x)K_{ix}(a)$	$\frac{1}{2}\pi(ay)^{\frac{1}{2}}(a+y)^{-1}\exp(-a-y)$
63	$x\sinh(bx)K_{ix}(a)$ $b < \pi$	$\frac{1}{2}\pi ay\,\sin b(a^2+y^2+2ay\,\cos b)^{-\frac{1}{2}}\cdot$ $\cdot K_1[(a^2+b^2+2ay\,\cos b)^{\frac{1}{2}}]$
64	$x(b^2+x^2)^{-1}\sinh(\pi x)K_{ix}(a)$	$\frac{1}{2}\pi^2 I_b(y)K_b(a) \qquad\qquad y < a$ $\frac{1}{2}\pi^2 I_b(a)K_b(y) \qquad\qquad y > a$
65	$x\sinh(\pi x)\cdot$ $\cdot\Gamma(\mu+\frac{1}{2}ix)\Gamma(\mu-\frac{1}{2}ix)\cdot$ $\cdot K_{ix}(a)$ $\mathrm{Re}\ \mu \ge 0$	$2^{1-2\mu}\pi^2(ayz^{-1})^{2\mu}K_{2\mu}(z)$ $z = (a^2+y^2)^{\frac{1}{2}}$

	$f(x)$	$g(y) = \int\limits_0^\infty f(x)K_{ix}(y)\,dx$
66	$x\sinh(2\pi x)\cdot$ $\cdot\Gamma(\mu+ix)\Gamma(\mu-ix)K_{ix}(a)$ $0 \le \mathrm{Re}\ \mu \le \tfrac12$	$\pi^{5/2}2^\mu[\Gamma(\tfrac12-\mu)]^{-1}\cdot$ $\cdot(|a^{-1}-y^{-1}|)K_\mu(|y-a|)$
67	$x\sinh(\pi x)\cdot$ $\cdot\Gamma(\mu+ix)\Gamma(\mu-ix)K_{ix}(a)$ $\mathrm{Re}\ \mu \ge 0$	$\pi^{3/2}2^{\mu-1}(a^{-1}+y^{-1})^{-\mu}\Gamma(\tfrac12+\mu)\cdot$ $\cdot K_\mu(a+y)$
68	$[\Gamma(\tfrac34+\tfrac12 ix)\Gamma(\tfrac34-\tfrac12 ix)]^{-1}\cdot$ $\cdot x\tanh(\pi x)K_{ix}(a)$	$\tfrac12(\pi az^{-1})^{\tfrac12}\exp[(a^2+y^2)^{\tfrac12}]$ $z = (a^2+y^2)^{\tfrac12}$
69	$x\sinh(\pi x)\cdot$ $\cdot P_{ix-\tfrac12}^\mu(z)K_{ix}(a)$ $z > 1$	$\pi^{3/2}2^{\mu-2}(z^2-1)^{\tfrac12\mu}c^{\tfrac12+\mu}\cdot$ $\cdot(z-\tau)^{-\tfrac12\mu-\tfrac14}J_{-\mu-\tfrac12}[c(z-\tau)^{\tfrac12}]$ $\qquad\qquad z > \tau$ $0\qquad\qquad z < \tau$ $c = (2ay)^{\tfrac12},\quad \tau = \tfrac12(ay^{-1}+ya^{-1})$
70	$x\tanh(\pi x)\cdot$ $\cdot P_{ix-\tfrac12}(z)K_{ix}(a)$ $z > 1$	$\tfrac12\pi(ay)^{\tfrac12}(a^2+y^2+2azy)^{-\tfrac12}\cdot$ $\cdot\exp[-(a^2+y^2+2azy)^{\tfrac12}]$

	$f(x)$	$g(y) = \int_0^\infty f(x) K_{ix}(y) dx$
71	$x \sinh(\pi x) \Gamma(\mu+ix) \cdot$ $\cdot \Gamma(\mu-ix) p_{ix-\frac{1}{2}}^{\frac{1}{2}-\mu}(z) \cdot$ $\cdot K_{ix}(a)$ $z > 1$	$2^{-\frac{1}{2}}\pi^{3/2} (ayb^{-1})^\mu (z^2-1)^{\frac{1}{2}\mu-\frac{1}{4}} K_\mu(b)$ $b = (a^2+y^2+2ayz)^{\frac{1}{2}}$
72	$x \tanh(\pi x) \operatorname{sech}(\pi x)$ $P_{ix-\frac{1}{2}}(a) K_{ix}(b)$	$\frac{1}{2}(by)^{\frac{1}{2}}c^{-1} [e^c Ei(-a-b-c) -$ $- e^{-c} Ei(-a-b+c)]$ $c = (a^2+y^2+2aby)^{\frac{1}{2}}$
73	$x \tanh(\pi x) K_{ix}(b) \cdot$ $\cdot [q_{ix-\frac{1}{2}}(a)+q_{-ix-\frac{1}{2}}(a)]$	$-\frac{1}{4}\pi(ab)^{\frac{1}{2}}z^{-1}[e^{-z}Ei(z-b-y) -$ $- e^z Ei(-z-b-y)]$ $z = (a^2+b^2+2aby)^{\frac{1}{2}}$
74	$x \begin{cases} \sinh(\frac{1}{2}\pi x) [J_{ix}(a)+J_{-ix}(a)] \\ \cdot i\cosh(\frac{1}{2}\pi x) [Y_{ix}(a)-Y_{-ix}(a)] \end{cases} \cdot$ $\cdot K_{ix}(a)$	$\frac{1}{2}\pi\sin(\frac{1}{2}a^2y^{-1})$
75	$x \begin{cases} \sinh(\frac{1}{2}\pi x) [Y_{ix}(a)+Y_{-ix}(a) \\ -i\cosh(\frac{1}{2}\pi x) [J_{ix}(a)-J_{-ix}(a) \end{cases} \cdot$ $\cdot K_{ix}(a)$	$-\frac{1}{2}\pi\cos(\frac{1}{2}a^2y^{-1})$

	$f(x)$	$g(y) = \int_0^\infty f(x) K_{ix}(y) dx$
76	$x[I_{-\frac{1}{2}ix}(a) I_{-\frac{1}{2}ix}(b) -$ $-I_{\frac{1}{2}ix}(a) I_{\frac{1}{2}ix}(b)]$	$2iy(4ab-y^2)^{-\frac{1}{2}} \cdot$ $\cdot \cosh\{\frac{1}{2}[(\frac{a}{b})^{\frac{1}{2}}-(\frac{b}{a})^{\frac{1}{2}}](4ab-y^2)^{\frac{1}{2}}\}$ $\qquad\qquad y < 2(ab)^{\frac{1}{2}}$ $\qquad 0 \quad y > 2(ab)^{\frac{1}{2}}$
77	$x \sinh(\frac{1}{2}\pi x) K_{\frac{1}{2}ix}(b)$ $[I_{\frac{1}{2}ix}(a)+I_{-\frac{1}{2}ix}(a)]$	$\pi y(4ab-y^2)^{-\frac{1}{2}} \cdot$ $\cdot \exp\{\frac{1}{2}[(\frac{a}{b})^{\frac{1}{2}}-(\frac{b}{a})^{\frac{1}{2}}](4ab-y^2)^{\frac{1}{2}}\}$ $\qquad\qquad y < 2(ab)^{\frac{1}{2}}$ $\pi y(y^2-4ab)^{-\frac{1}{2}} \cdot$ $\cdot \sin\{\frac{1}{2}[(\frac{a}{b})^{\frac{1}{2}}-(\frac{b}{a})^{\frac{1}{2}}](y^2-4ab)^{\frac{1}{2}}\}$ $\qquad\qquad y > 2(ab)^{\frac{1}{2}}$
78	$\cosh(\frac{1}{2}\pi x) [K_{\frac{1}{2}ix}(a)]^2$	$\frac{1}{2}\pi^2 K_0(z_1) K_0(z_2)$ $z_{\frac{1}{2}} = \frac{1}{2}[(y^2+4a^2)^{\frac{1}{2}} \mp y]$
79	$x \tanh(\pi x) K_{ix}(a) \cdot$ $\cdot [I_{ix}(a)-I_{-ix}(a)]$	$-\frac{1}{2}i\pi \exp(-y-\frac{1}{2}a^2y^{-1}) \mathrm{Erf}[ia(2y)^{-\frac{1}{2}}]$

	$f(x)$	$g(y) = \int_0^\infty f(x) K_{ix}(y)\,dx$
80	$I_{\mu-\frac{1}{2}ix}(a) K_{\mu+\frac{1}{2}ix}(a) +$ $+ I_{\mu+\frac{1}{2}ix}(a) K_{\mu-\frac{1}{2}ix}(a)$	$\pi I_\mu(\tfrac{1}{2}z_1) K_\mu(\tfrac{1}{2}z_2)$ $z_{\frac{1}{2}} = (4a^2+y^2)^{\frac{1}{2}} \mp y$
81	$x \sinh(\pi x) \cdot$ $\cdot [K_{i\frac{1}{2}x}(a)]^2$	$0 \qquad y < 2a$ $\pi^2 y (4a^2-y^2)^{-\frac{1}{2}} \quad y > 2a$
82	$x \sinh(\pi x) [K_{\frac{1}{2}+\frac{1}{2}ix}(a) \cdot$ $\cdot K_{\frac{1}{2}+\frac{1}{2}ix}(b) - K_{\frac{1}{2}-\frac{1}{2}ix}(a) \cdot$ $\cdot K_{\frac{1}{2}-\frac{1}{2}ix}(b)]$	$0 \qquad y < 2(ab)^{\frac{1}{2}}$ $2i\pi^2 (y^2-4ab)^{-\frac{1}{2}} \cdot$ $\cdot \cos\{\tfrac{1}{2}[(\tfrac{b}{a})^{\frac{1}{2}} - (\tfrac{a}{b})^{\frac{1}{2}}] (y^2-4ab)^{\frac{1}{2}}\}$ $y > 2(ab)^{\frac{1}{2}}$
83	$x \sinh(\tfrac{1}{2}\pi x) \cdot$ $\cdot K_{\frac{1}{2}ix}(ia) K_{\frac{1}{2}ix}(-ia)$	$\tfrac{1}{4}\pi^2 y (4a^2+y^2)^{-\frac{1}{2}}$
84	$x \sinh(\tfrac{1}{2}\pi x) \cdot$ $\cdot K_{\frac{1}{2}ix}(a) K_{\frac{1}{2}ix}(b)$	$\tfrac{1}{2}\pi^2 y \; z^{-1} \exp[-\tfrac{1}{2}(ab)^{-\frac{1}{2}}(az+bz)]$ $z = (y^2+4ab)^{\frac{1}{2}}$

	$f(x)$	$g(y) = \int_0^\infty f(x) K_{ix}(y) dx$
85	$x \sinh(\pi x) K_{\frac{1}{2}ix}(a) K_{\frac{1}{2}ix}(b)$	$0 \quad y < 2(ab)^{\frac{1}{2}}$ $\pi^2 y (y^2-4ab)^{-\frac{1}{2}} \cdot$ $\cdot \cos\{\frac{1}{2}[(\frac{a}{b})^{\frac{1}{2}}-(\frac{b}{a})^{\frac{1}{2}}](y^2-4ab)^{\frac{1}{2}}\}$ $y > 2(ab)^{\frac{1}{2}}$
86	$x \sinh(\pi x) K_{ix}(a) K_{ix}(b)$	$\frac{1}{4}\pi^2 \exp[-\frac{1}{2}y(\frac{a}{b} + \frac{b}{a} + ab y^{-2})]$
87	$x \tanh(\pi x) K_{ix}(a) K_{ix}(b)$	$\frac{1}{4}\pi^2 \exp[\frac{1}{2}(\frac{ay}{b} + \frac{by}{a} + \frac{ab}{y})] \cdot$ $\cdot \text{Erfc } 2^{-\frac{1}{2}}[(\frac{ay}{b})^{\frac{1}{2}} + (\frac{by}{a})^{\frac{1}{2}} + (\frac{ab}{y})^{\frac{1}{2}}]\}$
88	$x \sinh(\pi x) \cdot$ $\cdot K_{\mu+\frac{1}{2}ix}(a) K_{\mu-\frac{1}{2}ix}(a)$	$\pi^2 2^{-2\mu-1} a^{-2\mu} y(y^2-4a^2)^{-\frac{1}{2}} \cdot$ $\cdot \{[y+(y^2-4a^2)^{\frac{1}{2}}]^{2\mu}+[y-(y^2-4a^2)^{\frac{1}{2}}]^{2\mu}\}$ $y > 2a$ $0 \quad y < 2a$
89	$x \sinh(\pi x) \cdot$ $\cdot K_{\frac{1}{2}ix+ic}(a) K_{\frac{1}{2}ix-ic}(a)$	$0 \quad y < 2a$ $\pi^2 y(y^2-4a^2)^{-\frac{1}{2}} \cos\{2c\log[\frac{y}{2a}(\frac{y^2}{4a^2} -1)^{\frac{1}{2}}]\}$ $y > 2a$

	$f(x)$	$g(y) = \int\limits_0^\infty f(x) K_{ix}(y)\,dx$
90	$x \sinh(\tfrac{1}{2}\pi x) S_{0,ix}(a)$	$\tfrac{1}{2}\pi a y (a^2+y^2)^{-1}$
91	$x \tanh(\pi x) S_{0,2ix}(a)$	$-\tfrac{1}{8}(\tfrac{2y}{\pi})^{-\frac{1}{2}} \exp(\tfrac{1}{8}a^2 y^{-1} - y) \,\mathrm{Ei}(-\tfrac{1}{8}y^{-1})$
92	$x \sinh(\pi x)\,\Gamma(\tfrac{1}{2}-\mu+ix) \cdot$ $\cdot\,\Gamma(\tfrac{1}{2}-\mu-ix) S_{2\mu,2ix}(a)$ $\mathrm{Re}\ \mu \le \tfrac{1}{2}$	$\pi(\tfrac{2y}{\pi})^{-\frac{1}{2}} a 2^{2\mu-3}\Gamma(1-\mu) \cdot$ $\cdot\exp(\tfrac{1}{8}a^2 y^{-1}-y)\,\Gamma(\mu,\tfrac{1}{8}a^2 y^{-1})$
93	$x \tanh(\pi x) \cdot$ $\cdot\,[D_{-\frac{1}{2}+ix}(a) D_{-\frac{1}{2}-ix}(-a) +$ $+ D_{-\frac{1}{2}+ix}(-a) D_{-\frac{1}{2}-ix}(a)]$	$\pi y^{\frac{1}{2}} \cos[a(2y)^{\frac{1}{2}}]$
94	$x \sinh(\pi x)\,\Gamma(\tfrac{1}{2}-\mu+\tfrac{1}{2}ix) \cdot$ $\cdot\,\Gamma(\tfrac{1}{2}-\mu-\tfrac{1}{2}ix) W_{\mu,\frac{1}{2}ix}(a)$	$\pi^2 (4a)^\mu y^{1-2\mu} \exp(-\tfrac{1}{2}a - \tfrac{1}{4}y^2 a^{-1})$
95	$x \sinh(\pi x)\,\Gamma(\tfrac{1}{2}-\mu+ix) \cdot$ $\cdot\,\Gamma(\tfrac{1}{2}-\mu-ix) W_{\mu,ix}(2a)$ $\mathrm{Re}\ \mu \le \tfrac{1}{2}$	$\pi(\tfrac{1}{2}\pi)^{\frac{1}{2}} a\,\Gamma(1-\mu) y^{\frac{1}{2}-\mu}(a+y)^{\mu-1}\exp(-a-y)$

	$f(x)$	$g(y) = \int\limits_{0}^{\infty} f(x) K_{ix}(y)\,dx$
96	$x \sinh(\pi x)\,\Gamma(\tfrac{1}{2}-\mu+\tfrac{1}{2}ix)\cdot$ $\cdot\Gamma(\tfrac{1}{2}-\mu-\tfrac{1}{2}ix)\cdot$ $\cdot W_{\mu,\frac{1}{2}ix}(ia)W_{\mu,\frac{1}{2}ix}(-ia)$ $\mathrm{Re}\ \mu < \tfrac{1}{2}$	$\pi^2 ay^{1-2\mu}(a^2+y^2)^{-\frac{1}{2}}[(a^2+y^2)^{\frac{1}{2}}+a]^{2\mu}$
97	$x \sinh(\pi x)\cdot$ $\cdot W_{\mu,\frac{1}{2}ix}(a)W_{-\mu,\frac{1}{2}ix}(a)$	$0 \qquad y < a$ $\tfrac{1}{2}\pi ay^{1-2\mu}(y^2-a^2)^{-\frac{1}{2}}\cdot$ $\cdot\{[a+i(y^2-a^2)^{\frac{1}{2}}]^{2\mu}+[a-i(y^2-a^2)^{\frac{1}{2}}]^{2\mu}\}$ $=\pi ay(y^2-a^2)^{-\frac{1}{2}}\cos[2\mu\arccos(ay^{-1})]$ $\qquad y > a$

Chapter VI. Transforms with Lommel Functions as Kernel

Transforms with Lommel functions as kernel.

The pair of inversion formulas

(1) $g(y) = 2^{1-\mu} [\Gamma(\tfrac{1}{2}+\tfrac{1}{2}\mu-\tfrac{1}{2}\nu)\,\Gamma(\tfrac{1}{2}+\tfrac{1}{2}\mu+\tfrac{1}{2}\nu)]^{-1} \cdot$

$$\cdot \int_0^\infty f(x)\,(xy)^{\tfrac{1}{2}} s_{\mu,\nu}(xy)\,dx$$

(2) $f(x) = 2^{1-\mu} [\Gamma(\tfrac{1}{2}+\tfrac{1}{2}\mu-\tfrac{1}{2}\nu)\,\Gamma(\tfrac{1}{2}+\tfrac{1}{2}\mu+\tfrac{1}{2}\nu)]^{-1} \cdot$

$$\cdot \int_0^\infty g(y)\,(xy)^{\tfrac{1}{2}} [s_{\mu,\nu}(xy) - S_{\mu,\nu}(xy)]\,dy$$

where $s_{\mu,\nu}(z)$ and $S_{\mu,\nu}(z)$ denote the two Lommel functions is a generalization of cases previously considered.

For $\mu = \nu$

$$s_{\nu,\nu}(z) = \pi^{\tfrac{1}{2}} 2^{\nu-1} \Gamma(\tfrac{1}{2}+\nu) \mathbf{H}_\nu(z)$$

$$S_{\nu,\nu}(z) = \pi^{\tfrac{1}{2}} 2^{\nu-1} \Gamma(\tfrac{1}{2}+\nu) [\mathbf{H}_\nu(z) - Y_\nu(z)]$$

This leads to the inversion formulas occuring in Chapters III and IV. The special case $\mu = \nu - 1$ leads with the relations

$$\lim_{\mu \to \nu} \frac{s_{\mu-1,\nu}(z)}{\Gamma(\mu-\nu)} = 2^{\nu-1}\,\Gamma(\nu)\,J_\nu(z)$$

$$\lim_{\mu \to \nu} \frac{S_{\mu-1,\nu}(z)}{\Gamma(\mu-\nu)} = 0$$

to the pair of Hankel's inversion formulas (Chapter I).

References:

Cooke, R. G., 1925: Proc. London Math. Soc. 24, 381-420.

Erdélyi, A. et. al., 1953: Higher transcendental
                                functions, Vol. 1 p. 73.

Hardy, G. H., 1925: Proc. London Math. Soc. 23, XI.

	$f(x)$	$\int\limits_{0}^{\infty} f(x)\,(xy)^{\frac{1}{2}} s_{\mu,\nu}(xy)\,dx$
1	$x^{\nu+\frac{1}{2}}(a^2+x^2)^{-1}$ $1 > \mathrm{Re}(\mu+\nu) > -3$ $\mathrm{Re} < \frac{3}{2}$ $\mu+\nu \neq -1$ $\mu-\nu \neq -1, -3, -5, \cdots$	$\frac{1}{2}\pi a^{\nu} \sec[\frac{1}{2}\pi(\mu+\nu)] y^{\frac{1}{2}} \cdot$ $\cdot [2^{\mu-1}\Gamma(\frac{1}{2}+\frac{1}{2}\mu+\frac{1}{2}\nu)\Gamma(\frac{1}{2}+\frac{1}{2}\mu-\frac{1}{2}\nu) I_{-\nu}(ay) +$ $+ i e^{-i\frac{\pi}{2}\nu} s_{\mu,\nu}(iay)]$
2	$x^{-\frac{1}{2}}(a^2+x^2)^{-\frac{1}{2}} \cdot$ $\cdot [x+(a^2+x^2)^{\frac{1}{2}}]^{\mu+1}$ $-2 < \mathrm{Re}\ \mu < 0$	$\pi 2^{\mu-2}\Gamma(\frac{1}{2}+\frac{1}{2}\mu+\frac{1}{2}\nu)\Gamma(\frac{1}{2}+\frac{1}{2}\mu-\frac{1}{2}\nu) \cdot$ $\cdot a^{\mu+1}\csc(\pi\mu) y^{\frac{1}{2}} \cdot$ $\cdot [I_{\frac{1}{2}(\mu+\nu+1)}(\frac{1}{2}ay) I_{\frac{1}{2}(\mu-\nu+1)}(\frac{1}{2}ay) -$ $- I_{-\frac{1}{2}(\mu+\nu+1)}(\frac{1}{2}ay) I_{-\frac{1}{2}(\mu-\nu+1)}(\frac{1}{2}ay)]$
3	$x^{\frac{1}{2}-\nu}(x^2-a^2)^{\lambda} \qquad x > a$ $0 \qquad\qquad x < a$ $\mathrm{Re}(\lambda) > -1, \mathrm{Re}(\nu-2\lambda) > \frac{1}{2}$ $\mathrm{Re}(\nu-\mu-2\lambda) > \frac{1}{2}$ $\mu\pm\nu \neq -1, -3, -5, \cdots$	$\frac{1}{2}\Gamma(1+\lambda) a^{1+\lambda-\nu} y^{-\lambda-\frac{1}{2}} \{ \frac{\Gamma(\frac{1}{2}\nu-\frac{1}{2}\mu-\frac{1}{2}-\lambda)}{\Gamma(\frac{1}{2}\nu-\frac{1}{2}\mu+\frac{1}{2})} \cdot$ $\cdot S_{\lambda+1+\mu,\,\lambda+1-\nu}(ay) -$ $- 2^{\mu+\lambda}\Gamma(\frac{1}{2}\mu+\frac{1}{2}-\frac{1}{2}\nu)\Gamma(\frac{1}{2}\mu+\frac{1}{2}+\frac{1}{2}\nu) \cdot$ $\cdot [\sin(\frac{1}{2}\pi(\mu-\nu)) J_{\nu-\lambda-1}(ay) -$ $- \cos(\frac{1}{2}\pi(\mu-\nu)) Y_{\nu-\lambda-1}(ay)] \}$

	$f(x)$	$\int\limits_0^\infty f(x)(xy)^{\frac{1}{2}}s_{\mu,\nu}(xy)\,dx$
4	$x^{\lambda-\frac{1}{2}}e^{-ax}$ $\mathrm{Re}(\mu+\lambda) > -2$	$-\frac{1}{2}2^{\mu+\lambda+1}\Gamma(1+\frac{1}{2}\mu+\frac{1}{2}\lambda)\Gamma(\frac{3}{2}+\frac{1}{2}\mu+\frac{1}{2}\lambda)\cdot$ $\cdot[(\mu-\nu+1)(\mu+\nu+1)]^{-1}a^{\mu-\lambda-2}y^{\mu+\frac{3}{2}}\cdot$ $\cdot{}_3F_2(1,\frac{1}{2}\mu+\frac{1}{2}\lambda+1,\frac{1}{2}\mu+\frac{1}{2}\lambda+\frac{3}{2};\frac{1}{2}\mu-\frac{1}{2}\nu+\frac{3}{2},$ $\frac{1}{2}\mu+\frac{1}{2}\nu+\frac{3}{2};-y^2a^{-2})$
5	$x^{-\mu-\frac{3}{2}}\cos(ax)$ $\mathrm{Re}\,\mu > -\frac{3}{2}$	$\begin{array}{ll}0 & y < a\end{array}$ $\pi^{\frac{1}{2}}2^{\mu-\frac{1}{2}}\Gamma(\frac{1}{2}+\frac{1}{2}\mu+\frac{1}{2}\nu)\Gamma(\frac{1}{2}+\frac{1}{2}\mu-\frac{1}{2}\nu)\cdot$ $\cdot(y^2-a^2)^{\frac{1}{2}\mu+\frac{1}{4}}P^{-\mu-\frac{1}{2}}_{\nu-\frac{1}{2}}(ay^{-1})\qquad y > a$
6	$(a^2-x^2)^{-\frac{1}{2}}x^{-\frac{1}{2}}\cdot$ $\cdot\cos[(\mu+1)\arccos(xa^{-1})]$ $\qquad x < a$ $\qquad 0\qquad x > a$	$\pi2^{\mu-2}\Gamma(\frac{1}{2}+\frac{1}{2}\mu+\frac{1}{2}\nu)\Gamma(\frac{1}{2}+\frac{1}{2}\mu-\frac{1}{2}\nu)\cdot$ $\cdot y^{\frac{1}{2}}J_{\frac{1}{2}(\mu+\nu+1)}(\frac{1}{2}ay)J_{\frac{1}{2}(\mu-\nu+1)}(\frac{1}{2}ay)$
7	$x^{-\frac{1}{2}\nu-\frac{1}{2}\mu}J_{\frac{1}{2}(\mu-\nu+1)}(ax)$	$2^{\frac{1}{2}\mu-\frac{1}{2}\nu-\frac{1}{2}}\Gamma(\frac{1}{2}+\frac{1}{2}\mu-\frac{1}{2}\nu)a^{-\frac{1}{2}\mu+\frac{1}{2}\nu-\frac{1}{2}}\cdot$ $\cdot y^{\frac{1}{2}-\nu}(y^2-a^2)^{\frac{1}{2}\mu+\frac{1}{2}\nu-\frac{1}{2}}\qquad y > a$ $\qquad\qquad 0\qquad\qquad\qquad y < a$

	$f(x)$	$\int\limits_{0}^{\infty} f(x)(xy)^{\frac{1}{2}}s_{\mu,\nu}(xy)\,dx$
8	$x^{\frac{1}{2}-\frac{1}{2}\mu-\frac{1}{2}\nu}\cdot$ $\cdot Y_{\frac{1}{2}\mu-\frac{1}{2}\nu}(ax)$ $\mathrm{Re}(\mu+\nu)>0$ $-3<\mathrm{Re}(\mu-\nu)<1$	$\begin{aligned}&0\qquad\qquad y<a\\[4pt]&2^{\frac{1}{2}\mu-\frac{1}{2}\nu}\,[\Gamma(\tfrac{1}{2}\mu+\tfrac{1}{2}\nu)]^{-1}\Gamma(\tfrac{1}{2}+\tfrac{1}{2}\mu-\tfrac{1}{2}\nu)\cdot\\[4pt]&\cdot\,\Gamma(\tfrac{1}{2}+\tfrac{1}{2}\mu+\tfrac{1}{2}\nu)\,a^{\frac{1}{2}\nu-\frac{1}{2}\mu}y^{\frac{1}{2}-\nu}\cdot\\[4pt]&\cdot\,(y^2-a^2)^{\frac{1}{2}\mu+\frac{1}{2}\nu-1}\qquad\qquad y>a\end{aligned}$

Chapter VII. Divisor Transforms

The divisor transform kernel

$$k(u) = u^{\frac{1}{2}}[Y_0(u) - 2\pi^{-1}K_0(u)]$$

is a self reciprocal kernel. This means that the following
pair of inversion formulas hold

(1) $\quad g(y) = \int\limits_0^\infty f(x)(xy)^{\frac{1}{2}}[Y_0(xy) - 2\pi^{-1}K_0(xy)]dx$

(2) $\quad f(x) = \int\limits_0^\infty g(y)(xy)^{\frac{1}{2}}[Y_0(xy) - 2\pi^{-1}K_0(xy)]dy$

or also

(3) $\quad g(y) = \frac{1}{2}\int\limits_0^\infty f(x)\{2\pi^{-1}K_0[(xy)^{\frac{1}{2}}] - Y_0[(xy)^{\frac{1}{2}}]\}dx$

(4) $\quad f(x) = \frac{1}{2}\int\limits_0^\infty g(y)\{2\pi^{-1}K_0[(xy)^{\frac{1}{2}}] - Y_0[(xy)^{\frac{1}{2}}]\}dy$

This kernel occurs in Voronoï's Summation formula

$$-\frac{1}{4}f(0) + \sum_{n=1}^\infty d(n)f(n) = \int\limits_0^\infty (2\gamma + \log x)f(x)dx +$$

$$+ \sum_{n=1}^\infty d(n)\int\limits_0^\infty f(x)K(nx)dx$$

with

$$K(u) = 4K_0(4\pi u^{\frac{1}{2}}) - 2\pi Y_0(4\pi u^{\frac{1}{2}})$$

and $d(n)$ is the number of divisors of n. The tables in

this chapter give integrals of the form

$$g(y) = \tfrac{1}{2} \int_0^\infty f(x) \lambda_0 [(xy)^{\frac{1}{2}}] dx \qquad \text{with}$$

$$\lambda_0(u) = 2\pi^{-1} K_0(u) - Y_0(u) \ .$$

Then instead of (3), (4)

(5) $$g(y) = \tfrac{1}{2} \int_0^\infty \lambda_0 [(xy)^{\frac{1}{2}}] f(x) dx$$

(6) $$f(x) = \tfrac{1}{2} \int_0^\infty \lambda_0 [(xy)^{\frac{1}{2}}] g(y) dy$$

Examples for (5) can be obtained from the tables in Chapters II and III by substituting $y^{\frac{1}{2}}$ for y and $x^{\frac{1}{2}}$ for x. (For tables of the K_0 and Y_0 transform see also Chapters II and III.)

References:

Dixon, A. L., and W. L. Ferrar, 1932: Quarterly Journal of
    Mathematics 3, 43-59.

Dixon, A. L., and W. L. Ferrar, 1937: Quarterly Journal of
    Mathematics 8, 66-74.

Wilton, J. R., 1931: Proceedings Royal Society of London (A)
    134, 192-202.

Wilton, J. R., 1932: Quarterly Journal of Mathematics 3, 26-32.

Titchmarsh, E. C. 1937: Introduction to the theory of Fourier
    integrals, Ch. VIII. Oxford Press, 394 pp.

	$f(x)$	$\frac{1}{2}\int_0^\infty f(x)\,\lambda_0(\sqrt{xy})\,dx$
1	$1 \quad x < 1$ $0 \quad x > 1$	$-y^{-\frac{1}{2}}\{Y_1(\sqrt{y}) + \frac{2}{\pi}K_1(\sqrt{y})\}$
2	$x^{-\frac{1}{2}} \quad x < a$ $0 \quad x > a$	$a^{\frac{1}{2}}\{K_0[(ay)^{\frac{1}{2}}]\mathbf{L}_{-1}[(ay)^{\frac{1}{2}}] + \mathbf{L}_0[(ay)^{\frac{1}{2}}]K_1[(ay)^{\frac{1}{2}}]\}$ $-\frac{\pi}{2}a^{\frac{1}{2}}\{Y_0[(ay)^{\frac{1}{2}}]\mathbf{H}_{-1}[(ay)^{\frac{1}{2}}] + \mathbf{H}_0[(ay)^{\frac{1}{2}}]Y_1[(ay)^{\frac{1}{2}}]\}$
3	$x^\sigma \quad -1 < \mathrm{Re}\ \sigma < -\frac{1}{4}$	$\pi^{-1}2^{2\sigma+2}[\Gamma(1+\sigma)]^2[\sin(\frac{\pi}{2}\sigma)]^2 y^{-(\sigma+1)}$
4	$0 \quad x < a$ $(x-a)^{-\frac{1}{2}} \quad x > a$	$y^{-\frac{1}{2}}[e^{-a^{\frac{1}{2}}y^{\frac{1}{2}}} - \sin(a^{\frac{1}{2}}y^{\frac{1}{2}})]$
5	$0 \quad x < a$ $(x-a)^\mu \quad x > a$ $-1\ \mathrm{Re}\ \mu < -\frac{1}{4}$	$2^\mu a^{\frac{\mu+1}{2}} y^{-\frac{\mu+1}{2}} \Gamma(\mu+1)\{\frac{2}{\pi}K_{\mu+1}(a^{\frac{1}{2}}y^{\frac{1}{2}})$ $- Y_{-\mu-1}(a^{\frac{1}{2}}y^{\frac{1}{2}})\}$
6	$0 \quad x < a$ $x^{-\frac{1}{2}}(x-a)^{-\frac{1}{2}} \quad x > a$	$\frac{1}{\pi}[K_0(\frac{1}{2}a^{\frac{1}{2}}y^{\frac{1}{2}})]^2 + \frac{\pi}{4}[Y_0(\frac{1}{2}a^{\frac{1}{2}}y^{\frac{1}{2}})]^2$ $- \frac{\pi}{4}[J_0(\frac{1}{2}a^{\frac{1}{2}}y^{\frac{1}{2}})]^2$

	$f(x)$	$\frac{1}{2}\int_0^\infty f(x)\lambda_0(\sqrt{xy})\,dx$
7	$x^{-\frac{1}{2}}(a+x)^{-\frac{1}{2}}$ $\|\arg a\|<\pi$	$\frac{\pi}{4}[J_0(\frac{1}{2}a^{\frac{1}{2}}y^{\frac{1}{2}})]^2+\frac{\pi}{4}[Y_0(\frac{1}{2}a^{\frac{1}{2}}y^{\frac{1}{2}})]^2$ $+\frac{1}{\pi}[K_0(\frac{1}{2}a^{\frac{1}{2}}y^{\frac{1}{2}})]^2$
8	$x^{-\frac{1}{2}}(a+x)^{-1}$ $\|\arg a\|<\pi$	$\frac{\pi}{2}a^{-\frac{1}{2}}[H_0(a^{\frac{1}{2}}y^{\frac{1}{2}})-Y_0(a^{\frac{1}{2}}y^{\frac{1}{2}})]+a^{-\frac{1}{2}}K_0(a^{\frac{1}{2}}y^{\frac{1}{2}})$
9	$x^{-\frac{1}{2}}(a-x)^{-\frac{1}{2}}$ $x<a$ 0 $x>a$	$I_0(\frac{1}{2}a^{\frac{1}{2}}y^{\frac{1}{2}})K_0(\frac{1}{2}a^{\frac{1}{2}}y^{\frac{1}{2}})-\frac{\pi}{2}J_0(\frac{1}{2}a^{\frac{1}{2}}y^{\frac{1}{2}})Y_0(\frac{1}{2}a^{\frac{1}{2}}y^{\frac{1}{2}})$
10	$(x+a)^{-\frac{1}{2}}+(a-x)^{-\frac{1}{2}}$ $x<a$ $(x+a)^{-\frac{1}{2}}$ $x>a$	$y^{-\frac{1}{2}}\cos(a^{\frac{1}{2}}y^{\frac{1}{2}})$
11	$x^{\frac{\mu}{2}-\frac{1}{2}}(x^{\frac{1}{2}}+a^{\frac{1}{2}})^{-1}\,-1<\mathrm{Re}\ \mu<\frac{3}{2}$	$2^{\mu+1}(\pi)^{-1}[\Gamma(1+\frac{\pi}{2})]^2\{s_{-\mu-1,0}[(aye^{i\pi})^{\frac{1}{2}}]$ $+a^{\frac{\mu}{2}}\cos(\frac{\pi\mu}{2})S_{-\mu-1,0}[(ay)^{\frac{1}{2}}]\}$ $-2^{\mu}a^{\frac{1}{2}}\pi^{-1}[\Gamma(\frac{1+\mu}{2})]^2\{s_{-\mu,0}[(aye^{i\pi})^{\frac{1}{2}}]$ $-a^{\frac{\mu-1}{2}}\sin(\frac{\pi\mu}{2})S_{-\mu,0}[(ay)^{\frac{1}{2}}]\}$ $-2a^{\frac{\mu}{2}}\csc(\pi\mu)\{K_0[(ay)^{\frac{1}{2}}]$ $+\pi\cot(\pi\mu)I_0[(ay)^{\frac{1}{2}}]\}$

	$f(x)$	$\frac{1}{2}\int_0^\infty f(x)\,\lambda_0(\sqrt{xy})\,dx$
12	$x^{-\frac{1}{2}}(x+a)^{-\frac{1}{2}}\{[(x+a)^{\frac{1}{2}}+x^{\frac{1}{2}}]^{2\mu}$ $+[(x+a)^{\frac{1}{2}}-x^{\frac{1}{2}}]^{2\mu}\}$ $\|\text{Re }\mu\|<\frac{3}{4},\ \|\arg a\|<\pi$	$\frac{\pi}{2}a^{\mu}\{[J_{\mu}(\frac{1}{2}a^{\frac{1}{2}}y^{\frac{1}{2}})]^2+[Y_{\mu}(\frac{1}{2}a^{\frac{1}{2}}y^{\frac{1}{2}})]^2\}$ $+\frac{2}{\pi}a^{\mu}\cos(\pi\mu)[K_{\mu}(\frac{1}{2}a^{\frac{1}{2}}y^{\frac{1}{2}})]^2$
13	$\quad 0 \qquad\qquad x<a$ $x^{-\frac{1}{2}}(x-a)^{-\frac{1}{2}}\cdot$ $\cdot[\{x^{\frac{1}{2}}+(x-a)^{\frac{1}{2}}\}^{2\mu}+[x^{\frac{1}{2}}-(x-a)^{\frac{1}{2}}]^{2\mu}\}$ $\qquad\qquad x>a$ $\|\text{Re }\mu\|<\frac{3}{4}$	$\frac{2}{\pi}a^{\mu}[K_{\mu}(\frac{1}{2}a^{\frac{1}{2}}y^{\frac{1}{2}})]^2$ $-\frac{\pi}{2}a^{\mu}\{J_{\mu}(\frac{1}{2}a^{\frac{1}{2}}y^{\frac{1}{2}})J_{-\mu}(\frac{1}{2}a^{\frac{1}{2}}y^{\frac{1}{2}})$ $-Y_{\mu}(\frac{1}{2}a^{\frac{1}{2}}y^{\frac{1}{2}})\,Y_{-\mu}(\frac{1}{2}a^{\frac{1}{2}}y^{\frac{1}{2}})\}$
14	$(a+ix)^{-\nu-1}+(a-ix)^{-\nu-1}$ $\text{Re }\nu>-\frac{3}{4}$	$2^{-\nu}[\Gamma(\nu+1)]^{-1}y^{\frac{\nu}{2}}a^{-\frac{\nu}{2}}\{e^{(\frac{i\pi}{4}\nu)}K_{\nu}(e^{\frac{i\pi}{4}}a^{\frac{1}{2}}y^{\frac{1}{2}})$ $+e^{-i\frac{\pi}{4}\nu}K_{\nu}(e^{-i\frac{\pi}{4}}a^{\frac{1}{2}}y^{\frac{1}{2}})\}$
15	$(\alpha^2+x^2)^{-\frac{1}{2}}$	$-K_0[(\frac{\alpha y}{2})^{\frac{1}{2}}]\,Y_0[(\frac{\alpha y}{2})^{\frac{1}{2}}]$
16	$(\alpha^2+x^2)^{-1}$	$\alpha^{-1}\text{ker}(\alpha^{\frac{1}{2}}y^{\frac{1}{2}})$
17	$[(\alpha^2+x^2)(\beta^2+x^2)]^{-1}$	$[\alpha^2-\beta^2]^{-1}\{\frac{\text{ker}(\beta^{\frac{1}{2}}y^{\frac{1}{2}})}{\beta}-\frac{\text{ker}(\alpha^{\frac{1}{2}}y^{\frac{1}{2}})}{\alpha}\}$

	$f(x)$	$\frac{1}{2}\int_0^\infty f(x)\,\lambda_0(\sqrt{xy})\,dx$
18	$\begin{array}{ll}0 & x < a\\ x(x^2-a^2)^\mu & x > a\\ -1 < \text{Re}\,\mu < -\tfrac{5}{8}\end{array}$	$\frac{1}{4\sqrt{2}}a^{2\mu+2}(1+\mu)^{-1}\csc(\mu\pi)\,{}_0F_3(\tfrac{1}{2},\tfrac{1}{2},2+\mu;\frac{a^2y^2}{256})$ $+y^{-\mu-\frac{3}{2}}a^{\mu+\frac{1}{2}}2^{6\mu+5}\cdot[\frac{\Gamma(1+\mu)}{\Gamma(-\frac{1}{2}-\mu)}]^2\,{}_0F_3(-\mu-\tfrac{1}{2},-\mu-\tfrac{1}{2},-\mu;$ $\frac{a^2y^2}{256})$
19	$(\alpha^2+x^2)^{-\frac{1}{2}}\{(\alpha^2+x^2)^{\frac{1}{2}}+\alpha\}^{\frac{1}{2}}$ $\mid\arg\alpha\mid<\frac{\pi}{2}$	$y^{-\frac{1}{2}}e^{-(\frac{\alpha y}{2})^{\frac{1}{2}}}\{\cos[(\frac{\alpha y}{2})^{\frac{1}{2}}]-\sin[(\frac{\alpha y}{2})^{\frac{1}{2}}]\}$
20	$(\alpha^2+x^2)^{-\frac{1}{2}}\cdot$ $\cdot\{[(\alpha^2+x^2)^{\frac{1}{2}}+x]^\nu$ $+[(\alpha^2+x^2)^{\frac{1}{2}}-x]^\nu$ $\mid\text{Re}\,\nu\mid<\tfrac{3}{4}$	$-2\alpha\cos(\frac{\nu\pi}{2})K_\nu[(\frac{\alpha y}{2})^{\frac{1}{2}}]\{\sin(\frac{\nu\pi}{2})J_\nu(\frac{\alpha y}{2})^{\frac{1}{2}}]$ $+\cos(\frac{\nu\pi}{2})Y_\nu[(\frac{\alpha y}{2})^{\frac{1}{2}}]\}$
21	$e^{-\alpha x}\quad\text{Re}\,\alpha>0$	$-(2\pi\alpha)^{-1}\{e^{-\frac{y}{4\alpha}}\overline{Ei}(-\frac{y}{4\alpha})+e^{\frac{y}{4\alpha}}E_i(-\frac{y}{4\alpha})\}$
22	$x^{-\frac{1}{2}}e^{-\alpha x}\quad\text{Re}\,\alpha>0$	$(\pi\alpha)^{-\frac{1}{2}}K_0(\frac{y}{8\alpha})\cosh(\frac{y}{8\alpha})$

$f(x)$	$\frac{1}{2}\int\limits_{0}^{\infty} f(x)\,\lambda_0(\sqrt{xy})\,dx$
23 $x^{\mu-\frac{1}{2}}e^{-\alpha x}$ $\mathrm{Re}\ \mu > -\frac{1}{2}$ $\mathrm{Re}\ \alpha > 0$	$\pi^{-1}\alpha^{-\mu}y^{-\frac{1}{2}}[\Gamma(\mu+\frac{1}{2})]^2 \exp(\frac{y}{8\alpha}) W_{-\mu,0}(\frac{y}{4\alpha})$ $+\alpha^{-\mu}y^{-\frac{1}{2}}\sec(\pi\mu)\exp(\frac{-y}{8\alpha})\{W_{\mu,0}(\frac{y}{4\alpha})$ $-\ \Gamma(\mu+\frac{1}{2})\sin(\pi\mu) M_{\mu,0}(\frac{y}{4\alpha})\}$
24 $x^{-1}\exp[-\alpha^{\frac{1}{2}}x^{-\frac{1}{2}}]$ $|\arg\alpha| < \pi$	$4\pi^{-1}K_0[(4\alpha y e^{i\pi})^{\frac{1}{4}}] K_0[(4\alpha y\, e^{-i\pi})^{\frac{1}{4}}]$ $-2\ K_0[(4\alpha y)^{\frac{1}{4}}] Y_0[(4\alpha y)^{\frac{1}{4}}]$
25 $x^{-\frac{1}{2}}\exp[-(\alpha x)^{\frac{1}{2}}]$ $|\arg\alpha| < \pi$	$2\pi^{-1}\{(y+\alpha)^{-\frac{1}{2}}\log[(\frac{\alpha}{y})^{\frac{1}{2}} + (1+\frac{\alpha}{y})^{\frac{1}{2}}]$ $+\ (\alpha-y)^{-\frac{1}{2}}\log[\frac{\alpha}{y} + (\frac{\alpha^2}{y^2} - 1)^{\frac{1}{2}}]\}$
26 $x^{\frac{\mu-1}{2}}\exp[-(\alpha x)^{\frac{1}{2}}]$ $\mathrm{Re}\ \mu > -1$ $|\arg\alpha| < \pi$	$2\pi^{-\frac{1}{2}}(\alpha+y)^{-\frac{\mu+1}{2}}[\Gamma(\mu+1)]^2[\Gamma(\mu+\frac{3}{2})]^{-1}\ {}_2F_1(\mu+1,$ $\frac{1}{2};\ \mu+\frac{3}{2};\ \frac{\sqrt{\alpha}-\sqrt{y}}{\sqrt{\alpha}+\sqrt{y}})$ $+\ 2\pi^{-1}\Gamma(\mu+1)(y+\alpha)^{-\frac{\mu+1}{2}} Q_\mu[\frac{\alpha^{\frac{1}{2}}}{(\alpha+y)^{\frac{1}{2}}}]$
27 $e^{-ax^{\frac{1}{2}}}\cos(ax^{\frac{1}{2}})$ $|\arg a| < \frac{\pi}{4}$	$2^{\frac{1}{2}}a[4a^4+y^2]^{-3/4}\cos[3/2\ \arctan(\frac{1}{2}ya^{-2})]$

	$f(x)$	$\frac{1}{2}\int_0^\infty f(x)\,\lambda_0(\sqrt{xy})\,dx$
28	$x^{-\frac{1}{2}}\exp[-(ax)^{\frac{1}{2}}]\cos[a^{\frac{1}{2}}x^{\frac{1}{2}}+\frac{\pi}{4}]$ Re $a > 0$	$2^{-\frac{1}{2}}(4a^2+y^2)^{-\frac{1}{2}}\{(4a^2+y^2)^{\frac{1}{2}}+2a\}^{\frac{1}{2}}$
29	$x^{-3/4}\{\exp[-(4ax)^{\frac{1}{4}}]$ $-\exp[-(ax)^{\frac{1}{4}}]\sin[(ax)^{\frac{1}{4}}+\frac{\pi}{4}]\}$ $\lvert\arg a\rvert < \pi$	$\frac{\pi}{2}\,a^{\frac{1}{4}}\,y^{-\frac{1}{2}}[K_{\frac{1}{4}}(\tfrac{1}{4}a^{\frac{1}{2}}y^{-\frac{1}{2}})]^2$
30	$x^{-\frac{1}{2}}(x+a)^{-\frac{1}{2}}\exp[-b^{\frac{1}{2}}(x+a)^{\frac{1}{2}}]$ $\lvert\arg a\rvert < \pi$ $\lvert\arg b\rvert < \pi$	$\pi^{-1}\{K_0[\tfrac{1}{2}a^{\frac{1}{2}}(b^{\frac{1}{2}}+[b-y]^{\frac{1}{2}})]K_0[\tfrac{1}{2}a^{\frac{1}{2}}(b^{\frac{1}{2}}-[b-y]^{\frac{1}{2}})]$ $+K_0[\tfrac{1}{2}a^{\frac{1}{2}}([y+b]^{\frac{1}{2}}+b^{\frac{1}{2}})]K_0[\tfrac{1}{2}a^{\frac{1}{2}}([y+b]^{\frac{1}{2}}-b^{\frac{1}{2}})]\}$
31	$x^{-1}\exp(-\alpha^{\frac{1}{2}}x^{-\frac{1}{2}}-\beta^{\frac{1}{2}}x^{\frac{1}{2}})$ $\lvert\arg \alpha\rvert < \pi$ $\lvert\arg \beta\rvert < \pi$	$\frac{4}{\pi}K_0(z_1)K_0(z_2)-2Y_0(z_3)Y_0(z_4)$ $z_{\frac{1}{2}} = \alpha^{\frac{1}{4}}\{(\beta^{\frac{1}{2}}+y^{\frac{1}{2}})^{\frac{1}{2}}\pm(\beta^{\frac{1}{2}}-y^{\frac{1}{2}})^{\frac{1}{2}}\}$ $z_{\frac{3}{4}} = 2^{\frac{1}{2}}\alpha^{\frac{1}{4}}\{(\beta+y)^{\frac{1}{2}}\pm\beta^{\frac{1}{2}}\}^{\frac{1}{2}}$
32	$\log x \qquad 0 < x < 1$ $0 \qquad\quad 1 < x < \infty$	$-\frac{2}{y}\{Y_0(y^{\frac{1}{2}})+\frac{2}{\pi}K_0(y^{\frac{1}{2}})\}$

	$f(x)$	$\frac{1}{2}\int_0^\infty f(x)\lambda_0(\sqrt{xy})\,dx$
33	$x^{\nu-1}\log x$ $0 < \mathrm{Re}\ \nu < \tfrac{3}{4}$	$\pi^{-1}[2^\nu\cos(\tfrac{\nu\pi}{2})\Gamma(\nu)]^2 y^{-\nu}\{(1+\tfrac{\pi}{2})[\psi(\nu)-\tfrac{\pi}{2}\tan(\tfrac{\nu\pi}{2})]$ $+\ \pi\log(4)-\log y\}$
34	$\log(\lvert 1-a^2 x^{-2}\rvert)$	$\pi\,a^{\frac{1}{2}}y^{-\frac{1}{2}}J_1[(ay)^{\frac{1}{2}}]$
35	$\log(1+\alpha^2 x^{-2})$	$-2\alpha^{\frac{1}{2}}y^{-\frac{1}{2}}\{e^{-i\frac{\pi}{4}}K_1[(\alpha y e^{i\frac{\pi}{2}})^{\frac{1}{2}}]$ $+\ e^{i\frac{\pi}{4}}K_1[(\alpha y e^{-i\frac{\pi}{2}})^{\frac{1}{2}}]\}$
36	$[(x+\alpha)^{-1}+(\alpha-x)^{-1}]\log(\tfrac{\alpha}{x})$ $\lvert\arg\alpha\rvert < \pi$	$\pi K_0[(\alpha x)^{\frac{1}{2}}]$
37	$(a^2+x^2)^{-\frac{1}{2}}\log(\tfrac{a+(a^2+x^2)^{\frac{1}{2}}}{x})$ $\lvert\arg a\rvert < \pi$	$\tfrac{\pi^2}{2}\,J_0[(ax)^{\frac{1}{2}}]\,K_0[(ax)^{\frac{1}{2}}]$
38	$\dfrac{\sin ax}{x}$	$-\tfrac{1}{2}\,\mathrm{Ci}(\tfrac{y}{4a})$
39	$(\dfrac{\sin ax}{x})^2$	$-(\tfrac{a}{2})\,\mathrm{Ci}(\tfrac{y}{8a})-\tfrac{a}{2}\cos(\tfrac{y}{8a})-y\,\mathrm{si}(\tfrac{y}{8a})$
40	$\cos(ax)$	$(2a)^{-1}\cos(\tfrac{y}{4a})$

	$f(x)$	$\frac{1}{2}\int\limits_{0}^{\infty} f(x)\,\lambda_0(\sqrt{xy})\,dx$
41	$x^{-\frac{1}{2}}\sin(ax)$	$-\left(\frac{\pi}{8a}\right)^{\frac{1}{2}}\cos\left(\frac{y}{8a}\right)\{J_0\left(\frac{y}{8a}\right)+Y_0\left(\frac{y}{8a}\right)\}$
42	$x^{-\frac{1}{2}}\cos(ax)$	$\left(\frac{\pi}{8a}\right)^{\frac{1}{2}}\cos\left(\frac{y}{8a}\right)\{J_0\left(\frac{y}{8a}\right)-Y_0\left(\frac{y}{8a}\right)\}$
43	$x^{-\frac{1}{2}}\cos(ax-\frac{\pi}{4})$	$-\left(\frac{\pi}{4a}\right)^{\frac{1}{2}}\cos\left(\frac{y}{8a}\right)Y_0\left(\frac{y}{8a}\right)$
44	$x^{-\frac{1}{2}}\sin[(ax)^{\frac{1}{2}}]$	$2\pi^{-1}(y+a)^{-\frac{1}{2}}\log[\left(\frac{a}{y}\right)^{\frac{1}{2}}+(1+\frac{a}{y})^{\frac{1}{2}}]-(a-y)^{-\frac{1}{2}}$ $\times\log[\left(\frac{a}{y}\right)^{\frac{1}{2}}-\left(\frac{a}{y}-1\right)^{\frac{1}{2}}]\qquad y<a$ $\frac{2}{\pi}(y+a)^{-\frac{1}{2}}\log[\left(\frac{a}{y}\right)^{\frac{1}{2}}+(1+\frac{a}{y})^{\frac{1}{2}}]$ $-(y-a)^{-\frac{1}{2}}\arcsin\left(\frac{a}{y}\right)^{\frac{1}{2}}\qquad y>a$
45	$x^{-\frac{1}{2}}\cos[(ax)^{\frac{1}{2}}]$	$(y+a)^{-\frac{1}{2}}+(a-y)^{-\frac{1}{2}}\qquad y<a$ $(y+a)^{-\frac{1}{2}}\qquad\qquad\qquad y>a$
46	$(x^2+4\alpha^2)^{-3/4}$ $\cos[\frac{3}{2}\arctan(\frac{1}{2}xa^{-2})]$ $\|\arg\alpha\|<\frac{\pi}{4}$	$2^{-\frac{1}{2}}\alpha^{-1}e^{-\alpha y^{\frac{1}{2}}}\cos(\alpha y^{\frac{1}{2}})$

Appendix. List of Notations and Definitions

Abbreviations: ε_n = Neumann's number

$\varepsilon_0 = 1,\quad \varepsilon_n = 2,\quad n = 1, 2, 3,$

$\gamma = 0.57721\ldots$ Euler's constant

1. Elementary functions

Trigonometric and inverse trigonometric functions:

$\sin x,\quad \cos x,\quad \tan x = \dfrac{\sin x}{\cos x},\quad \cot x = \dfrac{\cos x}{\sin x}$

$\sec x = \dfrac{1}{\cos x},\quad \csc x = \dfrac{1}{\sin x},\quad \arcsin x,\quad \arccos x,$

$\arctan x,\quad \text{arccot} x.$

Hyperbolic functions:

$\sinh x = \tfrac{1}{2}(e^x - e^x),\quad \cosh x = \tfrac{1}{2}(e^x + e^{-x})$

$\tanh x = \dfrac{\sinh x}{\cosh x},\quad \coth x = \dfrac{\cosh x}{\sinh x},$

$\text{sech} x = \dfrac{1}{\cosh x},\quad \text{csch} x = \dfrac{1}{\sinh x}.$

2. Orthogonal polynomials

Legendre polynomials $P_n(x)$.

$$P_n(x) = 2^{-n}(n!)^{-1}\dfrac{d^n}{dx^n}(x^2-1)^n = {}_2F_1(-n,\ n+1;\ 1;\ \tfrac{1}{2}-\tfrac{1}{2}x)$$

Gegenbauer's polynomials $C_n^{\nu}(x)$

$$C_n^{\nu}(x) = [n!\,\Gamma(2\nu)]^{-1}\Gamma(2\nu+n)\,{}_2F_1(-n, 2\nu+n; \tfrac{1}{2}+\nu; \tfrac{1}{2}-\tfrac{1}{2}x)$$

Chebycheff polynomials $T_n(x),\quad U_n(x)$

$$T_n(x) = \cos(n\arccos x) = {}_2F_1(-n,n;\tfrac{1}{2};\tfrac{1}{2}-\tfrac{1}{2}x) = \tfrac{1}{2}n \lim_{\nu=0}\cdot\Gamma(\nu)C_n^\nu(x)$$

$$= \tfrac{1}{2}\{[x+i(1-x^2)^{\frac{1}{2}}]^n + [x-i(1-x^2)^{\frac{1}{2}}]^n\}$$

$$U_n(x) = (1-x^2)^{-\frac{1}{2}}\sin[(n+1)\arccos x] = C_n^1(x)$$

$$= x(n+1) \, {}_2F_1(\tfrac{1}{2}-\tfrac{1}{2}n, \tfrac{3}{2}+\tfrac{1}{2}n; \tfrac{3}{2};1-x^2)$$

$$= -\tfrac{1}{2}i(1-x^2)^{-\frac{1}{2}}\{[x+i(1-x^2)^{\frac{1}{2}}]^n-[x-i(1-x^2)^{\frac{1}{2}}]^n\}$$

Jacobi polynomials $P_n^{(\alpha,\beta)}(x)$

$$P_n^{(\alpha,\beta)}(x) = [n!\,\Gamma(1+\alpha)]^{-1}\Gamma(1+\alpha+n) \, {}_2F_1(-n,n+\alpha+\beta+1;\alpha+1;\tfrac{1}{2}-\tfrac{1}{2}x)$$

Laguerre polynomials

$$L_n^\alpha(x) = (n!)^{-1}x^{-\alpha}e^x \frac{d^n}{dx^n}(e^{-x}x^{n+\alpha})$$

$$= [n!\,\Gamma(1+\alpha)]^{-1}\Gamma(\alpha+1+n) \, {}_1F_1(-n;\alpha+1;x)$$

$$L_n^0(x) = L_n(x)$$

Hermite polynomials

$$He_n(x) = (-1)^n\exp(\tfrac{1}{2}x^2)\frac{d^n}{dx^n}\exp(-\tfrac{1}{2}x^2)$$

$$H_n(x) = (-1)^n e^{x^2}\frac{d^n}{dx^n}e^{-x^2} = 2^{\frac{1}{2}n}He_n(2^{\frac{1}{2}}x)$$

$$He_{2n}(x) = (-1)^n 2^{-n}(n!)^{-1}(2n)! \, {}_1F_1(-n;\tfrac{1}{2};\tfrac{1}{2}x^2)$$

$$He_{2n+1}(x) = x(-1)^n 2^{-n}(n!)^{-1}(2n+1)! {}_1F_1(-n;\tfrac{3}{2};\tfrac{1}{2}x^2)$$

3. The Gamma function and related functions

$$\Gamma(z) = \int_0^\infty e^{-t} t^{z-1} \, dt \quad , \qquad \text{Re } z > 0$$

ψ-function

$$\psi(z) = \frac{d}{dz} \log z = \frac{\Gamma'(z)}{\Gamma(z)}$$

Beta function $B(x,y)$

$$B(x,y) = \frac{\Gamma(x)\,\Gamma(y)}{\Gamma(x+y)}$$

4. Legendre functions

(Definition according to Hobson)

$$p_\nu^\mu(z) = [\Gamma(1-\mu)]^{-1}\left(\frac{z+1}{z-1}\right)^{\tfrac{1}{2}\mu} {}_2F_1(-\nu,\nu+1;1-\mu;\tfrac{1}{2}-\tfrac{1}{2}z)$$

$$e^{-i\pi\mu}q_\nu^\mu(z) = 2^{-\nu-1}[\Gamma(\tfrac{3}{2}+\nu)]^{-1}\pi^{\tfrac{1}{2}}\Gamma(1+\nu+\mu)\,z^{-\nu-\mu-1}(z^2-1)^{\tfrac{1}{2}\mu} \cdot$$

$$\cdot \; {}_2F_1(\tfrac{1}{2}\nu+\tfrac{1}{2}\mu+\tfrac{1}{2},\tfrac{1}{2}\nu+\tfrac{1}{2}\mu+1;\nu+\tfrac{3}{2};z^{-2})$$

z is a point in the complex z-plane cut along the real
z-axis from $-\infty$ to $+1$

$$(z^2-1)^{\tfrac{1}{2}\mu} = (z-1)^{\tfrac{1}{2}\mu}(z+1)^{\tfrac{1}{2}} \quad , \quad -\pi < \arg z < \pi, \quad -\pi < \arg(z\pm1) < \pi$$

$$P_\nu^\mu(x) = [\Gamma(1-\mu)]^{-1}\left(\frac{1+x}{1-x}\right)^{\tfrac{1}{2}\mu} {}_2F_1(-\nu,\nu+1;1-\mu;\tfrac{1}{2}-\tfrac{1}{2}x), \quad -1 < x < 1$$

$$Q_\nu^\mu(x) = \tfrac{1}{2}e^{-i\pi\mu}[e^{-\tfrac{1}{2}i\pi\mu}q_\nu^\mu(x+i0) + e^{\tfrac{1}{2}i\pi\mu}q_\nu^\mu(x-i0)], \quad -1 < x < 1$$

$$p_\nu^o(z) = p_\nu(z); \quad q_\nu^o(z) = q_\nu(z)$$

$$P_\nu^o(z) = P_\nu(z); \quad Q_\nu^o(z) = Q_\nu(z)$$

5. Bessel functions

$$J_\nu(z) = (\tfrac{1}{2}z)^\nu \sum_{n=0}^\infty \frac{(-1)^n (\tfrac{z}{2})^{2n}}{n!\,\Gamma(\nu+n+1)}$$

$$J_{-\nu}(z) = J_\nu(z)\cos(\pi\nu) - Y_\nu(z)\sin(\pi\nu);$$

$$Y_{-\nu}(z) = J_\nu(z)\sin(\pi\nu) + Y_\nu(z)\cos(\pi\nu);$$

$$Y_\nu(z) = \cot(\pi\nu)J_\nu(z) - \csc(\pi\nu)J_{-\nu}(z)$$

$$H_\nu^{(1)}(z) = J_\nu(z)+iY_\nu(z); \quad H_\nu^{(2)}(z) = J_\nu(z) - iY_\nu(z)$$

6. Modified Bessel functions

$$I_\nu(z) = e^{-\tfrac{1}{2}i\pi\nu}J_\nu(ze^{i\tfrac{1}{2}\pi}) = (\tfrac{1}{2}z)^\nu \sum_{n=0}^\infty \frac{(\tfrac{1}{2}z)^{2n}}{n!\,\Gamma(\nu+n+1)}$$

$$K_\nu(z) = \tfrac{1}{2}\pi \csc(\pi\nu)\,[I_{-\nu}(z) - I_\nu(z)]$$

$$= \tfrac{1}{2}i\pi e^{\tfrac{1}{2}i\pi\nu}H_\nu^{(1)}(ze^{i\tfrac{1}{2}\pi}) = -\tfrac{1}{2}i\pi e^{-i\tfrac{1}{2}\pi\nu}H_\nu^{(2)}(ze^{-i\tfrac{1}{2}\pi\nu})$$

7. Anger-Weber functions

$$\mathbf{J}_\nu(z) = \pi^{-1} \int_0^\pi \cos(z\,\sin t-\nu t)\,dt$$

$$\mathbf{E}_\nu(z) = -\pi^{-1} \int_0^\pi \sin(z\,\sin t-\nu t)\,dt$$

$$\mathbf{J}_n(z) = J_n(z), \qquad n = 0, 1, 2, \cdots ; \qquad \mathbf{E}_0(z) = -\mathbf{H}_0(z)$$

$$\mathbf{J}_{\frac{1}{2}}(z) = (\tfrac{1}{2}\pi z)^{-\frac{1}{2}}\{[C(z)-S(z)]\cos z + [C(z)+S(z)]\sin z\} = \mathbf{E}_{-\frac{1}{2}}(z)$$

$$\mathbf{J}_{-\frac{1}{2}}(z) = (\tfrac{1}{2}\pi z)^{-\frac{1}{2}}\ [C(z)+S(z)]\cos z - [C(z)-S(z)]\sin z\} = \mathbf{E}_{\frac{1}{2}}(z)$$

8. Struve functions

$$\mathbf{H}_\nu(z) = \sum_{n=0}^{\infty} \frac{(-1)^n (\tfrac{1}{2}z)^{\nu+2n+1}}{\Gamma(n+\tfrac{3}{2})\,\Gamma(\nu+n+\tfrac{3}{2})}$$

$$\mathbf{L}_\nu(z) = -ie^{-\frac{1}{2}i\pi\nu}\mathbf{H}_\nu(ze^{i\frac{1}{2}\pi}) = \sum_{n=0}^{\infty} \frac{(\tfrac{1}{2}z)^{\nu+2n+1}}{\Gamma(n+\tfrac{3}{2})\,\Gamma(\nu+n+\tfrac{3}{2})}$$

9. Lommel functions

$$s_{\mu,\nu}(z) = \frac{z^{\mu+1}}{(\mu-\nu+1)(\mu+\nu+1)}\ {}_1F_2(1;\tfrac{1}{2}\mu-\tfrac{1}{2}\nu+\tfrac{3}{2},\tfrac{1}{2}\mu+\tfrac{1}{2}\nu+\tfrac{3}{2};-\tfrac{1}{4}z^2)$$

$$\mu\pm\nu \neq -1,\ -2,\ -3,\ \cdots$$

$$S_{\mu,\nu}(z) = s_{\mu,\nu}(z) - 2^{\mu-1}\Gamma(\tfrac{1}{2}+\tfrac{1}{2}\mu-\tfrac{1}{2}\nu)\,\Gamma(\tfrac{1}{2}+\tfrac{1}{2}\mu+\tfrac{1}{2}\nu)\ \cdot$$

$$\cdot\ \{\sin[\tfrac{1}{2}\pi(\nu-\mu)]J_\nu(z) + \cos[\tfrac{1}{2}\pi(\nu-\mu)]Y_\nu(z)\}$$

$$s_{\nu,\mu}(z) = s_{\nu,-\mu}(z); \qquad S_{\mu,\nu}(z) = S_{\mu,-\nu}(z)$$

Special cases:

$$s_{\nu,\nu}(z) = \Pi^{\frac{1}{2}}2^{\nu-1}\Gamma(\tfrac{1}{2}+\nu)\mathbf{H}_\nu(z)$$

$$S_{\nu,\nu}(z) = \pi^{\frac{1}{2}}2^{\nu-1}\Gamma(\tfrac{1}{2}+\nu)\,[\mathbf{H}_\nu(z) - Y_\nu(z)]$$

$$s_{\nu+1,\nu}(z) = z^{\nu} - 2^{\nu}\Gamma(1+\nu)J_{\nu}(z); \qquad S_{\nu+1,\nu} = z^{\nu}$$

$$\lim_{\mu \to \nu} \frac{S_{\mu-1,\nu}(z)}{\Gamma(\mu-\nu)} = 2^{\nu-1}\Gamma(\nu)J_{\nu}(z)$$

$$s_{0,\nu}(z) = \tfrac{1}{2}\pi\csc(\pi\nu)[\mathbf{J}_{\nu}(z) - \mathbf{J}_{-\nu}(z)]$$

$$S_{0,\nu}(z) = \tfrac{1}{2}\pi\csc(\pi\nu)[\mathbf{J}_{\nu}(z) - \mathbf{J}_{-\nu}(z) - J_{\nu}(z) + J_{-\nu}(z)]$$

$$s_{-1,\nu}(z) = -\tfrac{1}{2}\pi\nu^{-1}\csc(\pi\nu)[\mathbf{J}_{\nu}(z) + \mathbf{J}_{-\nu}(z)]$$

$$S_{-1,\nu}(z) = \tfrac{1}{2}\pi\nu^{-1}\csc(\pi\nu)[J_{\nu}(z) + J_{-\nu}(z) - \mathbf{J}_{\nu}(z) - \mathbf{J}_{-\nu}(z)]$$

$$s_{1,\nu}(z) = 1 - \tfrac{1}{2}\pi\nu\csc(\pi\nu)[\mathbf{J}_{\nu}(z) + \mathbf{J}_{-\nu}(z)]$$

$$S_{1,\nu}(z) = 1 + \tfrac{1}{2}\pi\nu\csc(\pi\nu)[J_{\nu}(z) + J_{-\nu}(z) - \mathbf{J}_{\nu}(z) - \mathbf{J}_{-\nu}(z)]$$

$$S_{\frac{1}{2},\frac{1}{2}}(z) = z^{-\frac{1}{2}} \quad , \qquad S_{3/2,\frac{1}{2}}(z) = z^{\frac{1}{2}}$$

$$S_{-\frac{1}{2},\frac{1}{2}}(z) = z^{-\frac{1}{2}}[\sin z\; Ci(z) - \cos z\; si(z)]$$

$$S_{-3/2,\frac{1}{2}}(z) = -z^{-\frac{1}{2}}[\sin z\; si(z) + \cos z\; Ci(z)]$$

Lommel functions of two variables

$$U_{\nu}(w,z) = \sum_{n=0}^{\infty} (-1)^n \left(\frac{w}{z}\right)^{\nu+2n} J_{\nu+2n}(z)$$

$$V_{\nu}(w,z) = \cos(\tfrac{1}{2}w + \tfrac{1}{2}z^2 w^{-1} + \tfrac{1}{2}\pi\nu) + U_{2-\nu}(w,z)$$

Kelvin's functions

$$J_{\nu}(ze^{i\,3/4\,\pi}) = ber_{\nu}(z) + ibei_{\nu}(z)$$

$$J_{\nu}(ze^{-i\,3/4\,\pi}) = ber_{\nu}(z) - ibei_{\nu}(z)$$

$$K_\nu(ze^{i\frac{\pi}{4}}) = ker_\nu(z) + i\ kei_\nu(z)$$

$$K_\nu(ze^{-i\frac{\pi}{4}}) = ker_\nu(z) - i\ kei_\nu(z)$$

$$ber_0(z) = ber(z),\ bei_0(z) = bei(z),\ ker_0(z) = ker(z),$$

$$kei_0(z) = kei(z)$$

10. Gauss' hypergeometric series

$$_2F_1(a,b;c;z) = \frac{\Gamma(c)}{\Gamma(a)\Gamma(b)} \sum_{n=0}^{\infty} \frac{\Gamma(a+n)\Gamma(b+n)}{\Gamma(c+n)} \frac{z^n}{n!} \quad , \quad |z| < 1$$

11. Confluent hypergeometric functions

Kummer's functions

$$_1F_1(a;c;z) = \frac{\Gamma(c)}{\Gamma(a)} \sum_{n=0}^{\infty} \frac{\Gamma(a+n)}{\Gamma(c+n)} \frac{z^n}{n!} = z^{-\frac{1}{2}c}\ e^{\frac{1}{2}z}\ M_{\frac{1}{2}c-a,\frac{1}{2}c-\frac{1}{2}}(z)$$

$$_1F_1(a;c;z) = e^z\ _1F_1(c-a;c;-z)$$

Whittaker functions

$$M_{k,\mu}(z) = z^{\mu+\frac{1}{2}}e^{-\frac{1}{2}z}\ _1F_1(\tfrac{1}{2}+\mu-k;2\mu+1;z)$$

$$= z^{\mu+\frac{1}{2}}e^{\frac{1}{2}z}\ _1F_1(\tfrac{1}{2}+\mu+k;2\mu+1;-z)$$

$$W_{k,\mu}(z) = \frac{\Gamma(-2\mu)}{\Gamma(\frac{1}{2}-\mu-k)} M_{k,\mu}(z) + \frac{\Gamma(2\mu)}{\Gamma(\frac{1}{2}+\mu-k)} M_{k,-\mu}(z)$$

$$W_{k,-\mu}(z) = W_{k,\mu}(z)$$

Parabolic cylinder function

$$D_\alpha(z) = 2^{\frac{1}{4}+\frac{1}{2}\alpha} z^{-\frac{1}{2}} W_{\frac{1}{4}+\frac{1}{2}\alpha,\frac{1}{4}}(\tfrac{1}{2}z^2)$$

$$D_n(z) = e^{-\frac{1}{4}z^2} He_n(z), \qquad\qquad n = 0, 1, 2, \cdots$$

$$D_{-1}(z) = (\tfrac{1}{2}\pi)^{\frac{1}{2}} e^{\frac{1}{4}z^2} \ Erfc(2^{-\frac{1}{2}}z)$$

$$D_{-\frac{1}{2}}(z) = (2\pi z^{-1})^{-\frac{1}{2}} K_{\frac{1}{4}}(\tfrac{1}{4}z^2)$$

Error integrals

$$Erf(z) = 2\pi^{-\frac{1}{2}} \int_0^z e^{-t^2} dt = 2\pi^{-\frac{1}{2}} z \, {}_1F_1(\tfrac{1}{2};\tfrac{3}{2};-z^2) = 2(\pi z)^{-\frac{1}{2}} e^{-\frac{1}{2}z^2}$$

$$M_{-\frac{1}{4},\frac{1}{4}}(z^2)$$

$$Erfc(z) = 1-Erf(z) = 2\pi^{-\frac{1}{2}} \int_z^\infty e^{-t^2} dt = (\pi z)^{-\frac{1}{2}} e^{-\frac{1}{2}z^2} W_{-\frac{1}{4},\frac{1}{4}}(z^2)$$

$$Erf(x^{\frac{1}{2}} e^{i\frac{\pi}{4}}) = C(x) + S(x) + i[C(x) - S(x)]$$

$$Erfc(x^{\frac{1}{2}} e^{i\frac{\pi}{4}}) = 1-C(x)-S(x)+i[S(x)-C(x)]$$

Fresnel's integrals

$$C(x) = (2\pi)^{-\frac{1}{2}} \int_0^z t^{-\frac{1}{2}}\cos t \ dt; \ S(x) = (2\pi)^{-\frac{1}{2}} \int_0^z t^{-\frac{1}{2}}\sin t \ dt$$

Exponential integral

$$-Ei(-z) = \int_z^\infty t^{-1}e^{-t} dt = -\gamma-\log z - \sum_{u=1}^\infty \frac{(-z)^n}{n\cdot n!} = z^{-\frac{1}{2}} e^{-\frac{1}{2}z} W_{-\frac{1}{2},0}(z)$$

$$-\pi < arg \ z < \pi$$

$$\overline{Ei}(x) = \tfrac{1}{2}[Ei(x+i0)+Ei(x-i0)] = -P.V. \int_{-x}^{\infty} t^{-1}e^{-t}dt, \; x > 0$$

(P.V. = principle value)

$$\overline{Ei}(x) = \gamma + \log x + \sum_{n=1}^{\infty} \frac{x^n}{n \cdot n!} \; , \qquad x > 0$$

$$Ei(-ix) = Ci(x) - isi(x); \qquad \overline{Ei}(ix) = i\pi + Ci(x) + isi(x)$$

Sine and cosine integral

$$Si(x) = \int_{0}^{x} t^{-1} \sin t \; dt; \; si(x) = - \int_{x}^{\infty} t^{-1}\sin t \; dt = Si(x) - \frac{\pi}{2}$$

$$= \tfrac{1}{2}i[Ei(-ix) - Ei(ix)]$$

$$Ci(x) = - \int_{x}^{\infty} t^{-1}\cos t \; dt = \tfrac{1}{2}[Ei(-ix) + Ei(ix)]$$

$$- \gamma + \log x + \sum_{n=1}^{\infty} \frac{(-1)^n x^{2n}}{(2n)(2n)!}$$

Incomplete gamma function

$$\gamma(\nu,z) = \int_{0}^{z} t^{\nu-1} e^{-t} \; dt = \nu^{-1}z^{\nu} \; _1F_1(\nu, \nu+1; -z), \; \text{Re} \; \nu > 0$$

$$= \nu^{-1}z^{\frac{1}{2}\nu-\frac{1}{2}}e^{-\frac{1}{2}z} \; M_{\frac{1}{2}\nu-\frac{1}{2}, \frac{1}{2}\nu}(z)$$

$$\Gamma(\nu,z) = \Gamma(\nu) - \gamma(\nu,z) = \int_{z}^{\infty} t^{\nu-1} e^{-t}dt$$

$$= z^{\frac{1}{2}\nu-\frac{1}{2}}e^{-\frac{1}{2}z} \; W_{\frac{1}{2}\nu-\frac{1}{2}, \frac{1}{2}\nu}(z)$$

$$\Gamma(\tfrac{1}{2},z^2) = \pi^{\frac{1}{2}}\text{Erfc}(z); \qquad \Gamma(0,z) = -Ei(-z)$$

$$\gamma(\tfrac{1}{2},z^2) = \pi^{\frac{1}{2}}\text{Erf}(z); \qquad \gamma(1,z) = 1-e^{-z}$$

12. Generalized hypergeometric series

$$_mF_n(a_1,a_2,\cdots,a_m;\ b_1,b_2,\cdots b_n;\ z) =$$

$$= \frac{\Gamma(b_1)\cdots\Gamma(b_n)}{\Gamma(a_1)\cdots\Gamma(a_m)} \sum_{k=0}^{\infty} \frac{\Gamma(a_1+k)\cdots\Gamma(a_m+k)}{\Gamma(b_1+k)\cdots\Gamma(b_n+k)}\ \frac{z^k}{k!}$$

For $m = n+1$, $|z| < 1$ For $m < n+1$, $|z| < \infty$

   $m,\ n = 0,\ 1,\ 2\ \cdots$

13. Elliptic integrals

Complete elliptic integrals

$$\mathbf{K}(k) = \int_0^{\frac{1}{2}\pi} (1-k^2\sin^2 x)^{-\frac{1}{2}}dx = \tfrac{1}{2}\pi\ _2F_1(\tfrac{1}{2},\tfrac{1}{2};1;k^2)$$

$$\mathbf{E}(k) = \int_0^{\frac{1}{2}\pi} (1-k^2\sin^2 x)^{\frac{1}{2}}dx = \tfrac{1}{2}\pi\ _2F_1(-\tfrac{1}{2},\tfrac{1}{2};1;k^2)$$

14. Particular cases of Whittaker's functions

$$M_{-\frac{1}{4},\frac{1}{4}}(z) = \tfrac{1}{2}\pi^{\frac{1}{2}}z^{\frac{1}{4}}e^{\frac{1}{2}z}Erf(z^{\frac{1}{2}})$$

$$M_{\frac{1}{4},\frac{1}{4}}(z) = -i\tfrac{1}{2}\pi^{\frac{1}{2}}z^{\frac{1}{4}}e^{-\frac{1}{2}z}Erf(iz^{\frac{1}{2}})$$

$$M_{k,0}(z) = z^{\frac{1}{2}}e^{-\frac{1}{2}z}L_{k-\frac{1}{2}}(z)$$

$$M_{0,\mu}(z) = 2^{2\mu}\Gamma(1+\mu)\,z^{\frac{1}{2}}I_{\mu}(\tfrac{1}{2}z)$$

$$M_{\mu+\frac{1}{2},\mu}(z) = z^{\mu+\frac{1}{2}}e^{-\frac{1}{2}z}$$

$$M_{-\mu-\frac{1}{2},\mu}(z) \quad = \quad z^{\mu+\frac{1}{2}} e^{\frac{1}{2}z}$$

$$M_{k,k+\frac{1}{2}}(z) \quad = \quad (2k+1) z^{-k} e^{\frac{1}{2}z} \gamma(2k+1,z)$$

$$W_{-\frac{1}{4},\frac{1}{4}}(z) \quad = \quad \pi^{\frac{1}{2}} z^{\frac{1}{4}} e^{\frac{1}{2}z} \text{ Erfc}(z^{\frac{1}{2}})$$

$$W_{k,\frac{1}{4}}(z) \quad = \quad 2^{\frac{1}{2}-k} z^{\frac{1}{4}} D_{2k-\frac{1}{2}}[(2z)^{\frac{1}{2}}]$$

$$W_{0,\mu}(z) \quad = \quad \pi^{-\frac{1}{2}} z^{\frac{1}{2}} K_{\mu}(\tfrac{1}{2}z)$$

$$W_{-\frac{1}{2},0}(z) \quad = \quad -z^{\frac{1}{2}} e^{\frac{1}{2}z} \text{ Ei}(-z)$$

$$W_{k,\frac{1}{2}+k}(z) \quad = \quad z^{-k} e^{\frac{1}{2}z} \Gamma(2k+1,z)$$

$$W_{k,k-\frac{1}{2}}(z) \quad = \quad z^{\mu+\frac{1}{2}} e^{-\frac{1}{2}z}$$

List of Functions

Symbol	Name of the function	Listed under
$C(x)$	Fresnel's integral	11
$Ci(x)$	Cosine integral	11
$C_n^\nu(x)$	Gegenbauer's polynomial	2
$D_\nu(z)$	Parabolic cylinder function	11
$\mathbf{E}(k)$	Complete elliptic integral	13
$Ei(-x)$ $\overline{Ei}(x)$	} Exponential integrals	11
$Erf(z)$ $Erfc(z)$	} Error integrals	11
$\mathbf{E}_\nu(z)$	Anger-Weber functions	7
$_mF_n(z)$	Hypergeometric function	10, 11, 12
$He_n(x)$	Hermite's polynomial	2
$H_\nu^{(1),(2)}(z)$	Hankel's functions	5
$\mathbf{H}_\nu(z)$	Struve's function	8
$I_\nu(z)$	Modified Bessel function	6
$J_\nu(z)$	Bessel function	5
$\mathbf{J}_\nu(z)$	Anger-Weber function	7
$\mathbf{K}(k)$	Complete elliptic integral	13
$K_\nu(z)$	Modified Hankel function	6

Symbol	Name of the function	Listed under
$L_\nu(z)$	Laguerre's function	11
$L_n^\alpha(x)$	Laguerre's polynomial	2
$\mathbf{L}_\nu(z)$	Struve's functions	8
$M_{k,\mu}(z)$ $W_{k,\mu}(z)$	Whittaker's function	11
$P_n(x)$	Legendre's polynomials	2
$P_n^{(\alpha,\beta)}(x)$	Jacobi's polynomials	2
$p_\nu^\mu(z)$ $P_\nu^\mu(x)$ $q_\nu^\mu(z)$ $Q_\nu^\mu(x)$	Legendre functions	4
$S(x)$	Fresnel's integral	11
$\text{si}(x)$ $\text{Si}(x)$	Sine integrals	11
$s_{\mu,\nu}(z)$ $S_{\mu,\nu}(z)$	Lommel's functions	9